BEACHES OF THE VICTORIAN COAST
& PORT PHILLIP BAY

A guide to their nature, characteristics, surf and safety

ANDREW D SHORT

Coastal Studies Unit
School of Geosciences
University of Sydney
Sydney NSW 2006

Research officers and cartographers:

Christopher L Hogan
Katherine A McLeod
Surf Life Saving Australia

SYDNEY UNIVERSITY PRESS

Coastal Studies Unit   and   **Surf Life Saving Australia Ltd**
School of Geosciences F09   I Knott Ave
University of Sydney   Locked Bag 2
Sydney NSW 2006   Bondi Beach, NSW 2026

Short, Andrew D
     Beaches of the Victorian Coast and Port Phillip Bay   0 9586504 0 3
     A guide to their nature, characteristics, surf and safety   Published 1996
                                              Reprinted 2002, 2005

Other books in this series by Andrew D Short:
- *Beaches of the New South Wales Coast,* 1993   0-646-15055-3
- *Beaches of the Queensland Coast: Cooktown to Coolangatta,* 2000   0-9586504-1-1
- *Beaches of the South Australian Coast and Kangaroo Island,* 2001   0-9586504-2-X
- *Beaches of the Western Australian Coast: Eucla to Roebuck Bay,* 2005   0-9586504-3-8

Forthcoming books:
     Beaches of the Tasmania Coast and Islands (publication. 2006)   1-920898 12-3
     Beaches of the Northern Australian Coast: The Kimberley, Northern Territory and Cape York
                                            1-920898-16-6
     Beaches of the New South Wales Coast (2nd edition)   1-920898-15-8

Published by:
Sydney University Press
University of Sydney
www.sydney.edu.au /sup

Copies of all books in this series may be purchased online from the University of Sydney Press at:

http://purl.library.usyd.edu.au/sup/marine

Victorian beach database:
Inquiries about the Victorian beach database should be directed to Surf Life Saving Australia
at sls.com.au

Cover photograph: Venus Bay (A D Short)

Cover design by Stewart Irving

For Julia

TABLE OF CONTENTS

FOREWORD

BEACHES OF THE VICTORIAN COAST & PORT PHILLIP BAY
A guide to their nature, characteristics and safety

The coastline of Australia provides much pleasure to the recreating public. Millions of people experience the joys of surf bathing, surfing, bodyboarding, fishing, walking and boating activities each year. It is my pleasure to provide you with books like this, which aid people to understand and use their coastline a little better.

This book has been developed by a joint research program between Surf Life Saving Australia and the Coastal Studies nit of the niversity of Sydney. Experts in coastal geomorphology and dynamics, like Professor Andy Short, plus our own safety experts and staff have supplied information to ensure the validity of this book.

The Australian Beach Safety and Management Program is the first of its kind, not only in Australia but worldwide. In this book, the Program has identified coastal hazards that affect bathers along the Victorian coast, as well as rating all Victorian beaches as to the level of hazard they impose on the bather.

Research findings have shown that there is an ongoing need for beach safety education among both Australian residents (particularly those who live more than 50 kilometres from the coast) and international tourists alike.

Water safety programs and resources, such as this publication, are critical to ensure a greater awareness of beach safety and thereby to help reduce the number of beach related rescues and drownings.

I commend this publication to you in the interests of beach safety education.

Alan Whelpton, AM
President SLSA

PREFACE

This book had its origins in November 1986 at a meeting held at the niversity of Sydney, between Andrew May, National evelopment Officer with Surf Life Saving NSW (SLSNSW); Angus Gordon, Head of the NSW Public Works epartment (PW) Hydraulics Laboratory; and Professor Bruce Thom and myself from the Coastal Studies nit (CS). The meeting had been called by Angus Gordon to help address a problem brought to the PW by Andrew May; that of how to compile a database on the beaches of NSW and how to assess their safety. As a project of this nature was outside the role of the PW, the CS was approached to see if it could assist.

The result was the formation of the NSW Beach Safety Program, a joint project of the CS and SLSNSW. In 1990, with support from the Australian Research Council and in cooperation with Surf Life Saving Australia Ltd, the project was expanded to encompass the entire Australian coast.

The Coastal Studies nit commenced scientific investigations on the Victorian coast in 1981 at Eastern Beach, Lakes Entrance, followed by a major reconnaissance of the entire Ninety Mile Beach later that year. In 1986, studies focused on the multi-bar systems of eastern Port Phillip Bay. Investigations specifically for this project commenced in March, 1990. This required a number of field investigations on the ground, by boat and plane, to compile all the data required for the database and this book.

The NSW beach database and information systems were completed in 1993 and the first book in this series, 'Beaches of the New South Wales Coast' was published in that year. Now, three years later, both the Victorian beach database and this book are completed.

In compiling a book of this magnitude, there will be some errors and omissions, particularly with regard to the names of beaches, surfing spots, and some local factors. If you notice any, please communicate them to the author at the Coastal Studies nit, niversity of Sydney, NSW, 2006, phone (02) 9351 3625, fax (02) 9351 3644 or via Life Saving Victoria (03) 9567 0000 or Surf Life Saving Australia (02) 9300 4000. In this way we can ensure that future editions are up to date.

Andrew Short
Narrabeen Beach, April 1996

ACKNOWLEDGMENTS

This project, while based in Sydney, has relied upon and received tremendous support from Surf Life Saving Victoria. Nigel Taylor, the State Manager, always ensured the full cooperation of SLSV and its staff.

The first field investigations associated with the project were conducted between Nelson and Wilsons Promontory in March 1990. Those involved were Hamish erricks (SLSV), Andrew May and Chris Hogan (SLSNSW). Subsequent field investigations in eastern Victoria were conducted in January 1993 and 1994, with the assistance of Julia Short.

The Australian Beach Safety and Management Program has received significant financial support from funds provided to SLSA by the Australian Sports Commission. Scott erwin and arren Peters at SLSA have always encouraged the aims of the Program and have ensured the cooperation of the Australian Council, state centres and surf clubs in the pursuit of these aims.

The patrolling lifesavers of all 30 Victorian surf lifesaving clubs contributed by filling in beach maps and rescue forms.

At the niversity of Sydney, Nelson Cano supervised the beach sand analysis and Karen Lease provided the facilities for scanning of photographs.

The then Marine Models Laboratory, of the Port of Melbourne Authority, provided access to their aerial photograph library of the Victorian coast, and also assisted in a flight of much of the coast in March 1990. I am also indebted to my former colleague, Professor J L avies, who generously provided a complete set of black and white aerial photographs of the entire Victorian coast.

An additional flight of the coast was undertaken in January 1995 with the Wollongong Aerial Patrol.

Special thanks to Stewart Irving for designing the cover. Finally, Nigel Taylor (SLSV), Chris Hogan and Katherine McLeod (SLSA) took the time to read the entire draft of this book. Their comments have made a contribution to the finished product.

ABSTRACT

This book provides the first description of all Victorian ocean and Port Phillip Bay beaches. It is based on the results of the Victorian section of the Australian Beach Safety and Management Program. It has two aims firstly, to provide the public with general information on the origin and nature of Victorian beaches, including the contribution of geology, oceanography, climate and biota to the beaches, together with information on beach hazards and beach safety. The second aim is to provide a description of each beach emphasising its physical characteristics, including its names, location, access, facilities, dimensions and the character of the beach and surf zone. In particular, it comments on the suitability of the beach for bathing, surfing and fishing, with special emphasis on the natural hazards. Based on the physical hazards, all beaches are rated in terms of their public beach safety and are scaled accordingly from 1 (least hazardous) to 10 (most hazardous).

Keywords beaches, surf zone, rip currents, beach hazards, beach safety, Victoria, Port Phillip Bay

---

Australian Beach Safety and Management Program (ABSAMP)

Awards and Achievements

NSW Department of Sport, Recreation and Racing
Water Safety Award Research 1989
Water Safety Award Research 1991

Surf Life Saving Australia
Innovation Award 1993

International Life Saving
Commemorative Medal 1994

New Zealand Coastal Survey
In 1997 Surf Life Saving New ealand adopted and modified the ABSAMP in order to compile a similar database on New ealand beaches.

Great Britain Beach Hazard Assessment
In 2002 the Royal National Lifeboat Institution adopted and modified the ABSAMP in order to compile a similar database on the beaches of Great Britain.

Hawaiian Ocean Safety
In 2003 the Hawaiian Lifeguard Association adopted ABSAMP as the basis for their Ocean Safety survey and hazard assessment of all Hawaiian beaches.

Handbook on Drowning 2005
This handbook was a product of the inaugural World Congress on rowning held in Amsterdam in   2002. The handbook endorses the ABSAMP approach to assessing beach hazards as the international standard.

1 INTRODUCTION

The Victorian coast extends 1 230 km from Cape Howe in the east to the Glenelg River mouth in the west. Sixty-six percent of the coast, or 811 km, consists of 560 ocean beaches, together with another 132 beaches ringing Port Phillip Bay; in total 692 beaches. The beaches range in length from 176 km long Ninety Mile Beach, to small beaches only a few metres in length.

This book is about every beach on the Victorian coast, including those in Port Phillip Bay. It is based on the results of the Australian Beach Safety and Management Program, a cooperative project of the University of Sydney's Coastal Studies Unit, Surf Life Saving Australia and Surf Life Saving Victoria. It is part of the most comprehensive study ever undertaken of beaches on any part of the world's coast. In Victoria it has examined every beach on the ocean coast, as well as those around Port Phillip Bay and the Eastern Channel of Western Port.

In order to understand the nature of the hundreds of beaches that make up the coast, we must first examine the nature of the coast; as well as the processes that shape and change the beach and surf environment. These aspects are covered in the introductory chapters (2 and 3) on the Victorian coastal environment, its geology and coastal processes. Chapter 4 examines beaches in general and beach usage and hazards on the Victorian coast. The bulk of the book (Chapter 5) is however devoted to a description of every beach located on the open Victorian coast and in Port Phillip Bay (Table 1). Information is provided on each beach's name, location, access, facilities, physical characteristics and surf conditions. Specific comments are made regarding each beach's suitability for bathing, surfing and fishing, together with a beach hazard rating from 1 (safest) to 10 (least safe).

1.1   What is a beach?

A *beach* is a wave deposited accumulation of sediment, usually sand, but possibly cobbles and boulders. They extend from the upper limit of wave swash, as high as 5 m above sea level, out across the surf zone and seaward to the depth to which average waves can move sediment shoreward. On the high wave energy, open Victorian coast, this means they usually extend seaward to depths between 15 and 25 m and as much as 1 to 3 km offshore.

To most of us however, the beach is that part of the dry sand we sit on or cross to reach the shoreline and the adjacent surf zone. It is an area that has a wide variety of uses and users (Table 2). This book will focus on the dry or subaerial beach plus the surf zone or area of wave breaking, typical of the beaches illustrated in Figure 1.

Most Victorians have been to the beach. Many are frequent beach goers and have their favourite beach. The beaches most of us have been to, or go to, are close by our home or vacation area. They are usually at the end of a sealed road with a car park and other facilities. Often they are patrolled by lifesavers or professional lifeguards. These popular, developed beaches, however, represent only a minority of the state's beaches. In Victoria, 41 surf lifesaving clubs are located on 29 ocean beaches and 12 bay beaches, with an additional 15 Royal lifesaving clubs in Port Phillip Bay and one on the Bass coast. In addition, 29 beaches are patrolled by professional lifeguards (Table 3). In total, only 65 beaches (9%) are patrolled. While 364 of the state's beaches have sealed road access, another 233 have no vehicle access and, of these, only 127 can be reached on foot. What this means is that there are many beaches in a natural state. However, there are also many beaches (91%) where extra caution is required, because there is often no one available to assist bathers who might get into difficulties.

Table 1. Victorian coastal characteristics

	Ocean Coast	**Port Phillip Bay**	**Total**
Total length (km)	1 230	259	1 489
Rocky coast (km)	419	82	501
Sandy (beach) coast (km)	811	177	988
Number of beaches	560	132	692

Figure 1. Glenaire Beach, at the mouth of the Aire River, typifies the high energy beaches of the open Victorian coast. It is composed of fine to medium sand; a mixture of quartz and shell fragments. It has a wide, single bar surf zone dominated by rip currents and channels and it is bordered by cliffs and backed by both dunes and the river.

Table 2.   Types of beach users and their activities

Type	User	Location
Passive	sightseer, tourist	road, car park, lookout
Passive-active	sunbakers, picnickers beach sports	dry beach
Active	beachcombers, joggers	swash zone
Active	fishers, bathers	swash, inner surf zone
Active	surfers, water sports	breakers & surf zone
Active	skis, kayaks, windsurfers	breakers & beyond
Active	IRB, boats	beyond breakers

Table 3.   Victorian beaches - lifesaving facilities and access

	Ocean coast	Port Phillip Bay	Total
Surf Lifesaving Clubs	29 (5%)	12 (9%)	41
Royal Lifesaving Clubs	1 (<1%)	15 (11%)	16
Lifeguards (SLSV)	29	0	29
Sealed road access	260	104	364
Gravel road access	39	1	40
Dirt road access	56	5	61
Foot access only	127	0	127
Total beaches	**560**	**132**	**692**

2 THE COASTAL ENVIRONMENT

Beaches are a part of the coastal environment. They always occur at the shoreline, with part extending landward as the dry beach and part extending seaward as the surf zone and nearshore beach. For a beach to form however, a number of parameters and processes must combine. These include contributions from all the world's major spheres, the lithosphere or geology, the atmosphere, the hydrosphere or ocean and the biosphere (Figure 2). This chapter outlines the nature of the Victorian coastal environment and the contribution of each of these spheres to the evolution and nature of Victoria's beaches.

2.1 Geology

2.1.1 The coast

The beaches and geology of a region are inseparable. Simply put, without an underlying geological environment, there would be no beaches. The geology of a coast and often its hinterland, are essential for two reasons. Firstly, beaches are of a finite size; they have a certain width, height and depth below. This means they must rest on some other surface, and this surface is usually the bedrock geology.

On the Victorian coast, where beaches average about 1.3 km in length, many are bounded by prominent rock headlands. Others occupy drowned coastal river valleys, while many are fronted by rock reefs; and 98 rock islands fringe the coast. The bedrock or geology therefore provides what is called the coastal boundary within which, or on which, the beach rests. In addition, more than 500 km of the coast is rock, much of which is exposed on headlands, bluffs and cliffs.

Secondly, on many coasts and particularly in eastern and central Victoria, the beaches are predominantly composed of quartz sand grains, together with other minerals and rock fragments. These sediments originated in the coastal hinterland. Over many millennia they have been eroded from the cliffs and bluffs, or delivered by rivers and streams to the coast and continental shelf. Subsequently, they have been moved by waves on- or along-shore to be deposited as beaches and dunes. The regional geology therefore acts as the major source of sand for Victorian beaches, particularly east of Wilsons Promontory. West of Wilsons Promontory, carbonate detrital sediments derived from the continental shelf become the dominant sand source.

The beaches of any coast are therefore closely intertwined with the longer term geological history and setting of the coast and continental shelf. The geology provides the basic boundaries, shape and often the bulk of the beach sand, which, in combination with carbonate sediments, are acted upon by other processes to produce the beaches.

2.1.2 Geological history

The Victorian coast has a long and interesting geological history (Table 4). The rocks that form the headlands and cliffs range in age from 550 million to seven thousand years, while the coastline began taking on its present shape between 150 million and 70 million years ago. The evolution of the coast can be divided into four broad geological periods. These are summarised in Figure 3 and discussed in the next section.

2.1.3 Coastal geology

2.1.3.1 Ancient mountain building (450 to 280 million years ago)

The rocks that make up much of Victoria evolved during two ancient periods of mountain building. Firstly, during the Ordovician (450 million years ago), the Thompson Fold Belt, which extends from central Queensland to western Victoria, thrust up the sedimentary rocks (shales, slates and sandstones) that today compose the Grampians. None however are exposed at the coast. Secondly, the Lachlan Fold Belt formed between 450 and 400 million years ago. This resulted in the building of a chain of mountains all the way from the central Queensland coast down to northern Tasmania, including most of central and eastern Victoria and Bass Strait. These rocks are today found in the Victorian high country, along the coast at **Cape Liptrap** and in the massive granites at **Wilsons Promontory** and between **Point Ricardo** and **Cape Howe**.

2.1.3.2 Plate rifting and formation of the Otway, Bass and Gippsland basins (150 to 50 million years ago)

The next stage in the evolution of the coast occurred when the Australian plate began rifting away from Antarctica about 150 million years ago. The rifting defined the southern Australian coast, including Tasmania. It also resulted in buckling of the plate and the formation of the **Otway, Bass** and **Gippsland basins**, plus the Murray Basin in the north-west of the state. The rifting was also accompanied by uplift of the eastern highlands and volcanic activity, both of which supplied lavas, muds and sands that slowly filled the landlocked basins to depths of thousands of metres. This period of infilling continued into the Tertiary, when it was also accompanied by a series of marine transgressions (sea level rises, flooding the basins) and regressions (sea level falls, exposing the basins).

Table 4.   Age of Victorian coastal features

• rocks from	550 million years old
• shape of coast	150 million to 70 million years ago
• beaches from	120 000 years old
• most beaches less than	6 500 years old

The Coastal Environment and Processes

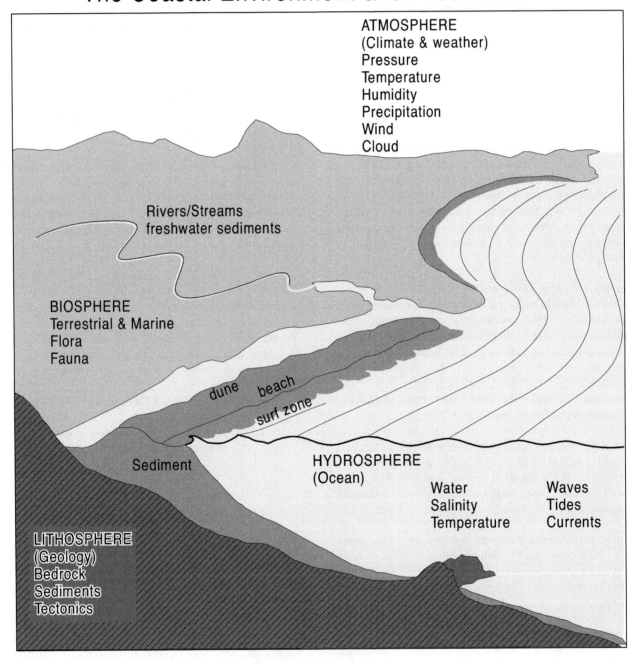

Figure 2. The coastal environment is the most dynamic part of the earth's surface. It contains elements of all four spheres that make up the earth, namely the atmosphere, hydrosphere or ocean, the lithosphere or geology and the biosphere. As the four spheres interact at the coast, they produce a wide spectrum of coastal environments, ranging from rocky coast to muddy tidal flats to sandy beaches.

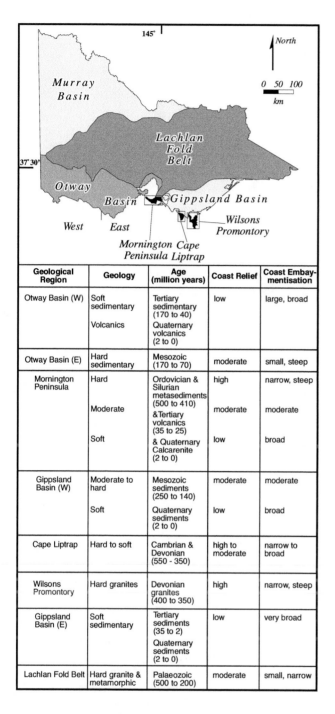

Geological Region	Geology	Age (million years)	Coast Relief	Coast Embay-mentisation
Otway Basin (W)	Soft sedimentary	Tertiary sedimentary (170 to 40)	low	large, broad
	Volcanics	Quaternary volcanics (2 to 0)		
Otway Basin (E)	Hard sedimentary	Mesozoic (170 to 70)	moderate	small, steep
Mornington Peninsula	Hard	Ordovician & Silurian metasediments (500 to 410)	high	narrow, steep
	Moderate	&Tertiary volcanics (35 to 25)	moderate	moderate
	Soft	& Quaternary Calcarenite (2 to 0)	low	broad
Gippsland Basin (W)	Moderate to hard	Mesozoic sediments (250 to 140)	moderate	moderate
	Soft	Quaternary sediments (2 to 0)	low	broad
Cape Liptrap	Hard to soft	Cambrian & Devonian (550 - 350)	high to moderate	narrow to broad
Wilsons Promontory	Hard granites	Devonian granites (400 to 350)	high	narrow, steep
Gippsland Basin (E)	Soft sedimentary	Tertiary sediments (35 to 2)	low	very broad
		Quaternary sediments (2 to 0)		
Lachlan Fold Belt	Hard granite & metamorphic	Palaeozoic (500 to 200)	moderate	small, narrow

Figure 3. The Victorian coast consists of eight geological provinces ranging in age from 550 million to 2 million years. The nature of the rocks and strata in each province, in association with the climate and erosion processes over millions of years, has largely shaped the bedrock relief of the coast. The map indicates the location and extent of each province, while the table indicates the name, geology, age, coastal relief and nature of the coastal bays.

The result is a series of marine (at high sea levels) and non-marine (at low sea levels) sediments, stacked in deposits thousands of metres thick over a period of 150 million to 50 million years ago.

2.1.3.3 Basin infilling and uplift (50 million years ago to present)

Otway Basin

The **Otway Basin** was filled with sandstones, mudstones and limestones which began uplifting episodically during the Cretaceous (130 million years ago), continuing during the Oligocene (40 million to 25 million years ago) to the present era. The rocks composing the Otway Ranges are Mesozoic (170 million to 70 million years old) arkose, sandstone and mudstone sedimentary rocks, while along the **Port Campbell** coast, the exposed limestones and marls that form the spectacular cliffs and sea stacks were deposited in Miocene seas between 70 million and 40 million years ago.

Gippsland Basin

The **Gippsland Basin** has been subject to episodes of erosion and deposition, coupled with lava flows, since the early Tertiary. This has produced a sequence of sedimentary rocks and basalts thousands of metres thick, including the rich Latrobe Valley brown coal fields and the oil and gas fields of Bass Strait. In eastern Gippsland, Tertiary sedimentary rocks only outcrop at the coast at **Red Bluff**. Most of the coastline, including the Gippsland Lakes, consists of Quaternary marine sediments.

In western Gippsland, the sedimentary rocks are more prominent between **Anderson Inlet** and **San Remo**, and the older Tertiary volcanics are found on **Phillip Island** and either side of **Flinders**.

2.1.3.4 Quaternary (past two million years)

Volcanics and lava flows

During the past two million years, continued uplift and tectonism have resulted in a series of major volcanic eruptions that blanketed the western plains with basalt lava flows. These newer Quaternary basalts are exposed along the western shores of **Port Phillip Bay** and on the ocean coast at **Ocean Grove** in the east and **Warrnambool, Port Fairy, Portland** and **Cape Bridgewater** in the west. The most recent eruption at Tower Hill occurred only 7 300 years ago, with the lava reaching the adjacent coast at Sisters Point.

Sea level transgressions and deposits

While ancient rock and more recent lavas form the hard core of the Victorian coast, the coastline itself is dominated by Quaternary marine deposits in the form of beaches and coastal dunes, together with extensive estuarine deposits.

These unconsolidated wave and wind deposits:

- dominate the coast between Cape Howe and Wilsons Promontory (92% of coast)
- compose the massive Gippsland Lakes barrier and lakes system (100%)
- fill the large Waratah, Venus and Kilcunda Bays (73%)
- blanket the valleys and cliffs of Phillip Island and the Mornington Peninsula (57%)
- fill the smaller valleys of the Otways and Port Campbell coast (46%)
- and fill the larger embayments of Warrnambool, Armstrong, Port Fairy, Portland, Bridgewater Bays and the large Discovery Bay (69%).

In all, 66% (811 km) of the coast is of Quaternary origin (less than 2 million years old). Most of this comprises Holocene marine deposits (beaches and dunes) associated with the most recent sea level rise, and is less than 7 000 years old.

2.1.4 Coastal evolution

The evolution of the southern and south-east coast of Australia and particularly the Victorian coastline goes back to the great supercontinent Gondwanaland, consisting of Antarctica, roughly in its present position, surrounded by South America, Africa, India and Australia-New Zealand. At that time the southern coast did not exist, as Australia was joined to Antarctica. The ensuing evolution of the Victorian coast consisted of three phases.

2.1.4.1 Australia separates from Antarctica - formation of Southern Ocean

The supercontinent began breaking up 180 million years ago. The Australian plate was the last to break off, separating from Antarctica 120 million years ago, and beginning its slow northward migration. It has moved north at about 5 to 6 cm per year, slowly opening up the Southern Ocean and forming the southern Australian coastline. The plate movements associated with this movement also buckled the coast and produced the three sedimentary depressions that continue to dominate the Victorian coast - the Gippsland, Bass and Otway basins (Figure 4).

2.1.4.2 Formation of Tasman Sea

Additional continental breakup occurred when what is called the Lord Howe Rise, including New Zealand, began separating from the south-east coast of Australia 75 million years ago (Figure 5). The spreading started in the Bass Strait region, opening up that part of the coast first and spreading northward.

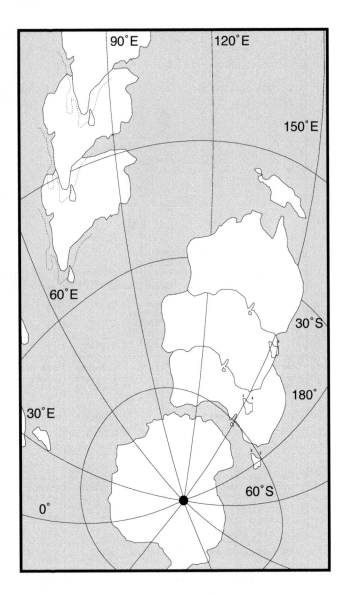

Figure 4. During the past 120 million years the Australian plate, containing Australia, New Guinea and half of New Zealand, has been moving northward at a rate of a few centimetres per year. This figure shows the movement of Australia and India as they have both moved northward away from Antarctica, the core of the once supercontinent - Gondwanaland.

As this separation continued, it not only formed the eastern Victorian, NSW and south-east Queensland coastline, but also opened up the Tasman and south Coral seas. Furthermore, the forces exerted on the south-east Australian coast resulted in the uplifting of the Eastern Highlands. By the time spreading ceased 20 million years later, the Rise, including Lord Howe Island and New Zealand, was located 1 500 km east of where it originated.

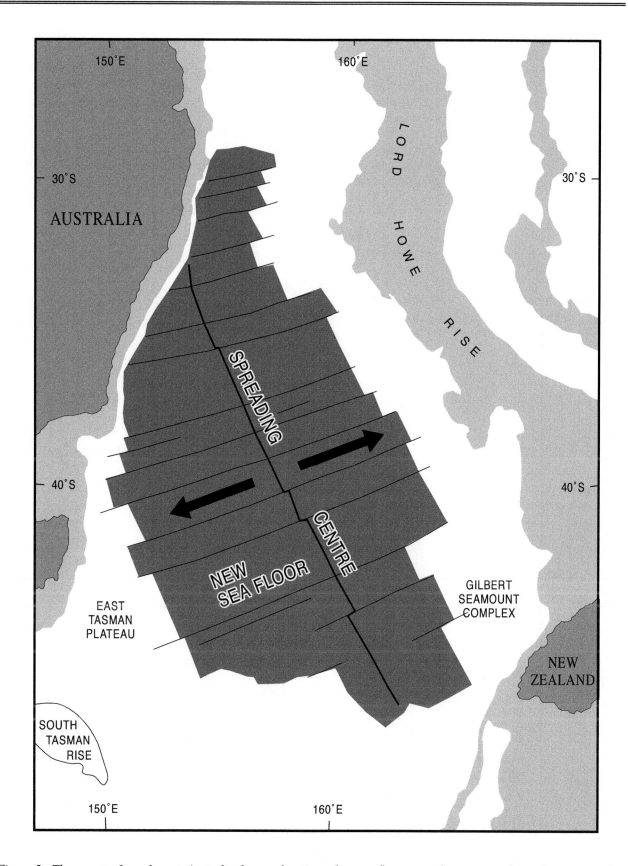

Figure 5. The coast of south-east Australia began forming when seafloor spreading occurred in what is now the southern Coral and Tasman seas. The spreading centre shown in the figure was active for 20 million years and caused the Lord Howe Rise, including New Zealand, to move up to 1 500 km east of the present eastern Victorian coast. At the same time, the movement of the oceanic plate under the east coast resulted in the uplift of Australia's eastern highlands, forming the Great Dividing Range.

2.1.4.3   Infilling of Gippsland, Bass and Otway basins

Following the formation of the Gippsland, Bass and Otway basins, they have been the focus of episodic, but continued, sedimentary deposition and tectonic activity, including volcanism, from 150 million years ago to the present. Today the Bass Basin is totally submerged by the high sea level, forming Bass Strait; while the Gippsland and Otway basins have uplifted portions forming large sections of the Victorian coast, as well as submerged sections on the continental shelf (Figure 3).

2.1.5   Geology and the present coast

The modern coast of Victoria is a product of its 550 million year old geological heritage, coupled with the impact of the Quaternary marine transgressions and regressions (sea level rise and fall). Each sea level rise has flooded the continental shelf (Figure 6), mobilising vast amounts of sand. This sand was reworked onshore by the energetic wave climate to form the extensive beaches and dune systems. We are now at a sea level maxima, that has flooded the continental shelf, including Bass Strait and drowned large bays like Port Phillip, Western Port and Corner Inlet (Figure 7).

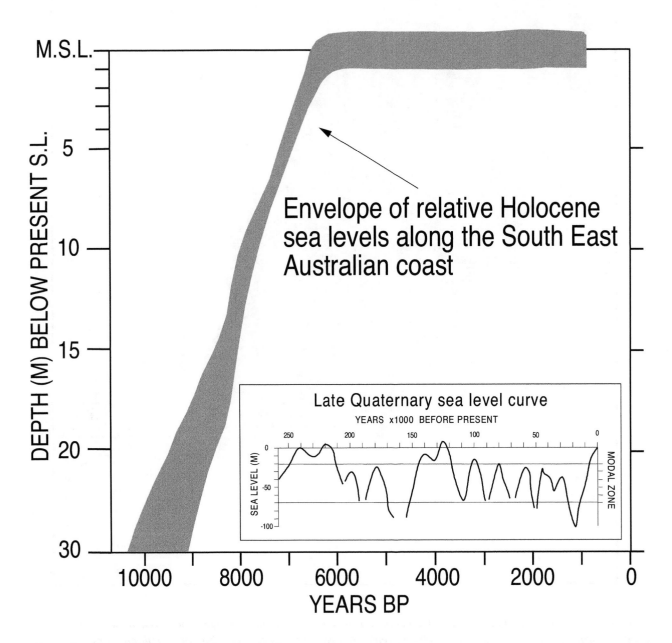

Figure 6. Two plots showing the recent rise in sea level, between 10 000 and 6 000 years ago, obtained from evidence along the south-east Australian coast; together with (insert) the regular oscillations in global sea level that have taken place over the past 250 000 years. The oscillations in sea level are a result of fluctuating earth surface temperature and its impact on the growth and decay of continental ice sheets or ice ages.

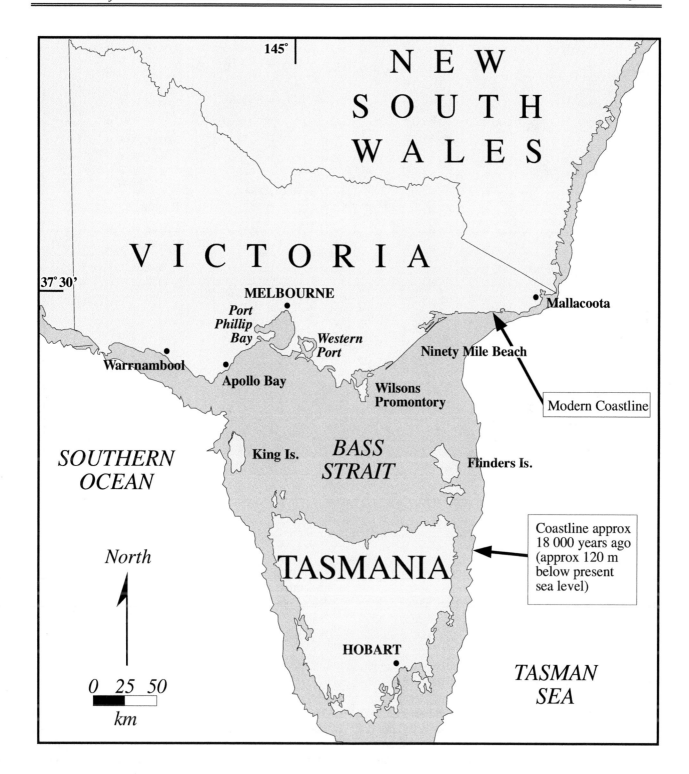

Figure 7. An outline of Australia showing the location of the present coast, together with the position of the coast at the peak of the last glacial maxima (ice age) 18 000 years ago, when sea level was up to 120 m below the present shoreline. At that time, Bass Strait was a dry coastal plain and Victoria was connected to Tasmania. Port Phillip was also a dry valley. As sea level rose between 18 000 and 6 000 years ago, it flooded the continental shelf including Bass Strait and Port Phillip and formed the present coast.

The coast is therefore a mixture of ancient rocks, more recent volcanics and the very recent marine sands. Based on the overriding impact of the bedrock geology, the present coast can be divided into nine coastal regions (Figure 8).

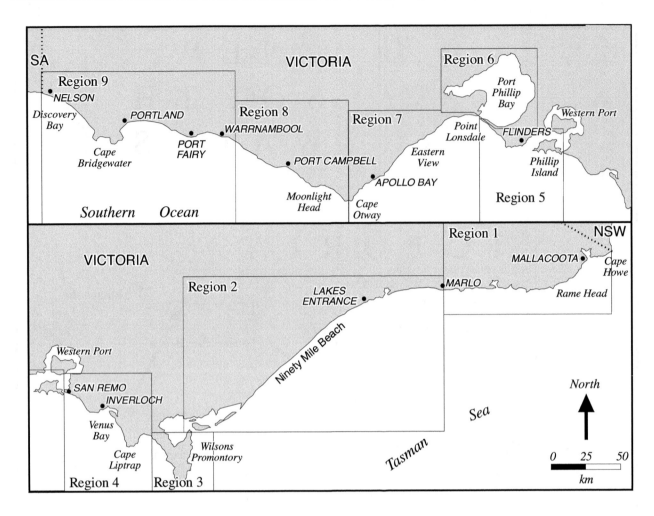

Figure 8. Map of Victoria showing the nine coastal regions used in this book.

2.1.5.1 Croajingolong (Cape Howe to Marlo)

This east trending 150 km section contains ancient (400 to 450 million year old) Paleozoic sedimentary and metasedimentary rocks, together with intrusive granites and granodiorites. The rocks are part of the Lachlan Fold Belt that extends north along the coast as far as Durras in New South Wales. They are generally hard, resilient rocks that form prominent headlands. However, their long period of exposure and erosion has also carved numerous valleys that have now been largely infilled with Quaternary marine sands and estuarine deposits (Figure 1). The strong westerly winds produce an energetic storm wave coast, whose waves have built numerous beaches, backed by extensive and often active transgressive dune systems. At Cape Howe the large active dune sheet is delivering sand across the rocky cape to New South Wales.

2.1.5.2 The Gippsland Lakes (Marlo to Corner Inlet)

Apart from some Miocene marine sandstones at Red Bluff, this section contains 50 km long Ewings Marsh, and the mighty Ninety Mile Beach which runs

uninterrupted for 127 km to Shoal Inlet, with beaches and inlets continuing on for a further 44 km to Snake Island and Corner Inlet; in all, 220 km of Holocene marine sands. Older Quaternary beach and barrier deposits back much of the modern barrier.

2.1.5.3 Wilsons Promontory (Corner Inlet to Darby River)

The promontory consists of a 400 million year old granitic core, formed by cooling magma deeper in the earth, but now exposed at the surface. The resilient granite rises to heights of 500 m and dominates the coast and adjoining granitic islands. Only in the smaller valleys have quartz rich sediments, derived from erosion of the granite, collected to form beaches and barriers. On the east coast, pure rounded quartz grains compose the beaches, while on the exposed west coast carbonate sands dominate. These sands are made up of coralline algae, echinoid spines, foraminifera and bryozoan fragments, derived from the inner continental shelf, that have been washed onshore by waves and blown further inland by westerly winds to blanket the isthmus with carbonate rich sediments. At Darby River, these Quaternary dune deposits have been partially cemented to form cliffs of aeolian calcarenite, commonly known as dunerock.

2.1.5.4   Waratah Bay to Point Nepean

Between Sandy Inlet and The Rip at Port Phillip Heads is 117 km of coast dominated by bedrock headlands. In the east these form the sweeping bays and beaches of Waratah, Venus and Kilcunda Bays, with generally smaller beaches between Kilcunda and Point Nepean. Cape Liptrap contains the state's oldest rocks, 550 million year old Cambrian greenstones, which lie adjacent to the 450 million year old Ordovician limestones and shales at Walkerville.

The bedrock coast between Anderson Inlet - Cape Paterson and Point Nepean consists of considerably younger 130 to 70 million year old Cretaceous sandstones and mudstones. These softer rocks have been eroded to form extensive shore platforms along the base of the cliffs. Between the rocky headlands are numerous beaches ranging in size from 5.2 km in length, down to small pockets of sand. Dunes of varying size back the exposed beaches with massive dunes extending several kilometres inland in Waratah Bay north of Darby River, in Venus Bay south of Anderson Inlet, and across parts of Phillip Island and Mornington Peninsula. The dunes at Darby River and particularly Mornington Peninsula, have been partially cemented to form more resilient dunerock, that has now been eroded to form many cliffs, arches and shore platforms along the coast all the way to Point Nepean.

2.1.5.5   Port Phillip

Once inside The Rip, both the geology and character of the bay coast changes dramatically. In the east, the uplifted Mornington Peninsula horst is composed of 200 to 500 million year old sedimentary and metasedimentary rocks. The bay itself and the western shore are a lower sunkland (depression), covered in the north by Miocene (25 million to 10 million year old) coarse volcanic sedimentary rocks; and in the west by Pliocene (10 million to 2 million year old) lava flows. To the south, the higher Bellarine Peninsula borders the sunkland and is composed of Tertiary sediments and volcanics. The sinking of the bay, combined with Quaternary sea level inundations, has permitted extensive Quaternary bay-marine and estuarine deposits in the lower reaches, particularly in the Seaford - Carrum area, and the Yarra, Werribee and Moorabool deltas.

2.1.5.6   The Otways and Port Campbell

The Otway Ranges are the crest of 3 000 m thick Mesozoic feldspathic sandstones and mudstones (180 to 70 million years old) that were uplifted episodically during the past 100 million years. These rocks dominate the coast from Eastern View down to Cape Otway and along to Moonlight Head. The range is surrounded by two younger phases of marine deposits; the first are (40 to 70 million year old) Paleocene - Eocene sediments, which outcrop along the eastern coast between Point Addis and Eastern View and the western coast between Moonlight Head and Princetown. The second are (40 to 10 million year old) Tertiary marine limestones and marls that occur in the east either side of Torquay, and in the west compose the famous cliffs and stacks between Princetown and Peterborough. Between the bedrock outcrops are numerous generally small beaches, barriers and estuaries, including extensive Quaternary coastal dune deposits between Cape Otway and Rotten Point.

2.1.5.7   Warrnambool to Glenelg River

Western Victoria contains a series of large bays infilled with extensive Quaternary marine deposits. The intervening headlands are composed of Quaternary volcanics that dominate south- western Victoria outcrop, often as low lava flows along the coast between Warrnambool and Port Fairy, and in the Portland - Cape Bridgewater area. However it is the Quaternary beach, barrier and dune deposits that occupy most of the 211 km of coast and blanket many of the headlands. In the west, these are increasingly dominated by calcareous sediments, which in turn have been partially cemented to form dunerock (Figure 9). This in turn covers some of the headlands, and outcrops on beaches in Bridgewater and Discovery Bays.

2.2   Coastal Processes

Coastal processes are the marine, atmospheric and biological activity that interact over time with the geology and sediments to produce a particular coastal system or environment.

The shoreline is the only part of the globe where the four great spheres, the atmosphere, the hydrosphere or ocean, the lithosphere or earth's surface and the biosphere all coexist. Consequently it is often the most dynamic part of the earth's surface. These dynamics are most evident on sandy beaches, particularly those exposed to high waves (Figure 10).

Sandy beaches consist of lithospheric elements, namely sand, lying ultimately on bedrock geology. They are acted on by the waves, tides and currents of the hydrosphere; the wind, rain and temperature of the atmosphere; and play host to the fauna of the biosphere. To understand beaches and how they form and change, we must first of all know about the processes that interact at the coast to produce beach systems.

Figure 9. The Craigs is a popular tourist stop just west of Port Fairy. This view shows the small platform beach surrounded by lithified Pleistocene sand dunes, called dune calcarenite. Similar calcarenite occurs on high energy beaches between Wilsons Promontory and the Victorian border.

2.3   Climate and Atmospheric Processes

Climate's contribution to beaches is in two main areas. Climate interacting with the geology and biology provides the geo- and bio-chemistry to weather the land surface, which together with the physical forces of rain, runoff, rivers and gravity erode and transport sediments to the coast. At the coast it is also climate, particularly winds, that interact with the ocean to generate waves and currents that are essential to move and build this sediment into beaches and dunes.

On a global scale, beaches can be classified by their climate. The **polar beaches** of the Arctic Ocean and those surrounding Antarctica are all dark in colour and composed of coarse sand, cobbles and even boulders. They receive only low waves, have no surf, but have steep swash zones and no dunes. All these characteristics are a product of the cold polar climate, which has low winds, hence low waves and no dunes. In fact, no waves occur for most of the year when the sea surface is frozen over. It also lacks the warm chemical weathering required to weather out the darker minerals that give rocks their colour, and helps break them down

into smaller fragments. Hence the sediment is dark and coarse. Finally, the coarse beach material and low waves result in steep beaches with no sand bars. This results in a barless beach with a shorebreak but no surf.

Tropical beaches also reside in areas of generally lower winds and consequently receive relatively low wave energy. This in turn builds steep beaches and few dunes. Their sediments however are often white, being composed of well weathered quartz sands, derived from plentiful tropical rivers, and bleached coral and algal fragments derived from coral and algal reefs.

The Victorian coast extends from 37°30' to 39°S latitude, placing it outside the tropics but still north of the cooler climates of the Southern Ocean. Victorian beaches have characteristics typical of those of the **temperate middle latitudes**. These beaches have sediments composed predominantly of well weathered quartz sand grains with variable amounts of shell fragments. In addition, wind and wave energy associated with the westerly wind stream is relatively high. The waves produce energetic surf zones, while the winds can build massive coastal dune systems.

Figure 10.   The Victorian coast is dominated by high waves, which in turn produce broad energetic beaches. This view of Portsea Beach shows the 200 m wide surf zone, which is dominated by strong persistent rips, including permanent rips against the calcarenite reefs at either end of the beach. The high waves and rips make this one of the state's most hazardous beaches. Immediately north are the contrasting protected and usually quiescent waters of Weroona Bay Beach in Port Phillip Bay.

2.3.1   High pressure systems

In terms of global weather systems, Victoria is dominated by both the subtropical high pressure system and the subpolar low pressure systems. The highs are centred around 36° S in summer and dominate the summer climate, bringing warm to hot and dry conditions. In winter the high moves north to 30° S, which permits the low pressure systems to move across Victoria. They are accompanied by strong westerly winds and cool, wet conditions (Figure 11). The day to day variation in weather conditions is caused by the continual movement of the highs from west to east across the Australian continent roughly once every ten days. When the highs are centred north of Victoria, warmer, calmer conditions prevail, whereas in between the highs, cooler cold fronts can penetrate the state, particularly in winter. Also in winter, the westerly winds are more dominant and are strongest along the southern coastline.

2.3.2   Sea breeze systems

Associated with the highs and the warmer weather they bring, are the coastal **sea and land breezes**. The sea

breeze results from the daily warming of the coastal land surface. The air above the land warms up and begins to rise. As the land gets hotter and more air rises, it is replaced by air moving in from above the adjacent, but relatively cooler, coastal ocean waters. Consequently a local circulation cell is initiated, resulting in the cooler sea breezes replacing the hotter land air. At night and in the early morning hours, the reverse can occur as the now cooler land surface and relatively warmer ocean causes the air over the ocean to rise, which is in turn replaced by the cooler land breeze. This produces a light offshore breeze, most commonly on clear mornings.

Around the Victorian coast the direction of the sea breeze is predominantly southerly, but may range from south-east to south-west depending on the orientation of the coastline. This is because of the earth's rotation, which produces what is called the Coriolis effect. In the southern hemisphere, it causes all large scale air and water movement to be deflected to the left as they move. Hence the onshore breeze is deflected to the left. This means that on south-east facing coasts it arrives from the east, on south facing coasts from the south-east, while on south-west to west facing coasts it will arrive from

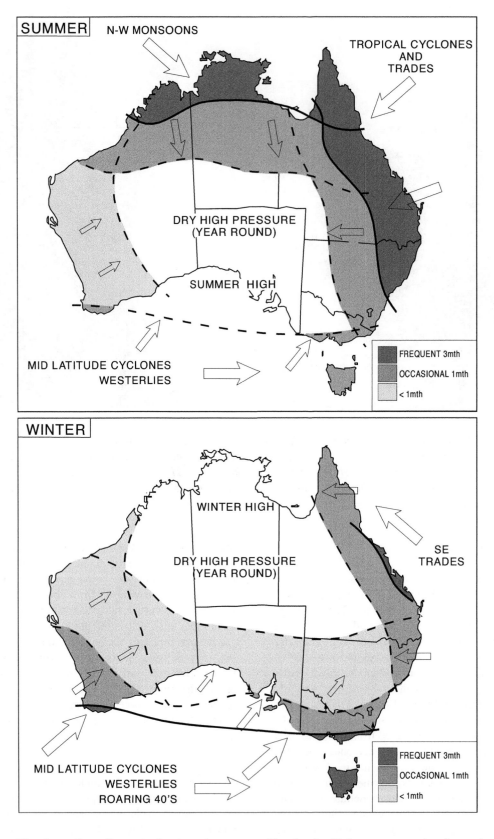

Figure 11. *The Australian climate is dominated year round by the dry high pressure anticyclones that sit over much of Australia. In summer, the equatorial low forms over northern Australia bringing the north-west monsoons (The Wet), while the dry subtropical high pressure systems dominate over southern Australia. In winter, the dry high moves north to dominate central and northern Australia, and permits mid latitude cyclones and their cold fronts to bring cool weather and rain across southern Australia. Consequently Victoria is seasonally influenced by warm to hot dry conditions in summer, while winter brings cooler, windier and wetter weather. This figure shows the source, season and extent of influence of the major air masses around the Australian coast.*

the south to south-west. The sea breezes are more frequent and intense in summer and consequently make an important contribution to the Victorian summer beach scene. Their full impact will be discussed in the section on waves.

2.3.3   Cyclonic systems

In addition to the high pressure systems, two types of low pressure cyclones can impact on the Victorian coast. The mid latitude cyclone is dominant, but in eastern Victoria the east coast cyclone can also have a major impact. These two are responsible for the strongest winds, heaviest rains, cooler weather and biggest seas.

2.3.3.1   East coast cyclones

East coast cyclones are a relatively poorly understood phenomena, yet they are the cyclones which wreak most havoc on the NSW and east Victorian coast. They can occur at any time of the year, but are more prevalent in early to mid-winter. While they occur each year, each lasting 4 to 5 days, they are highly variable in number, with years like 1977 having only one east coast cyclone,

while 1978 had seven such cyclones. The east coast cyclones generally form over the central coastal region of NSW and rapidly intensify, possibly reaching the strength of a tropical cyclone. They then meander in a south-west path across the Tasman Sea (Figure 12), while near the coast they produce the strongest winds, heaviest rainfall, and biggest sea and swell. They are responsible for episodically producing moderate to high easterly swell along the Gippsland coast.

2.3.3.2   Mid-latitude cyclones

Mid-latitude cyclones are part of the subpolar low pressure system which extends around the entire southern hemisphere, in a belt centred on 40° to 50° S latitude, the so called 'Roaring Forties'. It forms the southern boundary of the subtropical high pressure system. Like the high, it shifts with the seasons, moving closer to the southern Australian continent in winter and further south in summer. The lows or cyclones which are embedded in this belt are continually moving from west to east. On average, one passes south of Australia every three to four days.

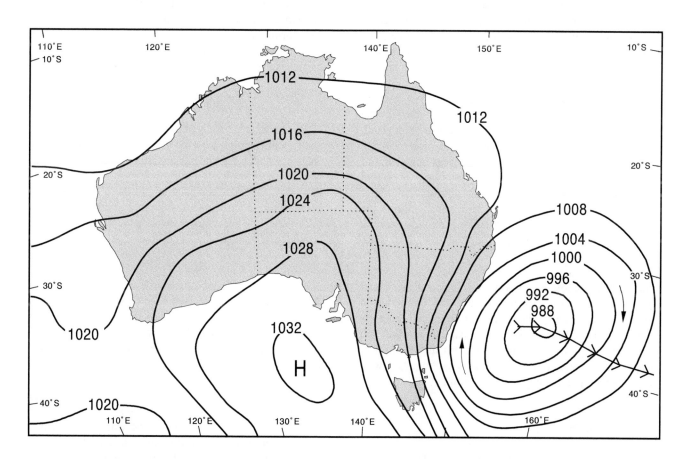

Figure 12.   An example of an East Coast Cyclone off the eastern Victorian coast. These cyclones can occur at any time of the year, usually originating over and just off the central NSW coast. They produce the strongest winds, heaviest rain, coastal flooding, and biggest waves and swell on the Gippsland coast. As they slowly meander off towards the south-west, the weather clears, however big swell continues to batter the coast. In this figure, the line tracks the cyclone with each arrow locating the centre of the cyclone on each day.

In summer, when they are kept south by the high pressure system over the southern half of the continent, they have limited impact on Victoria's weather. Only the top of the cyclones reach the coast, resulting in often strong north to north-westerly winds and hot winds ahead of a cooler change, but little rain (Figure 13). However, during winter when they move further north and when there are gaps between the highs, they can penetrate the continent with their effect being felt as far north as central Queensland. Their arrival is usually heralded by a cold front accompanied by strong to occasional gale force west through south winds and moderate to heavy rain. As the lows pass, the following highs take over as the winds lighten and tend more easterly.

2.3.4   Victoria's coastal climate

The **climate** of the Victorian coast is a composite of all the above weather systems averaged over a period of decades. The end result is a climate characterised by warm, dry summers with temperatures in the mid to high 20s and cool winters with temperatures in the low teens. Temperature extremes are produced by summer northerly winds bringing hot continental air to the coast, and by cold fronts pushing cooler subpolar air up along the coast, particularly in winter. At the coast the temperatures are also tempered by the westerly flow of cooler oceanic air, as well as the summer sea breezes.

The **rainfall** along the coast varies from a low of 500 mm around Geelong, to over 1 000 mm in the Otways, Wilsons Promontory and eastern Gippsland. Rainfall is highest to the south and on the higher coastal ranges, with pronounced rain shadow areas downwind of both the Otways and Wilsons Promontory. Western Victoria receives more of its rain during the winter months, while around Melbourne and along the Gippsland coast, the winter rainfall only slightly exceeds the summer total (Figure 14).

In summary, the Victorian coast has warm to hot summers and cool, wetter winters. Strong winds occur year round, but increase in intensity and frequency in winter when westerly gales are more frequent. There is considerable variation along the coast in rainfall and wind intensity, with the more southern and prominent Cape Otway and Wilsons Promontory, together with Gabo Island, receiving the fiercest weather.

2.4   Ocean Processes

Oceans occupy 71% of the world's surface. They therefore influence much of what happens on the remaining land surface. Nowhere is this more the case than at the coast, and nowhere are coasts more dynamic than on sandy beach systems. The oceans are the immediate source of most of the energy that drives coastal systems. Approximately half the energy arriving at the world's coastlines is derived from waves, much of the rest arrives as tides, with the remainder contributed by other forms of ocean and regional currents. In addition to supplying physical energy to build and reshape the coast, the ocean also influences beaches through its temperature, salinity and the rich biosphere that it hosts (Figure 2).

The Victorian coast is bordered by the southern Tasman Sea in the east, Bass Strait in the centre and the great Southern Ocean to the west. All three are swept by strong westerly winds, producing an easterly flow of water through Bass Strait and an energetic wave climate. However the size, fetch and orientation of each produce three distinct ocean-sea environments, which are described in Section 2.4.4.

There are seven types of ocean processes that impact the coast, namely: wind waves, tides, ocean currents, local wind driven currents, upwelling and downwelling, sea surface temperature and ocean biota (Table 5). Upwelling refers to the movement of deeper, cooler ocean waters to the coast and their arrival, or upwelling, at the shore, while downwelling refers to the movement of warmer, surface ocean water toward the coast and their turning down or downwelling at the shore. Ocean biota refers to all marine organisms at the shore and in the coastal ocean.

2.4.1   Ocean waves

There are many forms of waves in the ocean ranging from small ripples to wind waves, swell, tidal waves, tsunamis and long waves including standing and edge waves; lesser known but very important for beaches. In this book, the term 'waves' refers to the wind waves and swell, while other forms of waves are referred to by their full name. The major waves and their impact on beaches are discussed in the following sections.

2.4.1.1   Wind waves

Wind waves, or sea, are generated by wind blowing over the ocean. They are the waves that occur in what is called the area of wave generation; as such they are called '*sea*'. Five factors determine the size of wind waves:

- *Wind velocity* - wave height will increase exponentially as velocity increases;
- *Wind duration* - the longer the wind blows with a constant velocity and direction, the larger the waves will become until a fully arisen sea is reached; that is, the maximum size sea for a given velocity and duration;
- *Wind direction* will determine, together with the Coriolis force, the direction the waves will head ;

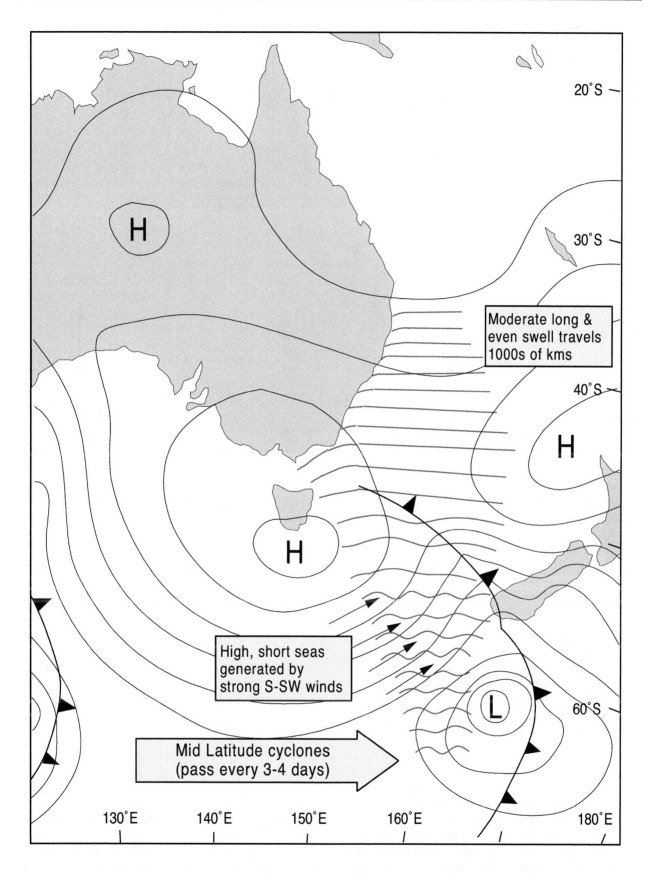

Figure 13.   An example of a Mid Latitude Cyclone and cold front passing south of Australia. As the cyclone traverses the Southern Ocean, its strong west to south-west winds, blowing over a long stretch of ocean, produce high seas. As the sea waves travel north of the cyclone, they become more regular swell. In summer when the lows are located further south of the continent, the swell can take one to two days to reach the coast. In winter the lows are closer to the coast and may even cross the coast, producing bigger seas and swell.

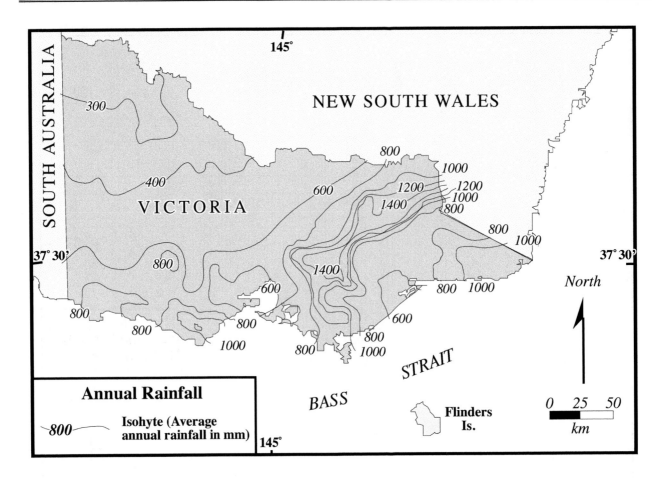

Figure 14.   Rainfall along the Victorian coast varies considerably in timing and amount. The winter cold fronts bring moderate rain to the entire coast, with precipitation increasing on the higher Otways and Wilsons Promontory. In addition, summer easterlies also bring rain to Gippsland.

Table 5.   *Major ocean processes impacting coast*

Process	Area of coastal impact	Type of impact
waves - sea & swell	shallow coast & beach	wave currents, breaking waves, wave bores
tides	shoreline & inlets	sea level, currents
ocean currents	continental shelf	currents
local wind currents	nearshore & shelf	currents
upwelling & downwelling	nearshore & shelf	currents & temperature
ocean temperature	entire coast	temperature
biota	entire coast	varies with environment

- *Fetch* - the sea or ocean surface is also important; the longer the stretch of water the wind can blow over, called the fetch, the larger the sea;
- *Water depth* is important as shallow seas will cause wave friction and possibly breaking, however this is not a problem in the deep ocean which averages 4.2 km in depth.

The **biggest seas** occur in those parts of the world where strong winds of a constant direction and long duration blow over a long stretch of ocean. The part of the globe where these factors occur most frequently is in the southern oceans between 40° and 50°S, where the Roaring Forties and their westerly gales prevail (Figure 13). Satellite sensing of all the world's oceans found that the world's biggest waves, averaging 6 m, and reaching up to 20 m, occurred most frequently in the Southern Ocean, south and west of Australia. For this reason, western Victoria receives some of the world's largest and most persistent waves.

2.4.1.2   Swell

Wind waves become swell when they leave the area of wave generation, by either travelling out of the area when the wind is blowing or when the wind stops blowing. Wind waves and swell are also called free waves or progressive waves (Figure 15). This means that once formed, they are free of their generating mechanism, the wind, and they can travel without it. They are also progressive, as they can move or progress unaided over great distances.

Once swell leaves the area of wave generation, the waves undergo a transformation that permits them to travel great distances with minimum loss of energy. Whereas in a sea the waves are highly variable in height and length, in swell the waves decrease in height, increase in length and become more uniform. As the speed of a wave is proportional to its length, they also increase in speed.

A quick and simple way to accurately calculate the speed of waves in deep water is to measure their period, that is, the time between two successive wave crests. The speed is equal to the wave period multiplied by 1.56 m. Therefore, a 10 second wave travels at 10 x 1.56 metres per second, which equals 15.6 m/s or 56 km per hour. In contrast, an 8 second wave travels at 36 km per hour and a 12 second wave at 67 km per hour. What this means is that as a swell propagates across the ocean, the longest and fastest waves arrive first.

Swell also travels in what surfers call 'sets' or more correctly *wave groups*, that is, groups of higher and lower waves. These wave groups are a source of long, low waves (the length of the groups) that become very important in the surf zone, as discussed in Section 2.4.2.

Swell and seas will move across the ocean surface through a process called orbital motion. This means the wave particles move in an orbital path as the wave crest and trough pass overhead. This is the reason the wave form moves while the water simply goes up and down, or more correctly around and around. However when waves enter water where the depth is less than 25% of their wave length (wave length equals wave period squared, multiplied by 1.56; for example, a 10 sec wave will be 10 x 10 sec x 1.56 = 156 m in length) they begin to transform into shallow water waves, a process that may ultimately end in wave breaking.

As waves move into shallower water and begin to interact with the seabed or feel the bottom, four processes take place, affecting the wave speed, length, height, energy and ultimately the type of wave breaking (Figure 16).

- *Wave speed* begins to decrease; the shallower the water, the slower the wave speed.

- Variable wave depth produces variable wave speed, causing the wave crest to travel at different speeds. At the coast this leads to *wave refraction*. This is a process which bends the wave crests, as that part of the wave moving faster in deeper water overtakes that part moving slower in shallower water.

- At the same time that the waves are refracting and slowing, they are interacting with the seabed, a process called *wave attenuation*. At the seafloor, some potential wave energy is transformed into kinetic energy as it interacts with the seabed, doing work such as moving sand. The loss of energy causes a decrease or attenuation in the overall energy and therefore height of the wave.

- Finally, as the water becomes increasingly shallow, the waves shoal, which causes them to slow further, decrease in length and increase in height. The speed and distance over which this takes place determines the type of *wave breaking*.

Wave types: sea and swell

Waves are generated by wind blowing over water surfaces.

Large waves require very strong winds, blowing for many hours to days, over long stretches of deep ocean.

Sea waves are still in the area of wave generation.

Swell are sea waves that have travelled out of the area of wave generation.

2.4.1.3   Wave breaking

There are three basic types of breaking wave:

- *Plunging or dumping waves,* which surfers know as a tubing wave, occur when shoaling takes place rapidly, such as when the waves hit a reef or a steep bar. As the trough slows, the following crest continues racing ahead and as it runs into the stalling trough, its forward momentum causes it to both move upward, increasing in height, and throw forward, producing a curl or tube.

- *Spilling breakers* on the other hand, occur when the seabed shoals gently, resulting in the wave breaking over a wide zone (Figure 17). As the wave slows and steepens, only the top of the crest breaks and spills down the face of the wave. Whereas a plunging wave may rise and break in a distance of a few metres, spilling waves may break over tens or even hundreds of metres.

Ocean Wave Generation, Transformation and Breaking

Wave type	Breaking	Shoaling	Swell	Sea
Environment	Shallow water - surf zone	Inner continental shelf	Deep water >> 100m	Deep water >>100m Long fetch = sea/ocean surface Wind velocity waves Wind duration waves Wind direction = wave direction
Distance travelled	100 m	1 km	100's to 1000's km	100's to 1000's km
Time required	seconds	minutes	hours to days	hours to days
Wave profile	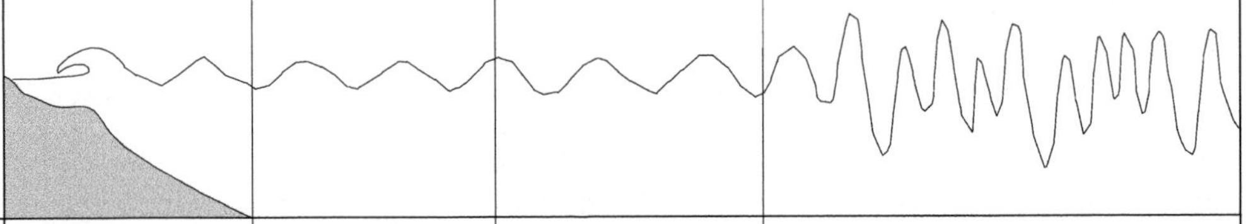			
Water depth	1.5 x water depth	< 100m	>> 100m	>> 100m
Wave character	wave breaks wave bore swash	higher shorter steeper same speed	regular lower longer flatter faster	variable height high short steep slow
Example; height (m) period (sec) length (m) speed (km/hr) distance travelled (km/day)	2.5 to 3m 12 sec 50 to 0m 15 to 0	2 to 2.5m 12 sec 220-50 m 66 to 15	2 m 12 sec 220 m 66 km/hr 1600 km/day	3 to 5 m 6 to 8 sec 50 to 100 m 33 to 45 km/hr 800 to 1100 km/day

Figure 15.   Ocean waves begin life as 'sea waves' produced by strong winds blowing over the surface of the deep ocean. When they leave the area of wave generation they transform into lower, longer, faster and more regular 'swell', which can travel for hundreds to thousands of kilometres.   As waves reach shallow water, they undergo a process called 'wave shoaling' which causes them to slow, shorten, steepen and finally break. This figure provides information on the characteristics of each type of wave.

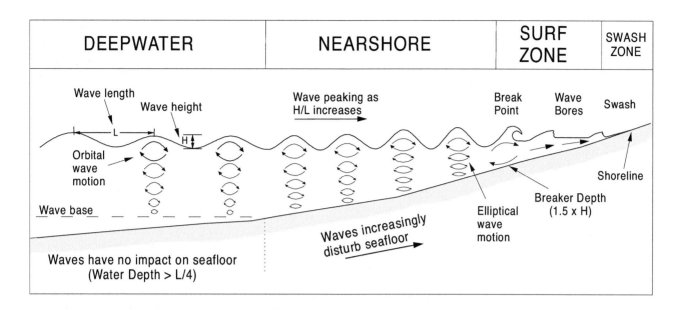

Figure 16.   A cross section of a beach showing what happens to deepwater sea waves and swell as they move into shallow coastal waters. On the surface they can be seen to slow, shorten, steepen, and perhaps increase in height. At the break point they break and move across the surf zone as wave bores (broken white water) and finally up the beach face as swash. Below the surface, the orbital wave currents are also interacting with the seabed, doing work by moving sand and ultimately building and forever changing the beach systems.

Figure 17.   Spilling waves breaking over the second bar at Seaford Pier. The 1 m high waves are peaking on the outer bar, spilling continuously across the second bar, reforming in the inner longshore trough before spilling again over the inner bar.

- The third type are *surging breakers* which occur when waves run up a steep slope without appearing to break. They transform from an unbroken wave to beach swash in the process of breaking. Such waves can be commonly observed on steeper beaches when waves are low, or after larger waves have broken offshore and reformed in the surf zone. They then may reach the shore as a lower wave, which finally surges up the beach face as swash.

Wave breaking and surf

- waves may break as plunging, spilling or surging waves
- as waves break they shorten, increase in height and slow down
- once broken, they become wave bores and finally swash

2.4.1.4 Broken waves

As waves break they are transformed from a progressive wave to a mass of water and foam called a *wave bore*. It is also called a wave of translation, as unlike the unbroken progressive wave, the water actually moves or translates shoreward. Boardriders, assisted by gravity, surf on the steep part of the breaking wave. Once the wave is broken, boards, bodies and whatever can be propelled shoreward with the leading edge of the wave bore.

2.4.2   Surf zone processes

Ocean waves can originate thousands of kilometres from a beach. They can travel as swell for days to reach their destination. However on reaching the coast, they can undergo wave shoaling and breaking in a matter of seconds. Once broken and heading for shore the wave has been transformed from a progressive wave containing potential energy, to a wave bore or wave of translation with kinetic energy, which can do work in the surf zone.

There are three major forms of wave energy in the surf zone - broken waves, surf zone currents and long waves (Table 6).

- *Broken waves* consist of wave bores and perhaps reformed or unbroken parts of waves. These move shoreward to finally run up and down the beach as swash and backwash. Some of the backwash may reflect out to sea as a reflected wave, albeit a much smaller version of the original source.

- *Surf zone currents* are generated by broken and unbroken waves, wave bores and swash. They include orbital wave motions under unbroken or reformed waves; shoreward moving wave bores; the up and down of the swash; the concentrated movement of the water along the beach as a longshore current; and where two converging longshore or rip feeder currents meet, as seaward moving rip currents.

- The third mechanism is a little more complex and relates to *long waves* produced by wave groups and, at times, other mechanisms. Long waves accompany sets of higher and lower waves. However, the long waves that accompany them are low (a few centimetres high), long (perhaps a few hundred metres) and invisible to the naked eye. As sets of higher and lower waves break, the accompanying long waves also go through a transformation. Like ocean waves they also become much shorter as they pile up in the surf zone, but unlike ocean waves, they do not break, but instead increase in height toward the shore. Their increase in height is due to what is called red shifting; a shift in wave energy to the red or lower frequency part of the wave energy field. These waves become very important in the surf zone, as their dimensions ultimately determine the number and spacing of bars and rips.

As waves arrive and break every few seconds, the energy they release at the break point diminishes shoreward, as the bores decrease in height toward the beach. The energy released from these bores goes into driving the surf zone currents and into building the long wave. The long wave crest attains its maximum height at the shoreline. Here it is visible to the naked eye in what is called *wave set-up* and *set-down*. These are low frequency, long wave motions, with periods in the order of several times the breaking wave period, that are

Table 6.   Waves in the surf zone

Wave form	Motion	Impact
Unbroken wave	orbital	stirs sea bed
Breaking wave	crest moves rapidly shoreward	wave collapses
Wave bore	all bore moves shoreward	shoreward moving turbulence
Surf zone currents	water flows shoreward, longshore and seaward (rips)	moves water and sediment in surf
Long waves	slow on-off shore	location of bars & rips

manifest as a periodic rise (set-up) and fall (set-down) in the water level at the shoreline. If you sit and watch the swash, particularly during high waves, you will notice that every minute or two the water level and maximum swash level rises then rapidly falls.

The height of wave set-up is a function of wave height, and also increases with larger waves and lower gradient beaches, to reach as much as one third to one half the height of the breaking waves. This means that if you have 1, 2, 5 and 10 m waves, the set-up could be as much as 0.3, 0.6, 1.5 and 3.0 m high, respectively. For this reason, wave set-up is a major hazard at the shoreline during high waves, particularly where the beach has a low gradient or slope.

Because the waves set up and set down in one place, the crest does not progress. They are therefore also referred to as a *standing wave*, one that stands in place with the crest simply moving up and down. These standing long waves are extremely important in the surf zone as they help determine the number and spacing of bars and rips. This interaction is discussed in Section 3 on beaches.

2.4.3   Wave measurements

While it is easy to see waves and to make an estimate of their height, period, length and direction, accurate measures of these statistics are more difficult. Yet we need to know just what type of waves are arriving at the coast, if we are to properly design for and manage the coast. Traditionally, wave measurements have been made by observers on ships and at lighthouses visually estimating wave height, length and direction. All of Victoria's lighthouses used to make such measurements, until they were progressively automated. Since the 1950s however, increasingly sophisticated electronic wave measuring devices have been developed and installed at some coastal locations. Regrettably, there are fewer electronic wave stations than there used to be manned lighthouses, so our record of Australian wave conditions has in fact diminished as lighthouses have been automated.

The present state-of-the-art on-site wave recording device is the Datawell *Waverider buoy* which was developed by the British Oceanographic Institute in the 1960s. It operates using an accelerometer housed in a water-tight buoy, about 50 cm in diameter. The buoy is chained to the sea floor, usually in about 80 m water depth. As the waves cause the buoy to rise and fall, the vertical displacement of the buoy is recorded by the accelerometer. This information is transmitted to a shore station and then by phone line to a central computer, where it is recorded.

The first Australian Waverider buoy was installed off the Gold Coast in 1968. Today, Queensland and New South Wales have a network of Waverider buoys stretching from Weipa to Eden; the most extensive in the world. Regrettably, Victoria has no permanent wave recording systems. Waveriders have been deployed periodically in Port Phillip Bay and waves are recorded on the Bass Strait oil rigs. However, for most of the Southern Ocean and Bass Strait coast, no reliable or long term wave data is available.

Today satellites, using laser beams, sense the height and direction of ocean waves on a global scale. This information is rapidly revolutionising our knowledge of ocean wave generation, dimensions and travel.

How to estimate wave height from shore

Wave height is the vertical distance from the trough to the crest of a wave. In making a visual measurement, it is easier with lower than high waves and easier if there are surfers in the water to give you a reference scale. If a surfer is standing up and the wave is waist height, then it's about 1 m; if as high as the surfer, it's about 1.5 m; if a little higher, then it's about 2m. For big waves just estimate how many surfers you could 'stack' on top of each other to get a general estimate, ie. two surfers about 4 m, three about 6 m and so on. Most surfers seem to underestimate wave height, either in ignorance or in some sort of false bravado.

2.4.4   Victorian wave climate

Wave climate refers to the seasonal variation in the size, character and direction of waves arriving at a location. Waves on the Victorian coast have four sources. Two are cyclonic; the mid-latitude and east coast cyclones; and two are associated with the high pressure systems, including the local sea breeze.

2.4.4.1   Wave sources

Mid-latitude cyclones move continuously across the Southern Ocean and generate most of the waves arriving on the western and west Bass Strait coasts of Victoria. Each month between five and nine cyclones cross south of Victoria, each producing waves which usually last a few days. During the year they generate waves on about 300 days. Because they have to travel up to 1 000 km before reaching the Victorian coast, they can take up to two days to reach the coast (Figure 13). In summer when the cyclones are located well south of Australia, the waves arrive as long, moderate to high swell (Figure 18). In winter (Figure 19) when the cyclones are closer to the coast, they can generate higher seas and swell, and accompanying onshore winds. On average these waves are 2 to 3 m in height, 12 to 14 seconds in period and arrive from the south-west. They are the dominant waves on the central-western Victorian coast and therefore are the waves that dominate the beaches in these areas.

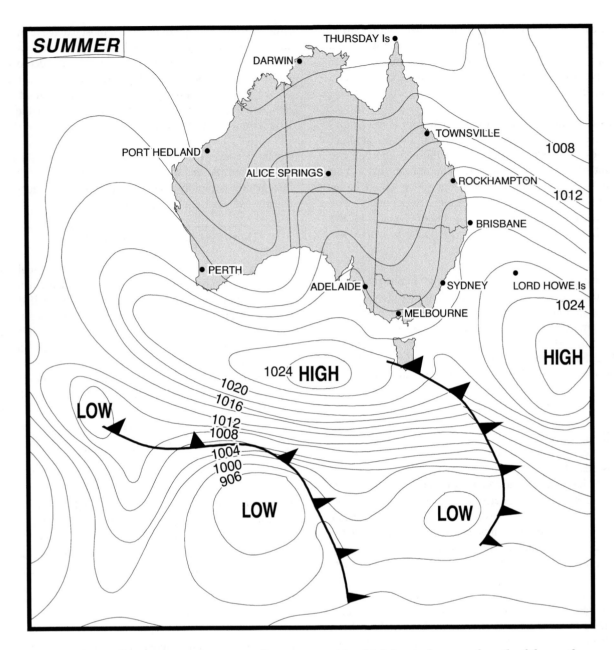

Figure 18.   A typical synoptic weather pattern for summer, with a high located over and south of the southern coast, and three low pressure mid-latitude cyclones located well south of the continent. These cyclones are responsible for the persistent swell that arrives on western Victorian and Bass Strait beaches.

East coast cyclones can occur in all months, with a preference for the early winter months of May, June and July. On average, three to four form each year and last for four to five days. Their waves only impact the Gippsland coast, where they arrive from the east to north-east (Figure 12) as sea and swell and average 2 to 3 m in the oil field, with wave periods between 10 and 12 seconds. These cyclones produce the biggest waves on the Gippsland coast and some of the biggest waves in the world.   In January 1978, a wave of 17 m was recorded on the Newcastle Waverider buoy.

2.4.4.2   Eastern Gippsland wave climate

The eastern Gippsland coast is dominated by westerly

wind waves that increase in strength toward Cape Howe, coupled with periodic east coast cyclones and high easterly swell (Figure 20).   The westerly winds blow offshore producing calms at the coast, with wave height increasing offshore. In the south they have no direct impact on the coast, however as the coast slowly trends to face more to the south, the westerly waves increase in size and, together with the wind, increasingly dominate the coast east of Cape Conran. Table 7 indicates the frequency of wave direction at Gabo Island. Even here, southerly and easterly waves dominate over the south-westerly waves. Table 8 gives a general indication of the height, period and direction of waves in eastern Gippsland.

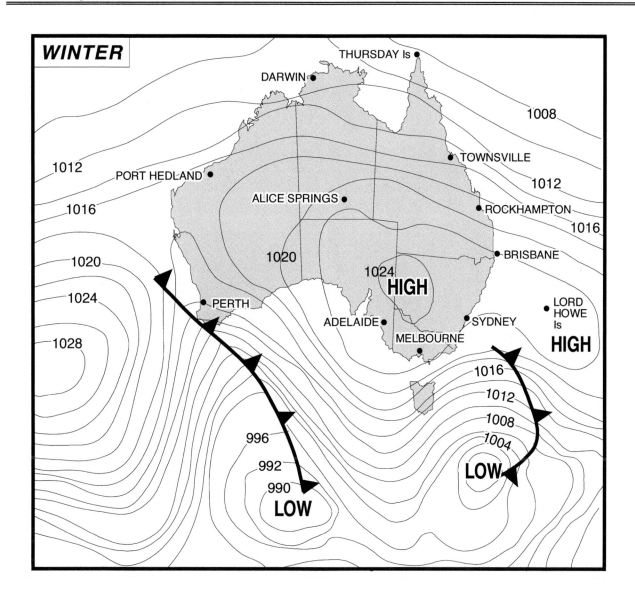

Figure 19.   During winter the high pressure system moves over the Australian continent, while low pressure cyclones track closer to the southern coast and commonly cross the coast. They generate higher waves at the coast and stronger accompanying onshore winds.

Table 7.   Swell direction frequency on the Victorian coast (%)

	NE	E	SE	S	SW	W	NW	N
Gabo Island	1	34	7	38	18	2	-	-
Wilsons Promontory	1	27	2	1	68	1	-	-
Port Phillip (sea)	-	-	-	30	19	8	41	2
Cape Northumberland	-	-	-	-	100	-	-	-

Table 8.   Typical wave characteristics on the Victorian coast

	Western Victoria	**Port Phillip**	**Eastern Victoria**
Average wave height (m)	1.5 - 2	0.5	1.5
Maximum wave height (m)	5	3	3
Average wave period (sec)	12-14	4-5	8-10
Direction	SW	SW-W-NW	SW, NE

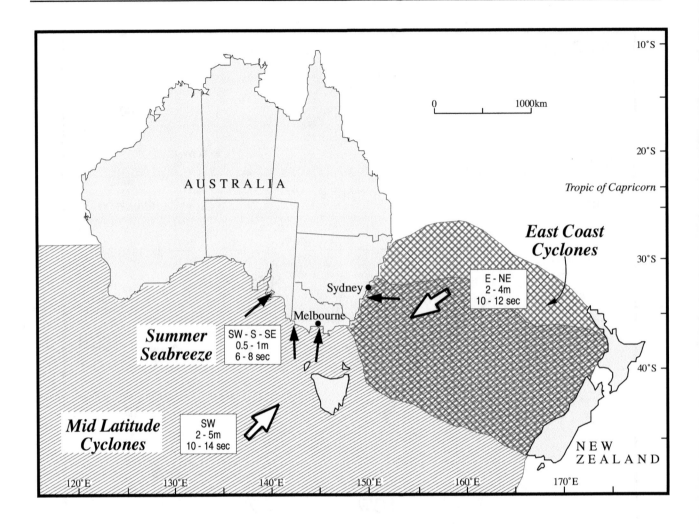

Figure 20.   Most waves reaching the southern Australian coast are generated in the shaded areas shown on this map. Across the entire southern coast, moderate to high swell waves generated by mid-latitude cyclones dominate year round. East of Wilsons Promontory, infrequent east coast cyclones can generate very large waves on the eastern Victorian and New South Wales coast. During the summer, locally generated sea breezes will also produce small wind chop along the coast. The boxes indicate the general direction and size of these waves on the Victorian coast.

2.4.4.3   Southern and western Victoria wave climate

West of Wilsons Promontory, the central and western Victorian coast is exposed to the persistent, moderate to high, south-west swell generated by the continuous stream of mid-latitude cyclones, which produce one of the world's highest energy coasts. Tables 7 and 8 indicate the dominance of the south-westerly waves, particularly in western Victoria. A summary of visual observations from Cape Northumberland, just across the border, indicates that south-west swell 0 to 2 m high occurs about 30% of the year; and 2 to 4 m high swell about 60 to 70%, with a slight peak during the winter months.

2.4.4.4   Port Phillip wave climate

Port Phillip has its own wave climate, as it receives no ocean waves. Rather, it is totally dependent on winds blowing over the bay. These tend to be dominated by westerly winds (north-west through to south-west) which, depending on the wind strength and fetch, can reach heights of 3 m and periods of 5 seconds. As a result, there is no background swell in Port Phillip; only calms when there is no wind, or offshore winds; low choppy waves less than 0.5 m under moderate winds; and only higher waves up to 3 m during occasional strong westerly winds. These tend to blow from the north-west in summer and south-west in winter.

The impact of these waves on the beaches of Victoria is examined in Section 3, on Beaches.

Freak waves, king waves, rogue waves, tidal waves

Freak waves do not exist.

All waves travel in wave groups or sets. A so-called 'freak', 'king' or 'rogue' wave is simply the largest wave or waves in a wave group.

However, unusually high waves are more likely in a sea than a swell. For this reason they are more likely to be encountered by yachtsmen than surfers or rock fishermen.

Tidal waves arrive on the Victorian coast twice each day. These are related to the predictable movement of the tides and not the damaging *tsunamis* with which they are commonly confused. Tidal waves are discussed in Section 2.4.5.

2.4.5   Tides

Tides are the periodic rise and fall in the ocean surface, due to the gravitational force of the moon and the sun acting on a rotating earth. The amount of force is a function of the size of each and their distance from the earth. While the sun is much larger than the moon, the moon exerts 2.16 times the force of the sun because it is much closer. Therefore, approximately 2/3 of our tidal forces are due to the moon and are called the lunar tides. The other 1/3 is due to the sun and these are called solar tides.

Because the rotation and orbit of the earth, and the orbit of the moon and sun are all rigidly fixed, the *lunar* tidal period, or time between successive high or low tides, is an exact 12.4 hours; while the *solar* period is 24.07 hours. Because these periods are out of phase, they progressively go in and out of phase. When they are in phase, their combined forces act together to produce higher than average tides, called *spring tides*. Fourteen days later they are completely out of phase and counteract each other to produce lower than average tides, called *neap tides*. The whole cycle takes 28 days and is called the lunar cycle over a lunar month.

The actual tide is in fact a wave, correctly called a *tidal wave,* not to be confused with tsunamis. They consist of a crest and trough, but are hundreds of kilometres in length. When the crest arrives it is called high tide, and the trough low tide. Ideally, the tidal waves would like to travel around the globe. However the varying size, shape and depth of the oceans, plus islands, continents, continental shelves and small seas complicate matters. The result is that the tide breaks down into a series of smaller tidal waves, that rotate around an area of zero tide called an amprodromic point. In the southern hemisphere, the Coriolis force causes the tidal waves to rotate in a clockwise direction, and anticlockwise in the northern hemisphere.

Tides in the deep ocean are zero at the amprodromic point and average less than 20 cm over much of the ocean. However, as the tidal waves move toward the coast and cross the relatively shallow continental shelves (< 150 m deep), they are amplified due to wave shoaling processes and increase in height (tide range) up to 1 - 3 m. In addition, certain large embayments can amplify the tide by a process of tidal wave resonance, which causes the tide to reach heights of several metres.

Tides are classified as being micro-tidal when having less than a 2 m range, meso-tidal when between 2 and 4 m, and macro-tidal when greater than 4 m.

2.4.5.1   Victorian tides

The Victorian coast has a micro- to meso-tide range, with a spring range that varies from 0.6 m inside Port Phillip to 2.3 m in neighbouring Western Port (Table 9).

Table 9.   Tidal characteristics of Victorian ports

Location	Spring tide range	Relative time of arrival 0 hr = Port Phillip Heads
Gabo Island	1.0 m	+5.5 (+ = earlier)
Waratah Bay	1.9	-0.05 (- = later)
Western Port	2.3	+0.6
Port Phillip Heads	**1.1**	**0 0 hr**
Geelong	0.8	-2.5
Melbourne	0.6	-3.0
Apollo Bay	1.3	-0.5
Port Campbell	0.6	+0.5
Warrnambool	0.5	+0.5
Portland	0.6	+0.5

The substantial variation in both the height of the tide around the coast and its time of arrival (Figure 21) is due to a number of factors:

- the tidal wave approaches from the east, reaching eastern Victoria first and Portland last;
- it is further slowed and refracted by Tasmania and enters Bass Strait from both sides;
- it is amplified by the shallow waters of Bass Strait;
- it is further amplified in Western Port (but not Port Phillip Bay);
- it is restricted by the narrow Rip entrance to Port Phillip Bay, which both lowers and delays the tide;
- it decreases in height west of Cape Otway.

Whereas all Victorian and southern Australian tides are less than 3 m in range; and most are less than 2 m; in northern Australia they generally exceed 2 - 3 m and reach heights of 9 m in Queensland's Broad Sound and

10 - 11 m in north-west Australia. The world's highest tide reaches 17 m in Canada's Bay of Fundy.

Note: Spring tides are also called 'king' tides and are highest around New Year and Christmas.

Spring or *king tides* are not responsible for beach erosion, unless they happen to coincide with large waves.

In Victoria, tides have no impact on ocean wave height; the arrival of both is purely coincidental. They will impact on breaking wave height, as they modify the depth of water immediately seaward of the beach.

Victorian tides: Gabo Island to Portland

Standard Tide Port: Port Phillip Heads

Range:	mean	1.1 m
	neap	0.7 m
	spring	1.3 m
	maximum	1.9 m

Range: The tidal range varies around the coast from 0.6 m at Melbourne to 2.3 m in Western Port (Figure 21a).

Time of tide: The tidal wave arrives from the east, first reaching Gabo Island and the Gippsland coast. It slows down on entering Bass Strait, while part of the wave moves around the bottom of Tasmania, reaching the western Victoria coast before Port Phillip. The two waves converge in Bass Strait, while on entering Port Phillip Bay it takes a further 3.5 hours to travel from The Rip to Port Melbourne, arriving 8 hours after it reached Gabo Island (Figure 21b).

2.4.6   Ocean Currents

Ocean currents refer to the continuous wind driven movement of the upper 100 to 200 m of the ocean. The major wind systems blowing over the ocean surface drive currents that move in large ocean gyres, spanning millions of square kilometres.

Victoria is impacted by ocean currents generated in both the Pacific and Southern Oceans (Figure 22). In Eastern Australia, the South Pacific ocean gyre moves in a giant anticlockwise circulation. Around Antarctica it is called the West Wind Drift, driven by the westerlies of the Roaring 40s. In the South-East Pacific, the drift is deflected equatorward by South America and becomes the northward moving Peru Current. It carries cool, polar water along the west coast of South America, before being deflected eastward in the sub-tropical latitudes (10 - 20° S) by the South-East Trade winds.

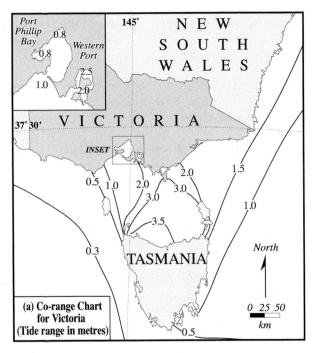

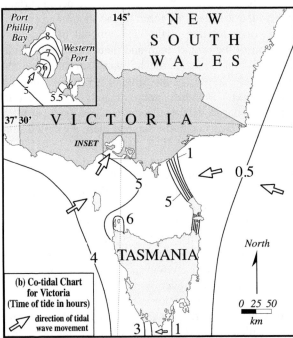

*Figure 21 a & b. Tidal co-range and co-tidal lines for the Victorian and Tasmanian coast. The **co-range** line (a) plots the height of the tidal wave or tide range along the coast, while the **co-tidal** line (b) plots the relative time of arrival of the wave. The tidal waves arrive from the east and move north and south of Tasmania. The northern part slows down on entering Bass Strait, while the south waves move around Tasmania to reach western Victoria and Port Phillip Bay, where they take another three hours to reach Melbourne. The tide is amplified in Bass Strait and Western Port, but remains low in eastern and western Victoria and in Port Phillip Bay.*

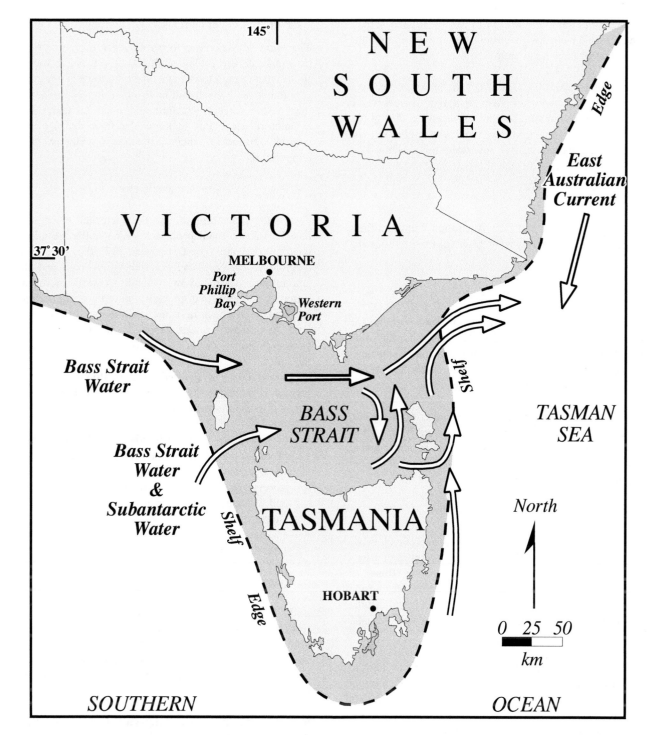

Figure 22.   The major ocean currents impacting the Victorian coast include the warm East Australian Current, cool but saline rich Bass Strait water, and cooler subantarctic waters.

The Trade Winds power the westerly moving Equatorial Drift, that travels for more than 10 000 km across the Pacific, warming as it goes. The drift pools in the Coral Sea as warm (25 -30° C) tropical water, and is deflected southward (toward Australia) by the land masses of New Guinea and northern Australia. As this warm water moves south, it is joined by equally warm water from the Great Barrier Reef lagoon, to form the warm *East Australian Current*. It flows south along the coast at speeds of 2 to 5 km per hour.

2.4.6.1   The East Australian Current

The East Australian Current parallels the Queensland and NSW continental shelf, occasionally producing currents that penetrate across the shelf to reach major promontories, such as Cape Byron. However, from Cape Byron south to Sugarloaf Point, the current at times veers seaward, sweeping into large meanders and counter-clockwise eddies. The eddies always have a warm centre (20-25° C); 4-5° warmer than the surrounding water and are up to 200 to 300 km across. Their counter-clockwise current can travel around the perimeter of the core at up to 6 to 8 km per hour.

What this all means is that along the NSW and eastern Victorian coast, the warm East Australian Current which flows as a continuous southerly current in the north, increasingly breaks into meanders and eddies to the south, with warmer eddies separated by cooler water. The eddies are also responsible for periodic and often rapid changes in water temperature, together with current direction and velocity.

In eastern Victoria, the East Australian Current brings warmer water with relatively high salinities, reflecting its tropical origin.

2.4.6.2   Bass Strait Water

The West Wind Drift that moves clockwise around the entire Southern Ocean is generally located too far south to directly impact the Victorian coastal waters. However, some of the cold water does reach the east and west areas of Bass Strait as sub-surface water.

At the surface, Bass Strait receives water flowing east from South Australia. Called Bass Strait Water, it is most dominant along the Victorian coast in winter and is characterised by low temperature (13° C) but higher salinities.

2.4.6.3   Local wind driven currents

In addition to the East Australian and Bass Strait currents, local winds will exert a force on the sea surface and produce local currents, particularly in times of strong winds. Along the Victorian coast, the strong westerly winds are the most prominent. They reinforce the Bass Strait current and generate westerly coastal currents along the Gippsland coast.

In general, the westerly winds tend to push water both to the east and toward the coast. This results in a slight increase in water level at the coast, which in turn causes the surface waters to sink at the coast and move down-slope across the inner continental shelf; a process called *downwelling*.

Northerly and easterly winds (particularly strong, hot summer northerlies) have the opposite effect. They push the surface waters seaward, resulting in a slight depression of the sea surface along the coast. To fill this depression, bottom water rises up from the inner continental shelf, replacing the warm surface water with cooler water. This process is called *upwelling*. It is most common in summer on the Gippsland and western Victorian coasts.

2.4.6.4   Other currents

There are several other forms of ocean currents driven by winds, density, tides, shelf waves and ocean waves. It is not uncommon to have several operating simultaneously. Each will have a measurable impact on the overall current structure and must be taken into account if one needs to know the finer detail of the coastal currents, their direction, velocity and temperature.

2.4.7   Sea surface temperature

The sea temperature along the Victorian coast is a product of two main processes. Firstly, the latitudinal location of the coast between 37°30' and 39°S determines the overhead position of the sun and the amount of solar radiation available to warm the ocean water. Secondly, two opposing currents bring warmer and cooler water to the coast. In the east, the warmer East Australian Current flows down the coast from warmer tropical waters and carries warmer water into the more temperate latitudes. Consequently the Gippsland coast is typified by a relatively warm sea surface (Figure 23). However, the actual temperature varies in response to four phenomena:

- the impact of El Niño Southern Oscillation, which produce pulses of relatively warmer and cooler water across the Pacific every few years;
- seasonally, in response to the movement of the sun and its warming influence on the East Australian Current;
- episodically, in response to local variation in the East Australian Current and its eddies;
- locally, the impact of wind driven upwelling (cooling) and downwelling (warming).

In the west, the Bass Strait and subantarctic water brings relatively cooler water to western Victoria and Bass Strait (Figure 23). This water flows through Bass Strait and at times cascades down the continental slope and under the warmer, more buoyant East Australian Current waters.

In Port Phillip, the temperature is influenced by the Bass Strait Water, plus the thermal warming and cooling of the shallow bay waters, warming to a maximum of 21° C in summer, while dropping to as low as 10° C in winter (Table 10).

Table 10. Victorian ocean temperatures

	Summer max. (°C)	Winter min. (°C)
Port Phillip Bay	21	10
Eastern Victoria	19 - 21	13 - 14
Western Victoria	18	13

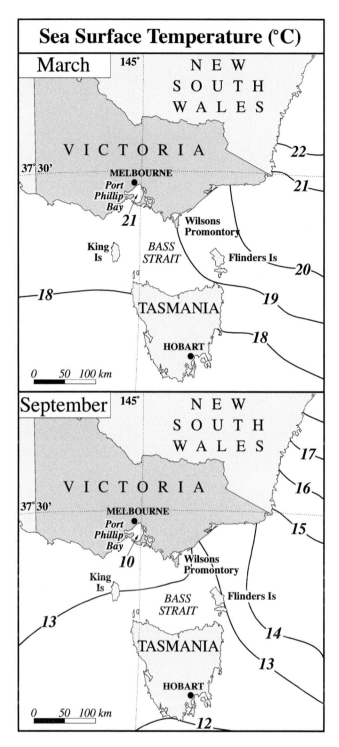

Figure 23.   Sea surface temperatures around the Victorian coast in the warmest month (March) and coolest (September).

2.4.8   Salinity

All oceans and seas contain dissolved salts derived from the erosion of land surfaces over hundreds of millions of years. Chlorine and sodium dominate and, together with several other minerals, account for the dissolved 'salt'.

The salts are well mixed and globally average 35 parts per thousand, increasing slightly into the dry sub-tropics and decreasing slightly in the wetter roaring 40s. Bass Strait water, derived from the arid south Australian coast and gulfs, is relatively high in salinity (but low in temperature), as are the warmer tropical waters of the East Australian Current.

2.5   Biological Processes

At first glance, sandy beaches resemble an arid, desert type landscape. While fish live in the sea and birds fly overhead, one might wonder what actually lives in and on the beach. Over the past decade our whole concept of beach ecology has changed dramatically, as scientists have looked at just what inhabits our beaches. As a result, we now know that beaches can be highly productive ecosystems, producing and exporting nutrients to the adjacent ocean; and contributing shell and algal fragments to beach sediments.

2.5.1   Beach ecology

The basis of the beach ecosystem are the microscopic diatoms that live in the water column (called phytoplankton) and microscopic meiofauna that live on and between sand grains; including bacteria, fungi, algae, protozoans and metazoans. Feeding on these are larger organisms that live in the sand, including meiofauna such as small worms and shrimp; and filter feeding benthos such as molluscs and worms. In the water column they are also preyed upon by zooplankton such as amphipods, isopods, mysids, prawns and crabs. At the top of the food chain are the fish and sea birds, and the occasional mammal such as dolphins, dugongs and even whales.

A number of hard bodied organisms also contribute their skeletal material to the beach in the form of sediment. When all organisms die, their internal and/or external skeletons can be washed onto beaches by waves. This material is composed of pure calcium carbonate. Common carbonate detritus found in Victorian sediments includes skeletal algae, barnacles, bivalve shells, bryozoa, small foraminifera shells, gastropods and echinoderm spikes. Along the Victorian coast there is a dramatic change in the contribution of carbonate material to beach sediments.   East of Wilsons Promontory, land-derived quartz sediments dominate. West of Wilsons Promontory, carbonate material derived from the inner continental shelf dominates increasingly to the west and right across southern Australia.   The percentage of carbonate beach sand commonly reaches between 60 and 100%.

2.5.2   Coastal biological hazards

A number of beach and ocean organisms pose particular problems to bathers, surfers and divers:

Blue-bottles (Portuguese Man O'Wars) and jelly fish are the most common cause of first aid on Victorian beaches. In summer, the warm tropical waters of the East Australian Current and onshore winds can deliver them in their thousands to the coast and beaches. The tentacles of the blue-bottle contain hundreds of minute, poisonous, pressure sensitive harpoons, which are fired and injected into the skin upon contact. They immediately appear as small blue spots, which soon swell into a red welt. The pain is felt instantly and usually lasts for about an hour, while the welts may remain for a few days. If stung by a blue bottle; swim to shore, remove any tentacles, seek first aid from the lifesavers and stay calm for the next hour. Recommended first aid is to pack ice on the stung area. Do not apply vinegar.

Sting rays are a common resident of the surf zone, where they usually lie hidden below a veneer of sand, as they feed on molluscs and crabs. If disturbed, they speed off in a cloud of sand. However, if you are unfortunate enough to stand directly on one, it might spear your leg with its sharp, serrated spear located below its tail. In extreme circumstances, the spear can lodge in the leg and require surgery to remove. Fortunately, sting ray spearings are rare.

Sharks are the most feared fish in the sea and in most years they attack one or two victims around the Australian coast. On the east coast, the frequency of attacks has been reduced since the introduction of meshing in 1937. However, as no Victorian beaches are meshed, caution should be used when swimming seaward of the breakers, particularly in remote areas.

3   BEACHES

Beaches throughout the world consist of wave deposited sediment and lie between the base of wave activity and the limit of wave run up, or swash. Only that part of the beach above the shoreline is clearly visible. Bars and channels are often present in the surf zone but are obscured below waves and surf. The shape of any surface is called its morphology, hence beach morphology refers to the shape of the beach and surf zone (Figure 24).

3.1   Beach Morphology

As all beaches are composed of sediment deposited by waves, beach morphology reflects the interaction between waves of a certain height, length and direction and the available sediment; whether it be sand, cobbles or boulders; together with any other object such as headlands, reefs and inlets.

Victorian beaches can be very generally divided into three groups: firstly, those on the open coast dominated by waves; secondly, open coast beaches where increasing tide range exerts considerable influence; and thirdly, the beaches in Port Phillip Bay. The wave dominated open coast beaches are exposed to persistent long period waves, and depending on the height of the waves and sand size, may be one of six types. In areas where wave height is low and tide range relatively large, such as parts of Western Port, the tide also plays an important role in determining the shape of the beach and can induce additional beach types. Port Phillip Bay beaches also receive waves, however they are considerably shorter, averaging less than 5 seconds in period. In addition, the waves only arrive when the wind is blowing onshore, and consequently during calms, light and/or offshore wind there are no waves, and the beach remains static. The following section briefly outlines the types of wave, tide and bay beaches that occur in Victoria.

Figure 24.   The nature and shape of beaches along the Victorian coast are products of both the inherited bedrock geology, which usually forms the boundary headlands bead rock reefs, and the waves and tides that have brought the sand onshore to build the ever changing beaches and surf zones. In this view of Wreck Beach, just west of Moonlight Head, the waves, sand and rocks all interact to form this series of crenulate beaches.

3.2   Wave dominated ocean and bay beaches

The simplest way to describe a beach is in two dimensions as shown in Figure 25. The beach consists of three zones: the subaerial beach, the surf zone and the nearshore zone.

3.2.1   Subaerial Beach

The *subaerial beach* is that part above sea level which is shaped by wave run up or swash. It starts at the shoreline and extends up the steeper swash zone or beach face. This may be backed by a flatter berm or cusps, which in turn may be backed by a runnel where the swash reaches at high tide. Behind the upper limit of spring tide and/or storm swash usually lies the beginning of the vegetated dunes. The subaerial, or dry, beach is

that part which most people go to and consider 'the beach'. However, the real beach is far more extensive, with the subaerial beach being literally the tip of the iceberg.

3.2.2   Surf Zone

The *surf zone* extends seaward of the shoreline and out to the area of wave breaking. This is one of the most dynamic places on earth. It is the zone where waves are continuously expending their energy and reshaping the seabed. It can be divided into two zones; firstly, the area of wave breaking, often underlain by a bar; and secondly, the area of wave translation where the wave bore moves toward the shoreline, transforming along the way into surf zone currents and at the shoreline, into swash.

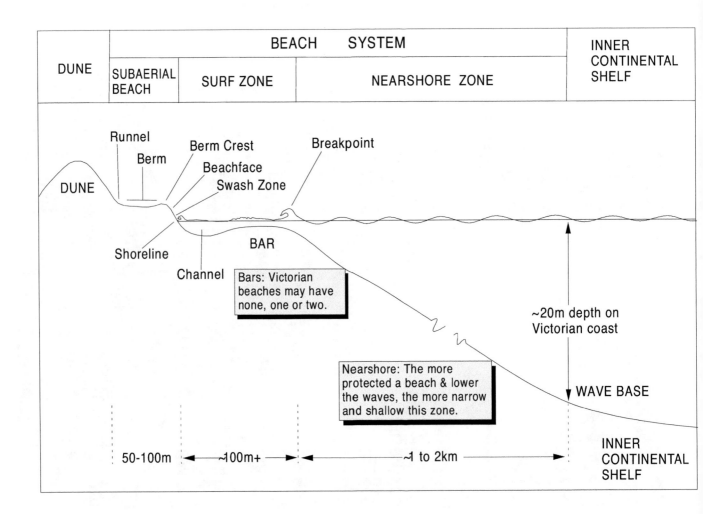

*Figure 25.   This cross section illustrates a typical Victorian beach system. It consists of the dry **subaerial beach** above the shoreline; the **surf zone** containing bars, troughs and breaking waves; and the **nearshore** zone which extends seaward of the breaker zone out to modal wave base. Wave base is the depth to which ocean waves can move beach sands. Seaward lies the **inner continental shelf.** The approximate width and depth of each zone are indicated.*

3.2.3   Nearshore Zone

The *nearshore zone* is the third and most extensive part of the beach, and the part most people never see. It extends seaward from the outer breakers, to the maximum depth at which average waves mobilise beach sediment and move it shoreward. The point where this begins is called the wave base, referring to the base of wave activity. On the Victorian coast it usually lies at a depth between 15 and 25 m and may extend 1, 2 or even 3 km out to sea, while in Port Phillip it extends to only 8 m depth on the higher energy, eastern shore beaches, and lies less than 1 km offshore. Hence the bulk of the beach lies hidden underwater.

3.2.4   Three dimensional beach morphology

In three dimensions, beaches become more complex. This is because most beaches are not uniform alongshore, but vary in a predictable manner.

Beaches vary longshore on two scales. On the Victorian coast, most beaches more than a kilometre long (and particularly those with prominent headlands) tend to curve toward their western end. This is because the shape or plan of the beach reflects the interaction of the waves with the seabed. On the Victorian coast, as the predominantly westerly waves reach the western headlands first, they begin to 'feel' the shallower seabed and slow down, while the part of the wave crest that is in deeper water races ahead. As a result, the waves arrive higher and straighter toward the northern end of beaches, but bend or refract toward the shore and decrease in height around southern headlands or reefs. The overall effect of the bending wave crests is to cause a spiral in the shape of the beach, so that the curvature increases to the more protected southern end. This bending of the wave crests is called *wave refraction*, while the loss of wave energy and height is called *wave attenuation*. The decrease in wave height down the beach results in lower breakers, a narrower and shallower surf zone, a narrower and often steeper swash zone and a lower and narrower subaerial beach.

The second scale of longshore beach variation relates to any and all undulations on the beach and in the surf zone, usually with scales in the order of tens of metres to as much as 500 m. Variable longshore forms produced at this scale include regular beach cusps located in the high tide swash zone and spaced between 15 - 40 m, and all variation in rips, bars, troughs and any undulations along the beach, usually with spacings of between 50 - 500 m. These features are associated with rip circulation and are known as rip channels, crescentic and transverse bars, and megacusp horns and embayments. Each of these is described in the section on Beach Types.

3.3   Beach Dynamics

Beach dynamics refers to the dynamic interaction between the breaking waves and currents and the sediments that compose the beach. This interaction not only builds all beaches, but, as waves change, causes continual changes in beach response and shape. Over a long period of time (10s - 1 000s of years) it can lead to a building out of the beach, called progradation or erosion, and perhaps ultimately complete removal of a beach. Over shorter periods of time (days to months) it causes all the changes in the shape of the beach and surf zone you usually see when you go to the beach.

Five factors determine the character of a beach. These are the size of the sediment, the height and length of the waves, the characteristics of any long waves present in the surf zone, and the tide range. The impact of each is briefly discussed below.

3.3.1   Beach sediment

The size of beach sediment determines its contribution to beach dynamics. Unlike in air where all objects fall at the same speed, sediment falls through water at a speed which is proportional to its size. Very fine sediment, like clay, will simply not sink but stay in suspension for days or weeks causing turbid, muddy water. Silt; sediments coarser than clay, but finer than sand; takes up to two hours to settle in a laboratory cylinder (Table 11).

Table 11. Sediment size and settling rates

Material	Size - diameter	Time to settle 1 m
clay	0.001 - 0.008 mm	hours to days
silt	0.008 - 0.063 mm	5 min to 2 hours
sand	0.063 - 2 mm	5 sec to 5 min
cobble	2 mm - 6.4 cm	1 to 5 sec
boulder	> 6.4 cm	< 1 sec

Sand which composes most beaches takes between a few seconds for coarse sand, to five minutes for very fine sand to settle through 1 m of water. For this reason it can be transported great distances by rivers to the sea, and then alongshore and onshore by waves, to build beaches.

Most Victorian beaches consist of sand. However, a few are composed of cobbles or even boulders. Cobbles and boulders require enormous energy to be lifted or moved, and then settle immediately. They are therefore very rarely moved, such as during extreme storms; and then only very slowly and over short distances. Consequently, beaches containing such coarse sediment always have a nearby source; usually an eroding cliff or bluff.

In the energetic breaking wave environment, anything as fine or finer than very fine sand stays in continual suspension and is flushed out of the beach system into deeper, quieter water. This is why ocean beaches never consist of silts or muds. Most beaches consist of sand because it can be transported in large quantities to the coast, while some beaches consist of cobbles and boulders if there is a nearby source of such material.

Depending on the nature of its sediment, each beach will inherit a number of characteristics. Firstly, the sediment will determine the mineralogy or composition of the beach; usually quartz sand or silica on eastern Victorian beaches, with significant quantities of carbonate (shell detritus etc.) west of Wilsons Promontory. Secondly, the size of the sediment will, along with waves, determine beach shape and dynamics. Fine sand produces a lower gradient (1 to 3°) swash zone, wider surf zone and potentially highly mobile sand. Medium to coarse sand beaches have a steeper gradient (5 to 10°), a narrower surf zone and less mobile sand. Cobble and boulder beaches are not only very steep, but have no surf zone and are almost immobile. Therefore, identical waves arriving at adjacent fine, medium and coarse sand beaches will interact to produce three distinctly different beaches.

Likewise, three beaches having identical sand size, but exposed to low, medium and high waves, will have three very different beach systems. Therefore, it is not just the sand or the waves, but the interaction of both, along with long waves and tides, that determine the nature of our beaches.

3.3.2   Wave energy - long term

Waves are the major source of energy that build and change beaches. Seaward of the breaker zone, waves interact with the sandy seabed to stir sand into suspension and, under normal conditions, slowly move it shoreward. The wave-by-wave stirring of sand across the nearshore zone and its shoreward transport has been responsible for the delivery of all the sand that presently composes the beaches and coastal sand dunes of Victoria.

Waves are therefore responsible for supplying the sand to build beaches. The higher the waves, the deeper the depth from which they can transport sand and the faster they can transport it. Consequently, the biggest volumes of sand that build our biggest beaches and dunes are all in areas of very high waves. In places, massive amounts up to 100 000 to 200 000 m³ of sand have been transported onshore for every metre of beach, amounting to 1 000 000 m³ for every few metres of beach. However, these same large waves can just as rapidly erode the very dynamic beaches they initially built.

Lower waves can only transport sand from shallow depths and at slower rates. Consequently, they build smaller barrier systems, usually delivering less than 10 000 m³ for every metre of beach. However, their beaches are less dynamic, more stable and are less likely to be eroded.

3.3.3   Wave energy - short term

Waves are not only responsible for the long term evolution of beaches, but also the continual changes and adjustments that take place as wave conditions vary from day to day. As noted above, a wave's first impact on a beach is felt as soon as the water is shallow enough for wave shoaling to commence, usually in less than 20 m water depth. As waves shoal and approach the break point, they undergo a rapid transformation which results in the waves becoming slower and shorter, but higher; and ultimately breaking, as the wave crest literally overtakes the trough.

As waves break, they release their kinetic energy; energy that may have been derived from the wind some hundreds or even thousands of kilometres away. This energy goes into turbulence, sound (the roar of the surf) and even heat. The turbulence stirs sand into suspension and carries it shoreward with the wave bore. The wave bore decreases in height shoreward, eventually turning into swash as it reaches the dry beach.

Breaking waves, wave bores and swash, together with unbroken and reformed waves, all contribute to a shoreward momentum in the surf zone. As these waves and currents move shoreward, much of their energy is transferred into two other major forms of surf zone currents, namely longshore, rip feeder and rip currents; and long waves and associated currents. These currents are responsible for returning the water seaward and play a major role in shaping the surf zone.

All these surf zone currents, which are described in Section 2.4.1 on Ocean Waves, are capable of moving sand. How much sand they move, for how long and how far and where, depends on the conditions operating at the time. These are discussed in Section 3.5 on Beach Change.

Needless to say, it is the variation in waves and sediment that produce the seemingly wide range of beaches present along the coast, ranging from the steep, narrow protected beaches, to the broad, low gradient beaches with wide surf zones, large rips and massive breakers. Yet every beach follows a predictable pattern of response, largely governed by its sediment size and prevailing wave height and length. The types of beaches that can be produced by waves and sand is discussed in the following section.

3.4   Beach types

Beach type refers to the prevailing nature of a beach, including the waves and currents, the extent of the nearshore zone, the width and shape of the surf zone including its bars and troughs, and the dry or subaerial beach. *Beach change* refers to the changing nature of a beach or beaches along a coast as conditions, waves, tides and sediment change.

The first comprehensive classification of beach types and change was developed by the Coastal Studies Unit (CSU), at the University of Sydney in the 1970s. This classification is now used internationally, wherever tide range is less than 2 m. In Australia this classification applies to most of the southern coast from Fraser Island in the east, around to Exmouth Peninsula in the west, but not the northern Australian coast where tides range from 3 -10 m; and parts of Victoria and South Australia where tide range increases above 2 m, as in Western Port Bay. In these areas, the larger tide range produces additional beach types, which have also recently been classified for the first time, again through research undertaken by the CSU.

3.4.1   Wave dominated beach types - ocean and bay beaches

The following section describes the types of beaches that are produced by waves in Victoria and in all parts of the world where tide range is low. The beaches consist of three types: reflective, intermediate or dissipative (Figures 26 & 27). Their overall occurrence in Victoria is listed in Table 12.

Wave dominated beaches occur both on the open coast and in Port Phillip Bay. However, there are two major differences between the two. The following descriptions apply to the open coast beaches. Those in Port Phillip Bay will differ in two ways. Firstly, the size of the bay causes waves to be limited in both their height and length. This directly impacts the dimensions of the beaches. Compared to the open coast, beaches of a similar type in the bay will have smaller dimensions, including the bar or bars closer to shore, shallower bars and more closely spaced rips. They will be a scaled down version of their open coast counterparts.

Secondly, because the waves in Port Phillip are totally dependent on wind blowing over the bay, the beaches tend to be maintained by periodic high waves associated with strong and gale force winds, and then remain near-dormant during calms and periods of low to moderate winds.

Table 12. Victorian beach types by number and length

Beach Type	No. of beaches		% of total beaches		Mean beach length (m)		Total beach length (km)		Proportion of length of beaches (%)		
	Ocean	Bay	Ocean	Bay	Ocean	Bay	Ocean	Bay	Ocean	Bay	Total
Wave dominated											
Dissipative	0 (2)	0 (2)	0	0	0	0	0	0	0	0	0
Longshore bar trough	0 (8)	0 (1)	0	0	0	0	0	0	0	0	0
Rhythmic bar beach	77	6	14	5	3589	603	276	4	34	2	28
Transverse bar rip	157	13	28	10	2058	2273	323	30	40	17	36
Low tide terrace	97	19	17	14	1130	777	110	15	14	8	13
Reflective	132	21	24	16	477	486	63	10	8	6	7
Wave + rock											
Reflective + platform	66	0	12	0	298	0	20	0	2	0	2
Wave + tide											
Reflective + LTT	31	1	5	1	621	200	19	0.2	2	0	2
Reflective + sand flat	0	72	0	55	0	1652	0	119	0	67	12
TOTAL	**560**	**132**	**100**	**100**	**1448**	**1343**	**811**	**177**	**100**	**100**	**100**

Beach state refers to inner bar only.   Parentheses indicate number of beaches with double bar morphology.

BEACH TYPE, HAZARDS AND HAZARD RATING

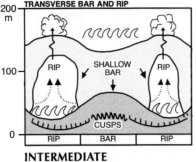

INTERMEDIATE
TRANSVERSE BAR AND RIP

CHARACTERISTICS
consists of attached bars, rip troughs and undulating beach, 1.0 - 1.5m breakers, distinct rip troughs separated by attached bars every 150 - 300 m

HAZARDS
pronounced changes in depth and current between bars and rips, safest bathing is on the bars

BEACH HAZARD RATING AND HINTS
5/10 (bathe on shallow bars adjacent to rips, however bathers can be washed off the bars into rips, inexperienced bathers may unknowingly enter rips)

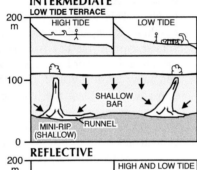

INTERMEDIATE
LOW TIDE TERRACE

CHARACTERISTICS
shallow bar or terrace often exposed at low tide, 0.5 - 1.0 m breakers

HAZARDS
safest bathing - safe at low tide, deeper water and weak rips at high tide

BEACH HAZARD RATING AND HINTS
3/10 (watch for plunging waves at low tide)

REFLECTIVE

CHARACTERISTICS
reflective - waves tend to reflect back off the beach, 0 - 1 m breakers, only occur on very low wave beaches and on harbour beaches

HAZARDS
safest bathing - safe apart from deep water close inshore and from shorebreak during higher waves, steep beach and abrupt drop off to deeper water can make access difficult for elderly and children

BEACH HAZARD RATING AND HINTS
2/10

WAVE HEIGHT (m)

PLEASE NOTE:
This model represents average wave conditions on these beach types in micro tidal (< 2 m tide range) regions of southern Australia (south Queensland, NSW, Victoria, Tasmania, South Australia and southern Western Australia).

BEACH SAFETY IS INFLUENCED BY:

HEADLANDS - rips usually occur and intensify adjacent to headlands, reefs and rocky outcrops.

OBLIQUE WAVES - stronger longshore currents, skewed and migratory waves.

HIGH TIDE - deeper water and in some cases stronger rips.

RISING SEAS - eroding bars, stronger currents, strong shifting rips, greater set up and set down.

HIGH TIDE AND RISING SEAS - more difficult to distinguish bars and troughs.

STRONG ONSHORE AND ALONGSHORE WINDS - reinforced downwind currents.

MEGARIPPLES - large migratory sand ripples common in rip troughs can produce unstable footing.

LOW TIDE - rips more visible but normally more intensified due to restricted channel.

CHANGING WAVE CONDITIONS - (rising, falling, change in direction or length) - produce a predictable change in beach topography and type; the reason why beaches are always changing.

*Figure 26.   A plan view of the **reflective**, **low tide terrace** and **transverse bar and rip** beach types. Note how as wave height increases between the reflective and transverse beaches, the surf zone and bar increase in width, rips form and increase in size, and the shoreline becomes crenulate. The physical characteristics and beach and surf hazards associated with each type are indicated.*

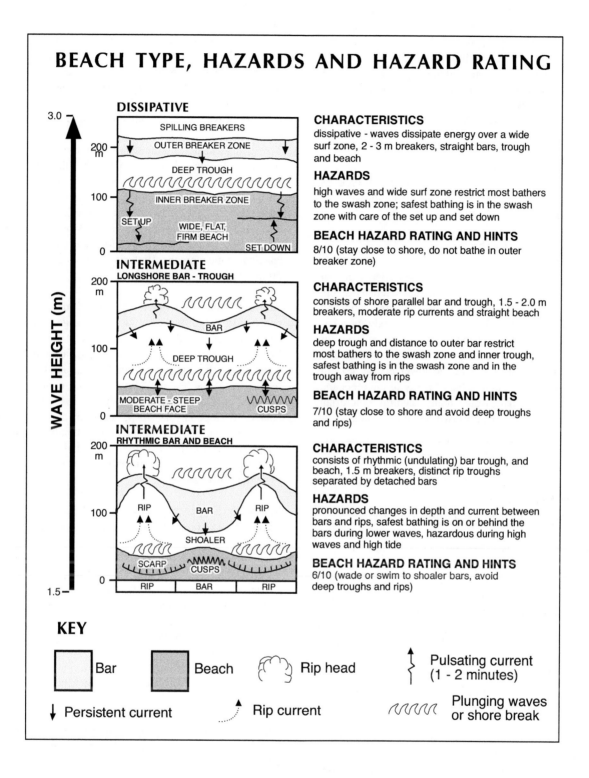

*Figure 27.   A plan view of the **rhythmic bar and beach, longshore bar and trough** and **dissipative** beach types. Note how as wave height increases between the rhythmic and dissipative beaches, the surf zone increases in width, rips initially increase and then are replaced by other currents, and the shoreline becomes straighter.   The physical characteristics and beach and surf hazards associated with each type are indicated.*

What this means is that while the bay beaches will assume the shape or morphology of the following wave dominated beach types, with bars, troughs and rip channels, they are only fully active during strong onshore winds and accompanying high waves. For much of the time the morphology is there, but the waves and currents are weak or absent. These low waves are not only too weak to drive the system, but also too weak to substantially modify it before the next period of high waves.

In the following sections each of the six beach types is described, together with examples and photographs of each beach type on the Victorian coast.

3.4.2   Reflective beaches

Reflective sandy beaches lie at the lower energy end of the beach spectrum. They are characterised by steep, narrow beaches usually composed of coarse sand and low waves. On the Victorian coast, sandy beaches require waves to be less than 0.5 m to be reflective. For this reason they are usually found inside the entrance to harbours and estuaries and at the lower energy end of some ocean beaches, and in lee of many of the reefs and rock platforms that front many central Victorian beaches. Figure 28 illustrates two reflective beaches.

Reflective ocean beaches:
Bells Beach, Inverloch, Henty Park, Cowes

Reflective bay beaches:
Black Rock, Weroona Bay, Ticonderoga Bay

In Victoria there are 251 reflective beaches (36%), however, many are located in lee of reefs and rock platforms and average only 298 m in length. Therefore, in terms of total beach length, they represent only 11% (112 km) of the Victorian sandy coast.

Reflective beaches are produced by coarse sand as well as low waves. Consequently, all beaches composed of gravel, cobble and boulders are always reflective, no matter what the wave height, as is the case with the energetic Bells Beach (Figure 28b).

Reflective beaches always have a steep, narrow beach and swash zone. Beach cusps are commonly present in the upper high tide swash zone. They also have no bar or surf zone as waves move unbroken to the shore, where they collapse or surge up the beach face. In this book they are referred to as 'steep barless beaches' and on figures by the abbreviation **'REF'**.

Their beach morphology is a product of four factors (Figure 29). Firstly, low waves will not break until they reach relatively shallow water (< 1 m); secondly, the coarse sand results in a steep gradient beach and

relatively deep nearshore zone (> 1 m); thirdly, because of the low waves and deep water, the waves do not break until they reach the base of the beach face; finally, because the waves break at the beach face, they must expend all their remaining energy over a very short distance. Much of the energy goes into the wave swash and backwash; the rest is reflected back out to sea as a reflected wave; hence the name reflective.

The strong swash, in conjunction with the usually coarse sediment, builds a high, steep beach face. The *cusps* which often reside on the upper part of the beach face are a product of sub-harmonic edge waves, meaning the waves have a period twice that of the incoming wave. The edge wave period and the beach slope determine the edge wave length, which in turn determines the cusp spacing, which on the Victorian coast can range from 20 to 40 m .

Another interesting phenomenon of most reflective beaches is that all those containing a range of sand sizes have what is called a *beach step*. The step is always located at the base of the beach face, around the low water mark. It consists of a band containing the coarsest material available, including rocks, cobbles, even boulders and often numerous shells. Because it is so coarse, its slope is very steep; hence the step-like shape. They are usually a few decimetres in height, reaching a maximum of perhaps a metre. Immediately seaward of the step, the sediments usually fine markedly, and assume a lower slope.

The reason for the step is twofold. The unbroken waves sweep the coarsest sediment continuously toward the beach and the step. The same waves break by surging over the step and up the beach face. However, the swash deposits the coarsest, heaviest material first, only carrying finer sand up onto the beach, then the backwash rolls any coarse material back down the beach. The coarsest material is therefore trapped at the base of the beach face by both the incoming wave and the swash and backwash.

3.4.2.1   Reflective platform beaches

On parts of the central Victorian coast, rock platforms and reefs front many beaches (Figure 30). Even on high energy exposed sections, the platforms will lower wave energy sufficiently so that a high tide reflective beach occurs behind the platform. These beaches have similar characteristics to the normal reflective beach, particularly at high tide. However, they also differ in two important ways. Firstly, the toe of the beach terminates at the usually horizontal rock platform; and secondly, they are stranded behind an often exposed rock platform at low tide.

Figure 28.   Reflective beaches are characterised by usually steep, often cusped beach faces fronted by relatively deep water, with no surf zone. Waves arrive unbroken and finally break by surging up the beach face. These photographs illustrate two contrasting reflective Victorian beaches. Refuge Cove (a) is a protected bay on the protected eastern side of Wilsons Promontory. Here the clear waters of the cove lap against a steep reflective beach. Other characteristics of this beach type are the relatively deep water right off the beach, no surf zone, with low waves surging against the beach face and, in this location, dense vegetation, including trees growing right behind the beach. The world famous Bells Beach (b) is renowned for its good surf, which breaks over the distant submerged rock reefs. The beach however, is composed of coarse sand and remains reflective even in big seas, when the surging waves, seen here, become a powerful shorebreak.

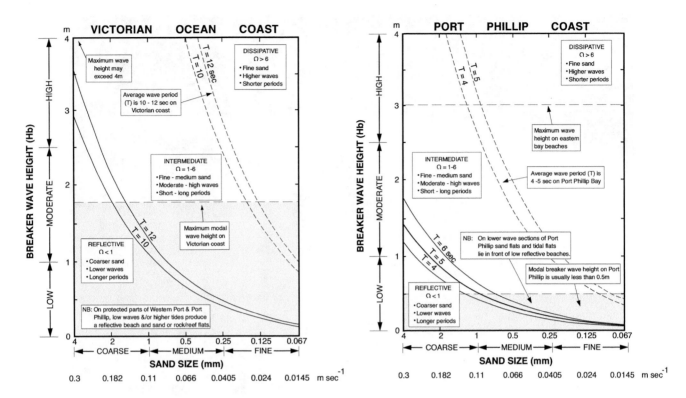

Figure 29.   A plot of breaker wave height versus sediment size, together with wave period, that can be used to determine approximate Ω and beach type.   To use the chart, determine the breaker wave height, period and grain size/fall velocity (mm or cm/sec).   Read off the wave height and grain size, then use the period to determine where the boundary of reflective/intermediate, or intermediate/dissipative beaches lie.   Ω = 1 along solid T lines, and 6 along dashed T lines.   Below the solid lines Ω < 1 and the beach is reflective; above the dashed lines Ω > 6 and the beach is dissipative; between the solid and dashed lines Ω is between 1 and 6 and the beach is intermediate.

Figure 30.   Sixty-six Victorian beaches are fronted by continuous intertidal rock platforms, that are exposed at low tide. In this view of the beach next to Eagle Nest (No. 148), the sandy beach is hemmed in between headlands, backed by a low foredune and vegetated bluffs and fronted by partly exposed shale and sandstone platforms.

Reflective plus platform beaches on ocean coast:
Eagle Nest, Cheviot, Twin Reefs

**Reflective plus platform beaches on bay coast:
Beaumaris**

3.4.2.2 Reflective beach hazards

The low waves and protected locations that characterise reflective beaches usually lead to relatively safe bathing locations. However, as with any water body, particularly one with waves and currents, there are hazards present that can produce problems for bathers and surfers.

Reflective beach hazards

- Steep beach face - may be a problem for toddlers, the elderly and disabled people.

- Relatively strong swash and backwash - may knock people off their feet.

- Step - causes a sudden drop off from shallow into deeper water.

- Deep water - absence of bar means deeper water close into shore, which can be a problem for non-swimmers and children.

- Surging waves and shorebreak - when waves exceed 0.5 m, they break increasingly heavily over the step and lower beach face. They can knock unsuspecting bathers over. If swimming seaward of the break, bathers may experience problems returning to shore through a high shorebreak.

- Most hazardous when waves exceed 1 m and shorebreak becomes increasingly powerful.

- Where fronted by a rock platform or reef, additional hazards are associated with the presence of the rock/reef.

- **Summary:** Relatively safe under low wave conditions, so long as you can swim. Watch children as deep water is close to shore. Hazardous shorebreak and strong surging swash under high waves (> 1 m).

3.4.3 Intermediate beaches

Intermediate beaches refer to those beach types that are intermediate between the lower energy reflective beaches and the highest energy beaches, called dissipative.

The most obvious characteristic of intermediate beaches is the presence of a surf zone with bars and rips. In

Victoria, 58% of all beaches are intermediate; in all, 401 beaches. In addition, they are usually longer than the reflective beaches and occupy 777 km or 79% of the Victorian sandy coast. This dominance is produced by a combination of fine to medium sand and waves between 0.5 and 2.5 m high, both of which are common on the Victorian coast.

However, between those beaches produced by 0.5 m waves and those by 2.5 m waves, there is quite a range in the shape and character of the beach. For this reason, intermediate beaches are classified into four beach states. The lowest energy one is called 'low tide terrace'; then as waves increase, the 'transverse bar and rip'; then the 'rhythmic bar and beach' and finally the 'longshore bar and trough'. Each of these beaches is described below.

3.4.4 Low tide terrace

Low tide terrace beaches are the lowest energy intermediate beach type. They occur on the Victorian coast where sand is fine to medium and wave height averages between 0.5 and 1 m. There are 97 modally low tide terrace beaches on the Victorian ocean coast, with another 19 in Port Phillip Bay, representing 17% of the beaches. These beaches have a total combined length of 125 km (13% of total beach length). They tend to occur toward the lower energy, more protected end of long beaches (Figure 31), in moderately protected embayments, and in more exposed locations where the sand is fine. They are referred to as low tide terrace beaches in this book and symbolised by **'LTT'** in figures and tables.

**Low tide terrace ocean beaches:
Cape Paterson, Woodside, Lorne**

**Low tide terrace bay beaches:
Hampton, Queenscliff**

Low tide terrace beaches are characterised by a moderately steep beach face, which is joined at the low tide level to an attached bar or terrace, hence the name - low tide terrace. The bar usually extends between 20 and 50 m seaward and continues alongshore, attached to the beach. It may be flat and featureless; have a slight central crest, called a ridge; and may be cut every several tens of metres by small shallow rip channels, called *mini rips*.

At high tide when waves are less than 1 m, they may pass right over the bar and not break until the beach face, behaving much like a reflective beach. However at spring low tide, the entire bar is usually exposed as a ridge or terrace running parallel to the beach. At this time, waves break by plunging heavily on the outer edge of the bar. At mid tide, waves usually break right across the shallow bar.

Figure 31.   Low tide terrace beaches consist of a steep, high tide beach, fronted by a wide, flat sand terrace or bar exposed at low tide. The beaches at Picnic Beach on Wilsons Promontory (a) and Cape Paterson (b) are typical of the 97 low tide terrace beaches around the Victorian coast.

Under typical mid-tide conditions, with waves breaking across the bar, a low 'friendly' surf zone is produced. Waves are less than 1 m and most water appears to head toward the shore. In fact it is also returned seaward, both by reflection off the beach face and via the mini rips, even if no rip channels are present. The rips however, are usually weak, ephemeral and shallow.

3.4.4.1   Low tide terrace hazards

Low tide terrace beaches are the safest of the intermediate beaches, because of their characteristically low waves and shallow terrace. However, changing wave and tide conditions do produce a number of hazards to bathers and surfers.

Low tide terrace beach hazards

- High tide - deep water close to shore; behaves like a reflective beach.

- Low tide - waves may plunge heavily on the outer edge of the bar, with deep water beyond. Take extreme care if body surfing or body boarding in plunging waves, as spinal injuries can result.

- Mid tide - more gently breaking waves and waist deep water, however weak mini rips return some water seaward.

- Higher waves - mini rips increase in strength and frequency and may be variable in location.

- Oblique waves - rips and currents are skewed and may shift along the beach, causing a longshore and seaward drag.

- Most hazardous at mid to high tide when waves exceed 1 m and are oblique to shore, such as during a strong summer north-east sea breeze.

- **Summary:** One of the safer beach types when waves are below 1 m high, at mid to high tide. Higher waves, however, generate dumping waves, strong currents and ephemeral rips; called 'side drag', 'side sweep' and 'flash' rips by lifesavers.

3.4.5   Transverse bar and rip

The *transverse bar and rip* beach type is the most common beach type on the Victorian ocean coast, with 157 beaches (28%) occupying 323 km (40%) of the total beach length. They occur on the Victorian coast where sand is fine to medium and waves average 1 to 1.5 m (Figure 32). As most Victorian beaches have such sand, and as waves average 1.6 m overall, they occur on all beaches when there is a slight reduction in deepwater

wave height. In this book they are referred to as 'transverse' beaches and are symbolised by the abbreviation **'TBR'** on figures and tables.

Transverse bar and rip ocean beaches:
Torquay, Bancoora, Ocean Grove, Apollo Bay

Transverse bar and rip bay beaches:
Half Moon Bay, South Melbourne, Elwood

Transverse beaches receive their name from the fact that as you walk along the beach, you will see bars transverse or perpendicular to, and attached to, the beach, separated by deeper rip channels and currents. The bars and rips are usually regularly spaced every 150 to 250 m, but can reach spacings of 500 m. Their surf zones range from 50 to 100 m wide.

Rip currents

Rip currents are a relatively narrow, seaward moving stream of water. They represent a mechanism for returning water back out to sea, that has been brought onshore by breaking waves. They originate close to shore as broken waves (wave bores) flow into longshore rip feeder troughs. This water moves along the beach as rip feeder currents. On normal beaches, two currents arriving from opposite directions usually converge, turn, and flow seaward through the surf zone. They commonly maintain a deeper rip feeder trough close to shore, and a deeper rip channel across the surf zone. As the confined rip current flows seaward of the outer breakers, it expands and may meander as a larger rip head. Its speed decreases and it will usually dissipate within a distance of two to three times the width of the surf zone. Rip currents will exist in some form on ALL beaches where there is a surf zone. Their spacing is usually regular and ranges from as close as 50 m on bay beaches, up to 500 m on high wave west coast beaches. Headlands and reefs in the surf will induce additional rips, called topographically controlled rips, and megarips can form during big seas.

Rip current spacing

- spacing approximately = surf zone width x 4
- on Victorian ocean coast, surf zone ranges from 30 to 250 m wide and rip spacing from 100 to 1 000 m; commonly 150 to 500 m apart
- in Port Phillip Bay, the rips on the inner bar average approximately 100 m spacing
- also a function of beach slope; the lower the slope (hence wider the surf zone), the wider the rip spacing

---

Figure 32. Surf zones dominated by alternating transverse bars, attached to the beach and separated by deep rip channels, occur on 157 Victorian ocean beaches. These three photographs show such bars and rips on Ninety Mile Beach (a), at Johanna Beach (b) and in the far west on a section of Discovery Bay (c). Wherever these beaches occur, they produce hazardous bathing conditions, with strong rip currents flowing seaward in the rip channels. They are easy to pick, owing to the highly crenulate nature of the beach, produced by the alternating rips and bars.

Transverse beaches differ from low tide terrace beaches in two obvious and important ways. Firstly, the beach and surf zone are discontinuous alongshore, being cut by prominent rips. Secondly, because of the alternation of shallow bars and deeper rip channels, there is a longshore variation in the way waves break across the surf zone. On the shallower bars waves break heavily, losing much of their energy. In the deeper rip channels they will break less, and possibly not at all, leaving more energy to be expended as a shorebreak at the beach face. Consequently, across the inner surf zone and at the beach face, there is an alternation of lower energy swash in lee of the bars, and higher energy swash/shorebreak in lee of the rips.

This longshore variation in wave breaking and swash causes the beach to be reworked, such that slight erosion usually occurs in lee of the rips, and slight deposition in lee of the bars. This results in a rhythmic shoreline, building a few metres seaward behind the attached bars as deposition occurs, and being scoured out and often scarped in lee of the rips. The rhythmic undulations are called megacusp horns (behind the bars) and embayments (behind the rips). Whenever you see such rhythmic features, which have a spacing identical to the bar and rips (150 to 500 m+), you know rips are present.

The transverse surf zone has a cellular circulation pattern. Waves tend to break more on the bars and move shoreward as wave bores. This water flows both directly into the adjacent rip channel and, closer to the beach, to the rip feeder channels located at the base of the beach.

The water in the rip feeder and rip channel then returns seaward in two stages. Firstly, water collects in the rip feeder channels and the inner part of the rip channel, building up an hydraulic head against the lower beach face. Once high enough, it pulses seaward as a relatively narrow accelerated flow; the rip. The water usually moves through the rip channel, out through the breakers, and seaward for a distance usually less than the width of the surf zone, that is, a few tens of metres.

The velocity of the rip currents varies tremendously. However, on a typical beach with waves less than 1.5 m, they peak at about 1 m per second, or 3.5 km per hour; about walking pace. However, under high waves they may double that speed. What this means is that under average conditions, a rip may carry someone out from the shore to beyond the breakers in 20 to 30 seconds. Even an Olympic swimmer would only be able to maintain their position when swimming against a strong rip.

Two other problems with rips and rip channels are their depth and their rippled seabed. They are usually 0.5 to 1 m deeper than the adjacent bar, reaching maximum depths of 3 m. Furthermore, the faster seaward flowing

water forms megaripples on the floor of the rip channel. These are sand ripples 1 to 2 m in length and 0.1 to 0.3 m high, that slowly migrate seaward. The effect of the depth and ripples on bathers is to provide both variable water depth in the rip channels and a soft sand bottom, compared to the more compact bar. As a result, it is more difficult to maintain your footing in the rip channel for three reasons: the water is deeper, the current is stronger and the channel floor is less compact. Also, someone standing on a megaripple crest who is suddenly washed or walks into the deeper trough, may think the bottom has 'collapsed'. This may be one source of the 'collapsing sand bar' myth; an event that can not, and does not, occur.

3.4.5.1   Transverse bar and rip beach hazards

Transverse beaches are one of the main reasons Victorian beaches have such good surf, as well as why there are over 1 000 rescues a year. The shallow bars tempt people into the surf, while lying to either side are the deeper, more treacherous rip channels and currents.

Transverse bar and rip beach hazards

- Bars - the centres of the attached bars are the safest place to bathe. They are shallow, furthest from the rip channels, and the wave bores move toward the shore.

- Rips - the rips are the cause of most rescues on the Victorian coast, so they are best avoided unless you are a very experienced surfer.

- Rip feeder channels - any channel close to shore has been carved by currents and is part of the surf zone circulation. It will be carrying water that is ultimately heading out to sea. Rip feeder channels usually run along behind and to the sides of the bar, adjacent to the base of the beach.

  In the rip embayment, the feeder currents converge and head out to sea. If you are not experienced, stay away from any channels, particularly if water is greater than waist depth and is moving.

- Children on floats must be very wary of feeder channels as they can drift from a seemingly calm, shallow, inner feeder channel located right next to the beach, rapidly out into a strong rip current.

- Breakers - waves will break more heavily on the bar at low tide, often as dangerous plunging waves or dumpers. In the rip embayment, the shorebreak will be stronger at high tide.

- Higher waves - when waves exceed 1 to 1.5 m, both wave breaking and rip currents will intensify.

- Oblique waves - these will skew both the bars and rips alongshore, and may make the rips more difficult to spot.

- Low tide - rip currents intensify at low tide, but are more confined to the rip channel.

- High tide - rip currents are weaker and may be partially replaced by a longshore current, even across the bar.

- **Summary:** This beach type is one of the main reasons why Victoria has surf lifesaving clubs and so many rescues. It is relatively safe on the bars during low to moderate waves, but beware, as many hazards, particularly rips, lurk for the young and inexperienced. Stay on the bar/s and well away from the rips and their side feeder currents.

3.4.6   Rhythmic bar and beach

The *rhythmic bar and beach* type is the highest energy beach type that commonly occurs on the Victorian ocean coast. In all, 77 beaches (14%) are this type. However, they tend to occur on the longer, more exposed beaches and consequently occupy 276 km (34%) of the total Victorian beach length. In this book they are referred to as 'rhythmic' beaches and are symbolised on figures and tables by the abbreviation **'RBB'**.

These energetic beaches require two primary ingredients for their formation: relatively fine sand and south-east exposure to the highest deepwater waves. On the Victorian coast, they occur where waves average at least 1.5 m and sand is more fine than medium (Figure 33). The energetic Portsea and Woolamai are good examples of rhythmic beaches.

> **Rhythmic bar and beach ocean beaches:**
> **Portsea, Woolamai, Jan Juc**
>
> **Rhythmic bar and beach bay beaches:**
> **Sandringham**

The rhythmic beach type has much in common with the transverse beach type, except for two important differences. Firstly, waves are usually higher owing to greater exposure to south-west waves. Secondly, the bars are detached from the beach. The transverse bars not only alternate with rips, but are also separated from the beach (megacusp horns) by continuous rip feeder channels. As the name implies, both the surf zone and bar are rhythmic as they weave between rips and bars, and the beach is rhythmic as it weaves between megacusp horns and embayments, as described for the transverse beach type.

Rhythmic beaches therefore consist of a rhythmic longshore bar that narrows where the rips cross the breakers, and in between broadens and approaches the shore. It does not, however, reach the shore, with a continuous rip feeder channel feeding the rips to either side of the bar. The beach face is usually rhythmic with a megacusp horn in lee of the detached bars and commonly scarped megacusp embayments behind the rips. The surf zone may be up to 100 to 250 m wide and the bar and rips are spaced every 250 to 1 000 m alongshore.

Waves break more heavily on the rhythmic bars and the water flows shoreward as a wave bore. The wave bore flows then move off the bar into the rip feeder channel. The water from both the wave bore and the swash piles up in the rip feeder channel and starts moving sideways toward the adjacent rip embayment, which may be several to more than 100 m alongshore. The feeder currents are weakest where they diverge behind the centre of the bar, but pick up in speed and intensity toward the rip. In addition, the rip feeder channels deepen toward the rip.

In the adjacent rip channels, waves break less or often not at all. They may move unbroken across the rip to finally break or surge up the steeper rip embayment swash zone. The strong swash often causes slight erosion of the beach face and cuts an erosion scarp.

In the rip embayment, the backwash returning down the beach face combines with flow from the adjacent rip feeder channels. This water builds up close to shore (called wave set-up), then pulses seaward as a strong, narrow rip current. The currents pulse every 30 to 90 seconds, depending on wave conditions. The rip current accelerates with each pulse and persists with lower velocities between pulses. Rip velocities are usually less than 1 m per second (3.5 km/h), but will increase up to 2 m per second under higher waves.

To identify this beach type, use the same procedure as for the transverse beaches, except that the bars are not attached to the shore. Usually a deeper trough, containing the rip feeder currents, separates the bar from the beach, with a current flowing longshore to the nearest rip embayment. Longshore beach rhythms are often very pronounced and readily observable.

3.4.6.1   Rhythmic bar and beach hazards

This is one of the most common beach types encountered on the Victorian coast and also one of the most hazardous. Most people are put off entering the surf by the deep longshore trough containing rips and their feeder currents. If you are bathing or surfing on a rhythmic beach, the following highlights some common hazards.

Figure 33.   Victoria's open coast has 77 beaches dominated by a rhythmic bar and beach. This view of Ninety Mile Beach shows such a system at Lake Tyers. The single bar is rhythmic alongshore, with bars underlying the areas of breaking waves. Troughs and rip channels occupy the deeper areas where waves are not breaking, with some of the channels cutting across the bar. The beach is also rhythmic, with the shoreline usually protruding seaward in lee of the bars, and cut away forming beach scarps in lee of the rips, as in Figure 32.

Rhythmic bar and beach hazards

- Bar - just to reach the bar requires crossing the rip feeder channel. This may be an easy wade at low tide or a difficult swim at high tide. Be very careful once water depth exceeds waist depth, particularly if a current is flowing. Also, as you reach the bar, water pouring off the bar may wash you back into the channel.

- The centre of the bar is relatively safe at low tide, but at high tide you run the risk off being washed into the rip feeder or rip channel.

- Rip feeder channel - depth varies with position and tide; both depth and velocity increase toward the rip.

- Rip - the rip channel is usually 2 to 3 m deep, with a continuous, but pulsating, rip current.

- High tide - deeper bar and channels, but weaker currents and rip.

- Low tide - waves break more heavily and may plunge dangerously; shallower bar and channels, but stronger currents and rip.

- Oblique waves - skew bar and rips alongshore.

- Higher waves - intensify wave breaking and strength of all currents.

- **Summary:** Caution is required by the young and inexperienced on rhythmic beaches, as the bar is separated from the beach by often deep channels and strong currents. Do not venture out to the bar unless you are between the red and yellow flags and you are a strong and experienced swimmer.

3.4.7   Longshore bar and trough

The *longshore bar and trough* beach type is rare on the Victorian coast, only occurring as outer bars on 8 beaches (1.4%), that occupy 76 km of the coast. They very rarely occur as the inner bar, where people bathe and swim. They are more common after periods of high seas, but with an average wave height on the coast of greater than 1.5 m, such beaches require full exposure to the south-west waves, coupled with fine sand. This beach type most commonly occurs as the outer bar on the longer, more exposed beaches of eastern Gippsland, including the Croajingolong region and Ninety Mile Beach (Figure 34). In this book they are called 'longshore' beaches and are symbolised by the abbreviation **'LBT'** in figures and tables.

Longshore bar and trough ocean beaches (outer bar): Ninety Mile, Port Fairy, Fairhaven

Longshore bar and trough bay beaches (outer bar): Mordialloc, Parkdale, Mentone

---

Longshore beaches are characterised by waves averaging 1.5 m or more, which break over a near continuous longshore bar located between 100 and 150 m seaward of the beach, with a 50 to 100 m wide, 2 to 3 m deep longshore trough separating it from the beach. The beach face, unlike the transverse and rhythmic beaches, is straight alongshore and usually has a lower gradient. While the bar, trough and beach may look straight and devoid of rips, the bar is usually crossed by rips every 250 to 500 m. The deep trough and the presence of less obvious rips make this a particularly hazardous bathing beach.

Higher waves tend to break continuously along the bar, with lower waves not breaking in the vicinity of the rip gaps. The wave bores cross the bar and enter the deeper longshore trough, where they quickly reform and continue shoreward as a lower wave to break or surge up the beach face. The water moving shoreward into the trough returns seaward using two mechanisms. Firstly, the water piles up along the beach face as wave set-up. As the water sets down, it moves both seaward as a standing wave, and longshore to feed the rip current. Secondly, as the converging feeder currents approach the rip, they accelerate, causing additional set-up in lee of the rip. As this set-up sets down, the rip pulses seaward. The dynamics that control these current flows are related to edge waves, described in the earlier Surf Zone Processes section.

Figure 34. The beach west of Red Bluff marks the beginning of the longest section of Ninety Mile Beach. Here it consists of a wide, gently curving beach, fronted by a single parallel longshore bar and trough. This type of beach can occur as a single bar, as in this photograph, but is more common as a second/outer bar. The trough is usually 2 to 3 m deep, and contains strong longshore and rip currents.

3.4.7.1 Longshore bar and trough hazards

Longshore beaches are hazardous. They require large waves to form and they have deep channels and troughs running continuously along the beach, which containlongshore moving feeder currents and periodic rip currents.

Longshore bar and trough beach hazards

- Bar - it's usually a long swim across a deep trough containing longshore and rip currents to reach the bar. At low tide, water pouring off the bar into the trough may make it difficult to get up onto the bar, or may wash you off the bar. Waves break more heavily on the bar at low tide.

- Trough - a wide, 2 to 3 m deep trough runs between the bar and beach and is occupied by longshore currents, which intensify toward the rips and seaward moving rip currents.

- High tide - deeper bar and troughs; weaker longshore currents and rips.

- Low tide - shallower channel, but often still greater than 2 m deep, and stronger longshore and rip currents; shallower bar with plunging waves.

- Higher waves - heavier waves breaking; stronger longshore and rip currents.

- Oblique waves - bias in longshore current in direction of waves; increase in velocity of current; plus skewed rip currents.

- **Summary**: Unless you are an experienced surfer stay between the flags, in the swash zone landward of the trough. As waves usually break first on the outer bar, they are lower at the beach face and can result in moderately safe conditions at the shore.

3.4.8 Dissipative

There are only two modally dissipative beaches in Victoria; in Bridgewater Bay and Discovery Bay (Figure 35). There are in fact very few in Australia, because to produce them on the open coast requires both long periods of waves greater than 2.5 m high, and fine sand. On the Victorian coast, the waves are not usually high or long enough. They do, however, occur on the above beaches because of their exposed location and fine beach sands. They also occur periodically following major periods of high waves, when some of the fine sand rhythmic and longshore beaches temporarily shift to a dissipative beach type. This also includes the multi-barred Frankston to Carrum Beach which occasionally goes dissipative during prolonged westerly gales. In this

book, they are referred to as 'dissipative' beaches and symbolised by the abbreviation **'D'** in figures and tables.

Dissipative ocean beaches:
Bridgewater and Discovery Bays

Dissipative bay beaches:
Frankston, Seaford, Carrum
(during prolonged westerly gale)

When *dissipative* beaches occur, the combination of high waves and fine sand ensures that they have wide surf zones and usually two to occasionally three shore parallel bars, separated by subdued troughs. The beach face is composed of fine sand and is always wide, low and firm; firm enough to support a 2WD drive car. The popular Goolwa Beach, south of Adelaide, is one of Australia's more accessible dissipative beaches, and on summer weekends it may be covered in cars.

Wave breaking begins as high spilling breakers on the outer bar, which reform to break again and perhaps again on the inner bar or bars. In this way they dissipate their energy across the surf zone, which may be up to 300 - 500 m wide. This is the origin of the name 'dissipative'.

In the process of continual breaking and re-breaking across the wide surf zone, the incident or regular waves decrease in height and may be indiscernible at the shoreline. The energy and water that commenced breaking in the original wave, is gradually transferred in crossing the surf zone to a lower frequency movement of water, called a standing wave. This is known as red shifting, where energy shifts to the lower frequency, or red end, of the energy spectrum.

At the shoreline, the standing wave is manifest as a periodic (every 60 to 120 seconds) rise in the water level (set-up), followed by a more rapid fall in the water level (set-down). As a rule of thumb, the height of the set-up is 0.3 to 0.5 times the height of the breaking waves (ie. 1 to 1.5 m for a 3 m wave). Because the wave is standing, the water moves with the wave in a seaward direction during set-down, with velocities between 1 to 2 m/sec closer to the seabed. As the water continues to set down, the next wave is building up in the inner surf zone, often to a substantial wave bore, 1 m+ high. The bore then flows across the low beach face and continues to rise, as more water moves shoreward and sets up. This process continuously repeats itself every one to two minutes.

Because of the fine sand and the large, low frequency standing wave, the beach is planed down to a wide, low gradient; with the high tide swash reaching to the back of the beach, often leaving no dry sand to sit on at high tide.

Figure 35.   Bridgewater Bay in the state's west is exposed to persistent high waves and is composed of fine sand. This combination produces a 300 m wide dissipative surf zone. In this view, 2 m high breakers are seen spilling over the outer bar, reforming and spilling again over the wider inner bar. Note the beach is wide and flat, as is typical of dissipative beaches.

Number of bars - none, one, two or more?

- Most ocean beaches in Victoria have one bar
- Some have none
- Some have two
- In Port Phillip Bay, the exposed Frankston - Carrum Beach has three bars (Figure 36).

- The number of bars depends on the slope of the beach, surf and nearshore zone and the wave height.
- Low waves and steep slopes (> 1.5°) result in no bar (reflective beaches).
- Moderate to high waves and moderate slopes (1 - 1.5°) produce one bar.
- High waves and low slopes (< 1°) produce two to three bars (the maximum on the ocean coast).
- High waves and short wave periods (< 5 sec) produce three bars in parts of Port Phillip.

- In Victoria, all longshore bars occur as the second or outer bar.
- In southern Australia, dissipative beaches always have two, and sometimes three, bars.

3.4.8.1   Dissipative beach hazards

The wide surf zones and high waves associated with dissipative beaches keep most bathers to the inner swash and surf zone. They are relatively safe close inshore, though not without some surprises; while the mid to outer surf zone is only for the fittest and most experienced surfers.

Dissipative beach hazards

- Most people do not venture far into dissipative surf zones as they are put off by their extremely wide surf and high outer breakers. However, if you do, this is what to watch out for:

- Outer surf zone - spilling breakers. Bigger sets break well seaward and catch surfers inside.

- Troughs - usually on/offshore currents, but chance of longshore and even rip currents, particularly under lower (< 1.5 m) wave conditions.

Figure 36.   Multi-bar beaches are a feature of the more exposed eastern Port Phillip Bay beaches.   This view of Seaford pier shows the rip dominated inner bar (transverse bar and rip), the straighter outer bar (longshore bar and trough), with a straight dissipative third bar (not visible) lying bayward of the pier.   The outer bars are only active during strong westerly winds and accompanying higher waves.

- Inner surf zone - watch for standing wave bores that can knock you over, fortunately shoreward. Set-down produces an often strong seaward flow, particularly closer to the sea bed, which may also drag children off their feet.

- Swash zone/beach face - this is where most bathers stay and where most get into trouble, owing to the set-up and set-down. Be aware that water level will vary considerably between set-up and down, and currents will reverse from onshore to offshore. At best you will be knocked over by the incoming bore, at worst you might be dragged seaward by the set-down. Children in particular are most at risk. Some young children, even babies in prams and parked cars, have been left on a seemingly safe part of the beach face, only to have a higher than usual set-up engulf them in water.

- **Summary:** Dissipative beaches are dangerous, however in Victoria they only occur in a few locations and when the seas are very big, so most people don't consider bathing, or at least not beyond the swash zone. Definitely for experienced surfers only out the back.

3.4.9   Wave and tide dominated beaches

In southern Australia, waves are the dominant process on most beaches. However, on parts of the ocean and bay shores, the occurrence of either very low waves and/or higher tide ranges increases the influence of tides to the extent that two additional beach types occur on the Victorian coast. The first type occurs in the higher tide ranges of Western Port Bay and parts of Port Phillip Bay; the second in the quieter reaches of Port Phillip Bay.

3.4.10   Reflective plus low tide terrace (with rips)

In areas where the tide range exceeds the average wave height by more than a factor of three, the reflective beaches are fronted by a low tide terrace, which is exposed at low tide. Rip channels may be present on the outer edge of the terrace. These occur in Western Port, on the northern shores of Phillip Island and from Sandy Point to Point Leo. In this area, not only is the tide range slightly higher at up to 2.3 m, but waves also average less than 0.5 m, making the tide up to four times higher than the waves.

What occurs on these beaches is that as the tide falls, the beach cannot drain all the water. As a result, an effluent line occurs on the beach at low tide, above which the surface of the beach is dry and below which it remains wet. The finer the sand, the higher on the beach the line will occur. Over time, the presence of the dry and wet areas impact the swash dynamics so that the upper dry beach is relatively steep and reflective, while the lower wet beach has a low gradient or slope and is seen as an exposed, but wet, low tide terrace at low tide (Figure 37).

Reflective + low tide terrace ocean beaches:
Point Leo, Port Campbell

Reflective + low tide terrace bay beaches:
Middle Brighton Baths

3.4.11   Reflective plus low tide terrace/sand flat

In Port Phillip Bay where waves are very low, the beaches tend to have a relatively steep reflective high tide beach, fronted by a broad, almost flat low tide terrace and/or sand flat. These beaches are a product of both the low waves which shape the beach and the tide range, which like the previous beach, determines (with sand size) the boundary between the dry and wet beach. This boundary forms the point of inflection, above which is the steeper, relatively high tide beach, and below which is the often broad, wet low tide terrace or

sand flat. These beaches occur in all the lower energy sections of Port Phillip (Figure 38), particularly on the western shores and in Corio Bay.

Reflective + sand flat bay beaches:
Rosebud, Sandridge, Williamstown, Altona

3.5   Beach change

As long as waves are breaking and the tide is rising and falling, beaches are always changing. The nature and extent of change, however, varies tremendously between beaches and over different periods of time. The changes may be short term in response to changing wave conditions, taking hours, days or weeks; or they may be long term in response to changes in sediment supply or climate, which can result in beach progradation or erosion.

3.5.1   Short term beach changes

As tide range is regular and persistent along the Victorian coast, most major beach changes are produced by the more variable and energetic waves.

To briefly explain how and why beaches change shape, erode and accrete, two sequences of beach change will be described. First, the accretionary sequence of beach change is presented, beginning with the high energy dissipative beaches, down through the six beach states to low wave reflective beaches. Then, beginning with a low wave reflective beach, the changes that occur as waves continue to increase in height and erode the beaches, will be described.

However, from the outset it must be understood that in nature most beaches never, and none regularly, move through the entire sequence from reflective to dissipative and vice versa.

Figure 37.   Summerland Bay is internationally famous for its penguin colony. The beach itself consists of a more reflective high tide beach, seen here with cusps (a) and a wide, exposed terrace at low tide (b). Such beaches are more common in the higher tide ranges around Western Port Bay.

Figure 38.   In the lower wave sections of Port Phillip Bay and parts of the open coast, the beaches consist of a steep reflective high tide beach fronted by tide dominated intertidal sand flats. In these photographs, (a) shows the beach at Clifton Springs at high tide, with shallow water evident over the tidal flats. In (b), an aerial view of the same location, the intricate tidal flats are more evident. Likewise in (c), the narrow high tide beach at Blairgowrie is fronted by a shallow, 200 m wide tidal flat dominated by very low ridges and runnels.

Rather, all have an average beach type and most oscillate up to one or two beach types either side, as wave conditions change.

3.5.2   Accretionary sequence - dissipative to reflective

A *fully dissipative* beach type, described on page 51, is produced by high waves (greater than 2.5 m) usually acting over a relatively fine sand beach for a period of weeks to months. The high waves acting across the wide surf zone produce shore parallel standing waves, which are in turn responsible for the location of the shore parallel bars and troughs (Figure 27).

If waves decrease to less than 2 m in height and stay low for a period of months, the following sequence of beach change would occur.

The standing wave dynamics of the dissipative surf zone would be replaced by (standing) edge waves. Edge waves stand perpendicular to shore and as they oscillate up and down, generate residual surf zone currents that cause water to move shoreward at regularly spaced intervals, while between these locations water moves seaward.

Once this circulation pattern is established, with the assistance of waves, it begins to move more sand shoreward under the shoreward currents, and some sand seaward under the offshore flows. The result is the formation of large scale rip circulation (cellular on- and off-shore currents); and the initiation of spatially segregated bars (onshore flows and sand) and rip channels (offshore flows and deeper bar). The beach remains straight (Figure 27). This is called the *longshore bar and trough* beach type described on page 49.

As the waves drop to 1.5 m, the bar continues to move shoreward. The most rapid bar movement is where the flow is shoreward; the least, or most retarded, is where the flow is seaward in the rips. Once the hydrodynamics has expressed itself in the morphology (bars and rips), a positive feed back occurs such that the morphology helps to reinforce and extenuate the dynamics. This is called morphodynamic coupling. What results is rapid movement of the bar toward the shore (possibly 1 to 2 m per day), assisted by shoreward moving wave bores and scouring of a deeper rip channel, occupied by intense seaward moving rip currents.

The bar and rip topography then causes a longshore variation in the degree of wave break and direction of wave approach across the surf zone. This results in a reworking of the beach face and the formation of megacusp horns in lee of the bars and megacusp embayments in lee of the rips, as described on page 48. At this stage, the beach is called *rhythmic bar and beach* after the crescentic bar and rips, and the rhythmic or undulating shoreline (Figure 27).

Continued low waves (< 1.5 m) will maintain the morphodynamic coupling and the bars will continue to move shoreward, separated by strong, well developed rips. Eventually, after a few days to a week or more, the bars will begin to reach the beach and attach to the base of the megacusp horns. At this stage, the beach becomes a *transverse bar and rip* beach type, as described on page 45. Both the bars and the rips are transverse or perpendicular to shore; the bars attached to the megacusp horns and the rips fronting the megacusp embayments (Figure 26). The megacusp horns and embayments ensure the beach face remains rhythmic.

As waves continue to fall to less than 1 m, the whole bar continues to move shoreward. In doing this, the surf zone decreases in width and the bars expand laterally at the expense of the rips. The rips and megacusp embayments stay in the same location, but begin to infill and shoal. As they decrease in size to narrow, shallow mini rips, separated by a shallow bar, the *low tide terrace* beach type is reached. It is characterised by a continuous attached bar and a relatively straight beach face with the mini rips eventually filling in and disappearing (Figure 26). This beach state is described in more detail, on page 43.

Finally, as waves decrease to less than 0.5 m, the bar or low tide terrace begins to be washed up onto the beach by the waves. This is usually manifest as a series of low, regularly spaced (20 to 40 m) beach cusps. As more sand moves onto the beach, the cusps increase in height and size. The cusp embayments may then completely infill to form a long, straight berm crest, backed by a slightly landward dipping berm; the rear of which is called a runnel, where swash collects at high tide. By this time, the bar has moved completely onto the beach face and relatively deep, barless water fronts the beach (Figure 26). This is the fully *reflective* beach type described on page 40.

3.5.3   Erosional sequence - reflective to dissipative

When an open coast *reflective* beach is exposed to an increase in wave height (greater than 0.5 m), the beach begins to rearrange itself to accommodate the larger waves. Essentially it does this using two complementary mechanisms. It lowers the beach gradient by eroding sand from the beach face and depositing it in the surf zone. Depending on wave height and state of the tide, this process may scarp or scour the beach face. The eroded sand shoals widen the surf zone and cause waves to break further seaward, thereby expending more energy crossing the wider surf zone. This process lowers wave height and energy at the shore, thereby protecting the beach from the higher waves.

As the steep reflective beach is initially eroded, the sand is deposited immediately seaward of, and attached to, the lower beach face, forming an *erosive low tide terrace*. During 1 m high wave erosive conditions, small ephemeral rips cross the terrace, helping to move more sand seaward.

As waves increase in height to greater than 1.5 m, the rips increase in size and intensity and move more sand seaward, thereby extending the bar further seaward and scouring deeper rip channels. While some parts of the bar are still attached to the shore, this is called the *erosional transverse bar and rip* beach type.

When the waves exceed 2 m, the rips not only grow in intensity, size and spacing, but they begin to link up, forming a continuous longshore trough, occupied by widely separated rips (300 to 600 m), joined by wide rip feeder channels lying behind the still rhythmic bar. This is the *erosional rhythmic bar and beach* type.

Increasing wave height above 2.5 m causes the rips to continue to increase in size and spacing and the bar to move still further seaward. Between the widely spaced rips (500 to 1 000 m), more and more water begins to return straight out to sea, instead of via the rips. This water is in the form of a standing wave. At this stage, the beach enters the *erosional longshore bar and trough* beach type. The bar lies 100 m or more seaward of the beach, with a continuous longshore trough, which has both widely spaced rips and on/offshore standing wave circulation.

Finally, if waves reach 3 m and maintain that height for several days, the *fully dissipative* beach type is reached. The bar is moved to its seaward-most limit, separated from the beach by a broad, moderately deep trough. The rips will increase in spacing to more than 2 to 3 km on long beaches and eventually disappear, being replaced by the far more efficient standing wave mechanism for seaward return of the water, described on page 22. On many of the longer, higher energy Victorian beaches, these conditions also produce a second bar seaward of the first; the first lying up to 100 m offshore, the second up to 250 m offshore.

On shorter embayed beaches, as along much of the central Victorian coast, or whenever beaches are less than about 4 km in length, they cannot reach this fully dissipative state. This is because the presence of headlands, reefs and/or the curvature of the beach will produce a longshore gradient in wave height. This means that rips will always persist. On these beaches, waves 3 m and above have produced some of the world's biggest rips called *megarips*. By definition they are a large scale, topographically controlled rip. What this means is that as the wave height increases, the location of the rip will be controlled by the wave height, period and direction as well as the topography of the beach system, particularly its bedrock. Beaches as long as 2 to 3 km will be drained by one huge megarip, fed by massive rip feeder currents, and flowing for up to 1 km seaward of the outer breakers.

Megarips are extremely hazardous and are responsible for the most severe erosion on any beach. On the Victorian coast, the most severe erosion usually occurs immediately behind such rips. This is the reason why on many embayed Victorian beaches, severe erosion is often localised and also why its location can vary from one storm to the next. It all depends on the location of the megarip. No one usually even considers bathing when the conditions that produce megarips are prevailing. However, some very experienced big wave surfers use them to gain access to the outer surf zone during big seas.

3.5.4 Beach change - long term

Long term beach change, particularly beach erosion, is not the subject of this book. However, as there is so much misunderstanding about the causes of beach erosion on the Victorian coast, it is worthwhile dispelling a few inaccuracies and identifying the main culprits.

Firstly, the most massive beach erosion in Victoria and right around the southern Australian coast, took place between 10 000 and 5 000 years ago. At that time we lost most of our larger, more energetic beaches. We know this because they left two major pieces of evidence. One of these sits up on top of many high exposed cliffs and headlands right round southern Australia, and are known as *cliff-top dunes* (Figure 39). The beaches that supplied and built these dunes have long since gone, but the dunes remain as evidence of their beach origin. Prominent cliff-top dunes occur

along the coast at Cape Bridgewater and Cape Nelson, on Cape Otway, along many parts of the Mornington Peninsula including Cape Schanck, on parts of western Wilsons Promontory and on all the major headlands between Cape Conran and Cape Howe.

Secondly, as these beaches were eroded, the sand had to go somewhere and as you would expect, it was carried longshore and immediately seaward and deposited in massive inner continental shelf sand bodies. These sand bodies were only discovered in the 1980s, initially off Sydney, but now along much of the NSW coast and off Gabo Island. They lie in water depths from 30 to 80 m and are found off most major exposed headlands on the NSW coast. It is expected that similar deposits should occur on parts of the central and western Victorian coast. They are the resting place of most of our original larger, more energetic beaches.

What this all means is that the beaches that are left are the smaller remnants of once larger systems; the more protected and sheltered leftovers. So as far as major beach erosion goes, it's all over.

In general, most of the 560 beaches remaining on the Victorian coast are in relatively good shape. This can be verified by the fact that they have already survived 6 500 years of wave attack. However, there is a range in the extent to which they can continue to survive. Basically, the main control is the sand budget of each beach or series of beaches. On the Victorian coast, most of the beaches are composed of sand that has been delivered to the continental shelf during lower sea levels, and is then reworked shoreward during each rise in sea level.

There was an abundance of sand on high energy sections of the coast as sea level neared its present level 7 000 year ago, as indicated by the cliff-top dunes. However, just as the high waves rapidly delivered this sand, they also removed it at a similar rate.

Beach sand budget

Sources:

- Ultimately from rivers and streams delivering sand to the continental shelf during lower sea levels
- Reworked onshore by waves during rising sea level
- Deposited as beaches and barriers and as tidal deltas in estuaries

Losses:

- Offshore to inner continental shelf sand bodies
- Longshore, particularly on long beaches
- Onshore to coastal dunes and into estuaries and river mouths
- Coastal dunes removed for sand aggregate

Figure 39.   Flaxman Hill Beach is bordered by limestone cliffs. In the centre of the beach, sand has climbed up over the backing 30 m high cliffs and moved more than 100 m inland as a cliff-top dune.

Elsewhere, in more moderate energy and protected sections of the coast, the sand from the inner shelf continued to arrive until 3 000 to 4 000 years ago on most beaches. These beaches experienced a seaward growth in their shorelines and some built 1, 2 and up to 3 km seaward. A small number of beaches have continued in this healthy state and appear to be still growing slowly seaward.

Most beaches however, have received all their sand supply and must make do with what they now have. Where no sand is being lost from the system, they remain stable. This is the case in most lower energy and protected, or deeply embayed beaches. The sand is trapped there by the combination of low waves and natural boundaries such as headlands.

However, whenever sand is lost by natural processes into coastal dunes, or estuaries, or longshore or

offshore; the beaches will erode. Along the Victorian coast, there is considerable longshore drift of sand from west to east. Where this drift is syphoned off naturally into dunes, estuaries or offshore, the sand budget decreases and erosion may result.

Finally, we can disturb the beach sand budget by either removing sand or interfering with its movement. Sand mining, even of dunes, removes the sand, while harbours and breakwaters will interfere with the longshore drift. This is the case at Portland, where the harbour construction has caused a major disruption to the longshore sand movement and erosion of the downdrift beach at Dutton Way. Also, in Port Phillip Bay, the major accumulation of sand in lee of Sandringham, Middle Brighton and St Kilda breakwaters, and the erosion of Hampton Beach, amongst others, are all evidence of our impact on natural coastal processes.

4   VICTORIAN BEACH USAGE AND HAZARDS

4.1   Beach usage and surf lifesaving

The Australian coast remained untouched until the arrival of the first Aboriginal Australians some tens of thousands of years ago. The coast and beaches they found, however, no longer exist. Probably crossing to Australia during one of the glacial low sea level periods, they not only reached a far larger, cooler, drier and windier continent, but one where the shoreline was some tens of metres below present sea level. Hence the coast they walked and fished lies out below sea level on the inner continental shelf.

The present Australian coast only obtained its position some 6 500 years ago, when sea level rose to approximately its present position. As it was rising, at about 1 m every 100 years, the locals no doubt followed its progress, by slowly moving inland and to higher ground. Therefore, we can assume that usage of the present Victorian beaches began as soon as they were formed some 6 000 years ago and continued in the traditional way until the 1830s.

As the long term Aboriginal inhabitants were removed and dispossessed, the European new-comers found the beaches to be of little use; except for running aground during storms, or loading goods and launching small boats in more sheltered locations. Fishermen however, did find (and often occupy) every anchorage suitable for small craft along the coast, within a few decades of settlement. However, the new settlers generally thought the beaches to be infertile and worthless, the waves a nuisance and hazard to all coastal sailors, and the dunes behind the beaches were just a waterless, barren wasteland.

As European settlement spread along the Victorian coast, the beaches were used as a more attractive route for foot and horse travel along the coast, compared to the heavily timbered and often rugged hinterland. The first great overland trek in Australia started in eastern Victoria. In 1790, the ship 'Sydney Cove' was wrecked off Flinders Island. Some of the crew set out for Sydney, only to be wrecked again on Ninety Mile Beach, in Bass Strait. They then walked along the east Victorian and south NSW coast to Port Hacking, where the surviving two crewmen were rescued by convict fishermen. The tradition of using the beaches for transport continued well into the 19th century, when many of the early coastal 'highways' followed beaches for parts of their routes.

The first profitable use of beaches was therefore for transport between suitable settlements and for loading of cargo in protected locations. However, as ports grew up along the coast and coastal shipping and roads improved, even these uses were overtaken by progress and once again beaches were left to the fishermen.

Australia celebrated its bicentenary in 1988 and the Surf Life Saving Association of Australia (now known as Surf Life Saving Australia) celebrates 100 years in 2006; however, for the first 114 years of European settlement, the beaches were unwanted, unpopulated and for a time it was even illegal to bathe in daylight hours.

Only in the early 1900s did beach and surf bathing become a popular recreational pastime. The popular beaches of the time were all either within walking distance of residential areas, or serviced by public transport. The public tended to surf at the nearest suitable beach, regardless of whether it was hazardous; as at Portsea; or relatively safe, as in Port Phillip Bay. The establishment of surf lifesaving clubs followed, as people got into difficulties and drownings occurred. In Port Phillip Bay, most of the lifesaving clubs were located within a short distance of a train station. Table 13 lists the date of establishment of Victorian surf lifesaving clubs.

Table 13. Date of establishment of Victorian surf lifesaving clubs

Year	Club	Year	Club
1910	Half Moon Bay	1952	Apollo Bay
1913	Hampton	1955	Seaspray
1914	St Kilda	1955	Point Leo
1917	Chelsea	1956	Lakes Entrance
1919	Mordialloc	1957	Fairhaven
1921	Mentone	1958	Wye River
1922	Williamstown	1959	Woolamai
1927	Altona	1960	Cape Paterson
1930	Warrnambool	1961	Venus Bay
1933	Bonbeach	1961	Barwon Heads
1937	Carrum	1961	Bancoora
1944	South Melbourne	1963	Jan Juc
1945	Torquay	1963	Kennett River
1947	Portland	1963	Waratah Beach
1947	Point Lonsdale	1963	Port Campbell
1947	Vic. State Centre	1966	Gunnamatta
1948	Lorne	1968	Woodside
1948	Ocean Grove	1986	Mount Martha
1949	Portsea	1998	Sorrento
1950	Port Fairy	1998	Inverloch
1952	Anglesea	2001	Mallacoota

4.1.1   Why the surf lifesaving clubs?

To understand the formation of the surf clubs and the broader Association is to understand two things. Firstly, in their rush to the surf, most beach-goers could not swim and had little or no knowledge about the surf and its dangers. Secondly, the coast of southern Australia, including Victoria, is as dangerous as it is inviting to the unprepared.

While people flocked to the beaches and the clubs were forming, all was still not happy with the law. It decreed that not only should both sexes wear neck-to-knee bathing costumes, but they should also cover their lower parts with a skirt. The neck-to-knees law was soon removed. However, four decades were to pass before men, after World War 2, were legally allowed to bathe topless; and another four decades before women were allowed the same rights in the 1980s.

The first surf lifesaving clubs and their Association had embarked upon the establishment and growth of an organisation that has become an integral part of Australian beach usage and culture, and through which it is so readily identified internationally. No country produces surfers and surf lifesavers like Australia.

Now, 100 years after the initial rush to the beaches and the foundation of the early surf lifesaving clubs, both beach usage and the 280 surf lifesaving clubs around the coast are accepted as an integral part of Australian beaches and beach life. However, as we begin the 21st century, beach usage is undergoing yet another surge as the Australian population and visiting tourists increasingly concentrate on the coast (Figure 40). This is resulting in more beaches being used, more of the time by more people, many of whom are unfamiliar with beaches and their dangers. In addition, most of the newly used beaches have no surf lifesaving clubs or patrols.

There is therefore a greater need than ever to maintain public safety on these beaches, a service that has been provided on patrolled beaches by volunteer lifesavers and professional lifeguards. This book is the result of a joint Surf Life Saving Australia, Surf Life Saving Victoria and the Coastal Studies Unit project that is addressing this problem. The book is designed to provide all information on each and every ocean beach in Victoria (including Port Phillip Bay) and its suitability for bathing and surfing. In this way, bathers may be forewarned of potential hazards before they get to the beach and consequently bathe more safely.

Figure 40.   A wide variety of people use beaches for a range of reasons and activities.   In Port Phillip Bay and Western Port Bay, besides bathing and surfing, many beaches are used for launching sailing boats.   Both ocean and bay beaches are also popular for fishing and windsurfing.

Oldest and Youngest Clubs

The oldest surf lifesaving club in Victoria is Half Moon Bay in Port Phillip Bay, formed in 1910 (Warrnambool was the first on the ocean coast, formed in 1930). The youngest is Mallacoota SLSC in Far East Gippsland, formed in 2001.

4.1.2   Surf Life Saving Australia

If you are interested in joining a surf lifesaving club or learning more about surf lifesaving, contact Surf Life Saving Australia, your state centre or nearest surf lifesaving club.

Surf Life Saving Australia
Surf House
Level 1, 1 Notts Ave
Bondi Beach NSW 2026
Phone (02) 9130 7370   info@slsa.asn.au
Fax   (02) 9130 8312   www.slsa.asn.au

Surf Life Saving Victoria
A W Walker House
Beaconsfield Parade
St Kilda VIC 3182
Phone (03) 9534 8201   slsv@slsv.asn.au
Fax   (03) 9534 0311   www.surflifesaver.com.au

Surf Life Saving New South Wales
PO Box 430
Narrabeen NSW 2101
Phone (02) 9984 7188   experts@surflifesaving.com.au
Fax   (02) 9984 7199   www.surflifesaving.com.au

Surf Life Saving Queensland
PO Box 3747
South Brisbane QLD 4101
Phone (07) 3846 8000   slsq@lifesaving.com.au
Fax   (07) 3846 8008   www.lifesaving.com.au

Surf Life Saving South Australia
PO Box 108
Torrensville SA 5031
Phone (08) 8354 6900   surflifesaving@surfrescue.com.au
Fax   (08) 8354 6999   www.surfrescue.com.au

Surf Life Saving Western Australia
PO Box 1048
Osborne Park Business Centre WA 6916
Phone (08) 9244 1222   slswa@slswa.asn.au
Fax   (08) 9244 1225   www.slswa.asn.au

Surf Life Saving Tasmania
GPO Box 1745
Hobart TAS 7001
Phone (03) 6272 7788   slst@slst.asn.au
Fax   (03) 6272 6500

Surf Life Saving Northern Territory
PO Box 43096
Casuarina NT 0811
Phone (08) 8941 3501   slsnt@austarnet.com.au
Fax   (08) 8981 3890

4.2   Beach safety

4.2.1   Beach hazards

Beach hazards are elements of the beach-surf environment that expose the public to danger or harm. Beach safety is the mitigation of such hazards and requires a combination of common sense, swimming ability and beach-surf knowledge and experience. The following section highlights the major physical hazards encountered in the surf, with hints on how to spot, avoid or escape from such hazards.

There are six major physical hazards on Victorian beaches:

1.   water depth (deep and shallow)

2.   breaking waves

3.   surf zone currents (particularly rip currents)

4.   strong winds

5.   rocks and headlands

6.   water temperature

In the surf zone, three or four hazards usually occur together, as indicated in Figure 41. In order to bathe safely, it is simply a matter of avoiding the above when they constitute a hazard to you, your friends or children.

4.2.1.1   Water depth

Any depth of water is potentially a hazard. Water that is too deep can drown the non-swimmer or inexperienced, while water too shallow can cause injury to the unwary.

Knee depth water can be a problem to a toddler or young child. Chest depth is hazardous to non-swimmers, as well as to panicking swimmers. In the presence of a current, it is only possible to wade against the current when water is below chest depth. Be very careful when water begins to exceed waist depth, particularly if younger or smaller people are present.

Shallow water is also a hazard when people are diving in the surf or catching waves. Both can result in spinal injury if people hit the sand head first.

Figure 41. Beach hazards include deep water, breaking waves, currents and rocks and reefs. In this series of photographs, the beach at Cathedral Rocks (a) shows all these hazards, with a strong rip current flowing out through the deep water in the centre of the beach. At Walkerville South (b) is a lower energy beach, which still has hazardous rocks that are submerged at high tide. The high, steep crumbling cliffs at Anglesea (c) are another type of hazard on the many cliff-backed beaches on the open coast and in Port Phillip Bay.

Water Depth

- Safest: knee deep - can walk against a strong rip current
- Moderately safe: waist deep - can maintain footing in rip current
- Unsafe: chest deep - unable to maintain footing in rip current

Remember: what is shallow and safe for an adult can be deep and distressing for a child.

Shallow water hazards

Spinal injuries are usually caused by people surfing dumping waves in shallow water, or by people running and diving into shallow water.
To avoid these:
- Always check the water depth prior to entering the surf.
- If unsure do not run and dive into the surf. Walk; do not run.
- Only dive under waves when water is at least waist deep.
- Always dive with your hands outstretched.
Also
- Do not surf dumping waves.
- Do not surf in shallow water.

4.2.1.2   Breaking waves

As waves break, they generate turbulence and currents which can knock people over, drag and hold them under water, and dump them on the sand bar or shore. If you do not know how to handle breaking waves (as most people don't), stay away from them; stay close to shore and on the inner part of the bar.

If you are knocked over by a wave, remember two points - it usually holds you under for only 2 to 3 seconds (though it may seem like much longer), therefore do not fight the wave; you will only waste energy. Rather, let the wave turbulence subside, then return to the surface. The best place to be when a big wave is breaking on you is as close to the seafloor as possible. Experienced surfers will actually 'hold on' by digging their hands into the sand as the wave passes overhead, then kick off the seabed to speed their return to the surface.

If a wave does happen to gather you up in its wave bore (white water) it will only take you towards the shore and will quickly weaken, so you can at least reach the surface after 2 to 3 seconds, usually leaving you in a safer location than where you started.

Breaking waves and wave energy

Surging waves - safe when low

- break by surging up beach face; usually less than 50 cm high
- can be a problem for children and elderly, who are more easily knocked over
- become increasingly strong and dangerous when over 50 cm high

Spilling waves - relatively safe

- break slowly over a wide surf zone
- are good body surfing waves

Plunging (dumping) waves - the most dangerous wave

- break by plunging heavily onto sand bar
- strong wave bore (white water) can knock bathers over
- very dangerous at low tide or where water is shallow
- waves can dump surfers onto sandbar, causing injury
- most spinal injuries are caused by people body surfing or body boarding on dumping waves
- to avoid spinal injury, do not surf dumping waves or in shallow water; if caught by a wave do not let it dump you head first; turn sideways and cover your head with your arms
- only very experienced surfers should attempt to catch plunging waves

---

Wave energy ≈ square of the wave height

wave energy represents the amount of power in a wave of a particular height

0.3 m wave	=	1 unit	wave energy/power
1.0 m wave	=	11 units	
1.5 m wave	=	25 units	
2.0 m wave	=	44 units	
2.5 m wave	=	70 units	
3.0 m wave	=	100 units	

Therefore, a 3 m wave is 10 times more powerful than a 1 m wave and 100 times more powerful than a 0.3 m wave.

4.2.1.3   Surf zone currents and rip currents

Surf zone currents and particularly *rip currents* are the biggest hazards to most bathers. They are the hardest for the inexperienced to spot and when bathers are caught by them, they can generate panic. See the following section on rips, for how to identify and escape from rip currents.

The problem with currents, particularly rip currents, is that they can move you unwillingly around the surf zone and ultimately seaward (Figure 42). In moving seaward they will also take you into deeper water and possibly toward and beyond the breakers. As mentioned earlier, currents are manageable when the water is below waist level, but as water depth reaches chest height they will sweep you off your feet.

Rip and surf zone current velocity

Breaking waves are travelling at 3 - 4 m/sec (10 - 15 km/hr)

Wave bores (white water) travel at 1 - 2 m/sec (3 - 7 km/hr)

Rip feeder and longshore currents travel at 0.5 - 1.5 m/sec (2 - 5 km/hr)

Rip currents under average wave conditions (< 1.5 m high) attain maximum velocities of 1.5 m/sec = 5.4 km/hr (Note: Olympic swimmers swim at 7 km/hr)

An average rip in a surf zone 50 m wide can carry you outside the breakers in as little as 30 seconds.

4.2.1.4   Strong winds

Strong winds can be a major hazard on exposed beaches.

Onshore winds will help pile more water onto the beach and increase the water level at the shore. They also produce a more irregular surf, which makes it more difficult to spot rips and currents.

Longshore winds will cause wind waves to run along the beach with accompanying longshore and rip currents also running along the beach. The waves and currents can very quickly sweep a person along the beach and into deeper rip channels and stronger currents.

Offshore winds tend to clean up the surf. However, if you are floating on a surfboard, bodyboard, ski or wind surfer, it also means it will blow the board offshore. In very strong offshore winds, it may be difficult or impossible for some people to paddle against this wind.

To accommodate these factors in the beach hazard grading of Victorian beaches, a wind hazard factor has been developed and is shown in Figure 43a, and on all individual beach maps.

Figure 42. There are approximately 720 rips operating at any one time along the Victorian coast. These three photographs show how to spot rip channels and currents. At Johanna Beach (a), waves are breaking over a shallow bar with deep water to either side. The water moving onshore with the breaking waves will move sideways against the shore (small arrows) and flow seaward as rip currents in the deeper water (large arrows). At Logans Beach (b) near Warrnambool, the walkway leads straight to a deep longshore trough (small arrows), with a deeper rip channel and current breaching the bar (large arrow). Near the Powlett River mouth (c), a rip leaves the beach next to some submerged rocks, then turns and runs out through the surf zone in the deeper rip channel (large arrows).

(a) Victorian ocean coast

Wind Hazards (Add to Beach Hazard Rating)				
Direction	Light	Mod	Strong	Gale
Longshore	0	1	3	4
Onshore	0	1	2	3
Offshore	0	1	1	2

(b) Port Phillip Bay

Onshore Wind Strength	Max Wave Height (m)
Light	0.3
Moderate	0.5
Strong	1.0
Gale	3.0

Figure 43. (a) Wind hazard rating for Victorian beaches, based on wind direction and strength. Winds blowing on and alongshore will intensify wave breaking and surf zone currents, with strong longshore winds capable of producing a strong longshore drag. Their impact on surf zone hazards and beach safety is indicated by the relevant hazard number, which should be added to the prevailing beach hazard rating. (b) Port Phillip Bay wave height based on strength of onshore wind. Maximum wave height in the bay is about 3 m. In addition, bay waves tend to be short and steep, compared to ocean waves of the same height.

In large enclosed bays such as Port Phillip Bay, the beaches are dependent on wind blowing across the bay to generate waves, as ocean swell is blocked by the narrow bay entrance. In Port Phillip Bay, calms result in no waves; light winds build low waves up to 30 cm high; moderate winds build waves up to 50 cm; strong winds build waves up to 1 m and more; and gale force winds build waves up to 3 m and more (Figure 43b). The winds must be blowing on or alongshore to produce waves at the shore; offshore winds will result in calms. However, what is an offshore wind on one side of the bay is an onshore wind on the other side. As Port Phillip's strongest winds come from the north-west, west and south-west, the highest waves occur along the west-facing, eastern shore beaches, with usually low waves year round on the east-facing, western shore beaches.

4.2.1.5   Rocks, reefs and headlands

Many Victorian beaches have some rocks, rock reefs and headlands. These pose more problems because they cause additional wave breaking; generate more (and stronger) rips; and have hard and often dangerous surfaces.

Rocks, reefs and headlands

- if there is surf against rocks or a headland, there will usually be a rip channel and current next to the rocks
- rocks and reefs can be hidden by waves and tides, so be wary
- do not dive or surf near rocks, as they generate greater wave turbulence and stronger currents
- rocks often have sharp, shelled organisms growing on their surface which can inflict additional injury
- if walking or fishing from rocks, be wary of being washed off by sets of larger waves

4.2.1.6   Safe bathing

The safest place to bathe is on a patrolled beach, between the red and yellow flags. If there are no flags then stay in the shallow inshore or toward the centre of attached bars, or close to shore, if deep. However, remember that rip feeder currents are strongest close to shore, and rip currents depart from the shore. The least safe is in or near rips and/or rocks, outside the flags, or on unpatrolled beaches and alone.

Remember these points:

- **Do** bathe on patrolled beaches.

- **Do** bathe between the red and yellow flags.

- **Do** observe and obey the instructions of the lifesavers or lifeguards.

- **Do** bathe close to shore, on the shallow inshore and/or on sand bars.

- **Always** have at least one experienced surf swimmer in your group.

- **Never** bathe alone.

- **Do** bathe under supervision if uncertain of conditions.

- **Do not** enter the surf if you cannot swim or are a poor swimmer.

- **Do not** bathe or swim in rips, troughs, channels or near rocks.

- **Do not** enter the surf, if you are at all unsure where to swim or where the rips are.

- **Be aware** of hypothermia caused by exposure to cold air and water, particularly on bare skin, and with small children. **Wind** will add to the chill factor.

Patrolled beaches

- bathe between the red and yellow flags

- obey the signs and instructions of the lifesavers or lifeguards

- still keep a check on all the above, as over 1 200 people are rescued from patrolled beaches in Victoria each year

---

Unpatrolled beaches

- always look first and check out the surf, bars and rips

- select the safest place to bathe, do not just go to the point in front of your car or the beach access track

- try to identify any rips that may be present

- select a spot behind a bar and away from rips and rocks

- on entering the surf, check for any side currents (these are rip feeder currents) or seaward moving currents (rip currents)

- if they are present, look for a safer spot

- it's generally safer to swim at low tide, if you avoid the rips

Children

- **never** let them out of your sight

- **advise** them on where to bathe and why

- **always** accompany young children or inexperienced children and teenagers into the surf

- **remember** they are small, inexperienced and usually poor swimmers and can get into difficulty at a much faster rate than an adult

---

4.2.2   Beach Hazard Rating

The *beach hazard rating* refers to the scaling of a beach according to the hazards associated with its beach type together with any local hazards. It ranges from a low, least hazardous rating of 1 to a high, most hazardous rating of 10.

Figure 44 illustrates each of the six beach types under average wave conditions for that type, together with their characteristics, hazards and beach hazard rating. This rating is called the average or *modal beach hazard rating*.

As the modal rating refers to average wave conditions, any change in wave height, length or direction will change the surf conditions and accompanying hazards. This will in turn change the beach's hazard rating. Figure 45 provides a method for calculating the actual or *prevailing beach hazard rating*. It assumes you know the beach type and can estimate the breaker wave height. With these two factors, the prevailing beach hazard rating can be read off the chart. This figure also describes the general hazards associated with each beach type as wave conditions rise and fall.

Two other factors must also be considered in calculating the prevailing beach hazard. They are additional local hazards such as rocks, reefs, inlets and headlands, and the presence of strong winds. Figure 43 indicates the impact of strong winds on beach conditions and associated wind generated hazards.

What this means for any particular beach is that as waves rise and fall, the surf conditions, hazards and hazard rating will change accordingly. The changes that occur due to a change in either wave height or beach type are plotted on Figure 45, which also indicates the accompanying change in beach hazards and hazard rating.

To illustrate how beaches with different average wave heights will have both different beach types and hazard ratings, Figure 46 shows Jan Juc, Torquay, Front and Fishermans beaches. The beaches on either side of Point Danger receive variable wave heights, resulting in a range of beach types from a low tide terrace to transverse bar and rip, and beach hazard ratings from a low of 3 to a high of 7.

Finally, to show how one beach can change both beach type and beach hazard rating as a result of changing wave height and/or wave direction, Figure 47 shows Eastern Beach at Lakes Entrance under three different combinations of wave height and direction. Each set of wave conditions produces a different set of beach types along the beach, together with a different level and range in beach hazard ratings. The result is a change in both beach type and beach hazard rating, both along the beach and over time.

BEACH TYPE AND HAZARD RATING

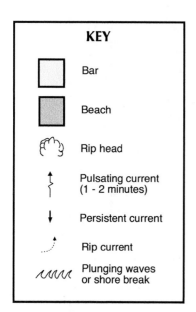

KEY

☐ Bar

▨ Beach

🖐 Rip head

↕ Pulsating current (1 - 2 minutes)

↓ Persistent current

⤶ Rip current

〰 Plunging waves or shore break

PLEASE NOTE:

This model represents average wave conditions on these beach types in micro tidal (< 2 m tide range) regions of southern Australia (south Queensland, NSW, Victoria, Tasmania, South Australia and southern Western Australia).

BEACH SAFETY IS INFLUENCED BY:

HEADLANDS - rips usually occur and intensify adjacent to headlands, reefs and rocky outcrops.

OBLIQUE WAVES - stronger longshore currents, skewed and migratory waves.

HIGH TIDE - deeper water and in some cases stronger rips.

LOW TIDE - rips more visible but normally more intensified due to restricted channel.

RISING SEAS - eroding bars, stronger currents, strong shifting rips, greater set up and set down.

HIGH TIDE AND RISING SEAS - more difficult to distinguish bars and troughs.

STRONG ONSHORE AND ALONGSHORE WINDS - reinforced downwind currents.

MEGARIPPLES - large migratory sand ripples common in rip troughs can produce unstable footing.

CHANGING WAVE CONDITIONS - (rising, falling, change in direction or length) - produce a predictable change in beach topography and type; the reason why beaches are always changing.

BIBLIOGRAPHY
Short 1979
Short 1985
Short and Hogan 1990
Short and Hogan 1991
Short and Wright 1981
Wright and Short 1984

INCREASING WAVE HEIGHT

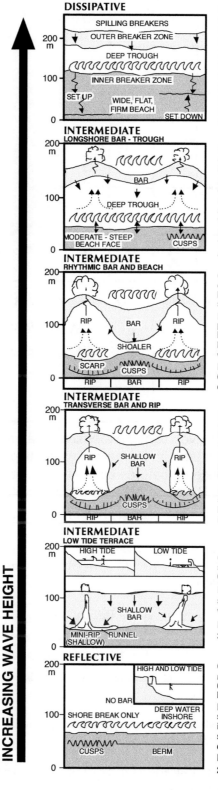

DISSIPATIVE

CHARACTERISTICS
dissipative - waves dissipate energy over a wide surf zone, 2 - 3 m breakers, straight bars, trough and beach

HAZARDS
high waves and wide surf zone restrict most bathers to the swash zone; safest bathing is in the swash zone with care of the set up and set down

BEACH HAZARD RATING AND HINTS
8/10 (stay close to shore, do not bathe in outer breaker zone)

INTERMEDIATE
LONGSHORE BAR - TROUGH

CHARACTERISTICS
consists of shore parallel bar and trough, 1.5 - 2.0 m breakers, moderate rip currents and straight beach

HAZARDS
deep trough and distance to outer bar restrict most bathers to the swash zone and inner trough, safest bathing is in the swash zone and in the trough away from rips

BEACH HAZARD RATING AND HINTS
7/10 (stay close to shore and avoid deep troughs and rips)

INTERMEDIATE
RHYTHMIC BAR AND BEACH

CHARACTERISTICS
consists of rhythmic (undulating) bar trough, and beach, 1.5 m breakers, distinct rip troughs separated by detached bars

HAZARDS
pronounced changes in depth and current between bars and rips, safest bathing is on or behind the bars during lower waves, hazardous during high waves and high tide

BEACH HAZARD RATING AND HINTS
6/10 (wade or swim to shoaler bars, avoid deep troughs and rips)

INTERMEDIATE
TRANSVERSE BAR AND RIP

CHARACTERISTICS
consists of attached bars, rip troughs and undulating beach, 1.0 - 1.5m breakers, distinct rip troughs separated by attached bars every 150 - 300 m

HAZARDS
pronounced changes in depth and current between bars and rips, safest bathing is on the bars

BEACH HAZARD RATING AND HINTS
5/10 (bathe on shallow bars adjacent to rips, however bathers can be washed off the bars into rips, inexperienced bathers may unknowingly enter rips)

INTERMEDIATE
LOW TIDE TERRACE

CHARACTERISTICS
shallow bar or terrace often exposed at low tide, 0.5 - 1.0 m breakers

HAZARDS
safest bathing - safe at low tide, deeper water and weak rips at high tide

BEACH HAZARD RATING AND HINTS
3/10 (watch for plunging waves at low tide)

REFLECTIVE

CHARACTERISTICS
reflective - waves tend to reflect back off the beach, 0 - 1 m breakers, only occur on very low wave beaches and on harbour beaches

HAZARDS
safest bathing - safe apart from deep water close inshore and from shorebreak during higher waves, steep beach and abrupt drop off to deeper water can make access difficult for elderly and children

BEACH HAZARD RATING
2/10

Figure 44. This figure illustrates the six beach types, their general characteristics and patterns of wave breaking, bars, troughs and currents. The characteristics of each, together with their hazards are noted, together with the influence of other factors that affect beach safety.

BEACH TYPE \ WAVE HEIGHT	< 0.5 (m)	0.5 (m)	1.0 (m)	1.5 (m)	2.0 (m)	2.5 (m)	3.0 (m)	> 3.0 (m)
Dissipative	4	5	6	7	8	9	10	10
Long Shore Bar Trough	4	5	6	7	7	8	9	10
Rhythmic Bar Beach	4	5	6	6	7	8	9	10
Transverse Bar Rip	4	4	5	6	7	8	9	10
Low Tide Terrace	3	3	4	5	6	7	8	10
Reflective	2	3	4	5	6	7	8	10

BEACH HAZARD RATING

Safest: 1 - 3
Moderately safe: 4 - 6
Low safety: 7 - 8
Least safe: 9 - 10

KEY TO HAZARDS

- Water depth and/or weak currents
- Shorebreak
- Rips and surfzone currents
- Rips, currents and large breakers

NOTE: All hazard level ratings are based on a bather being in the surf zone and will increase with increasing wave height and/or wind strength or with the presence of features such as inlet, headland or reef induced rips and currents. Rips also become stronger with the falling tide.

BOLD gradings indicate hazard level under modal wave conditions

Figure 45.   A table for calculating the beach hazard rating for any Victorian wave dominated beach, based on the prevailing wave or breaker height and beach type. Note that the rating for individual beaches may also include additional hazards such as winds, rocks, reefs, etc.

Beach Hazard Ratings

1 - least hazardous beach
10 - most hazardous beach

Beach hazard rating is the scaling of a beach according to the hazards associated with its beach type and local beach and surf environment.

Modal beach hazard rating is based on the beach type prevailing under average or modal wave conditions, for a particular beach type (Figures 44 & 45) or beach (Figure 46).

Prevailing beach hazard rating refers to the level of beach hazards associated with the prevailing wave and beach conditions on a particular day or time (Figure 47).

In addition, if moderate to strong winds are blowing, then the *Wind Hazard Rating* (Figure 43) must also be added to the prevailing beach hazard rating. In this book, all Victorian beaches are rated by their *modal beach hazard rating*. Table 14 summarises the rating for all Victorian beaches.

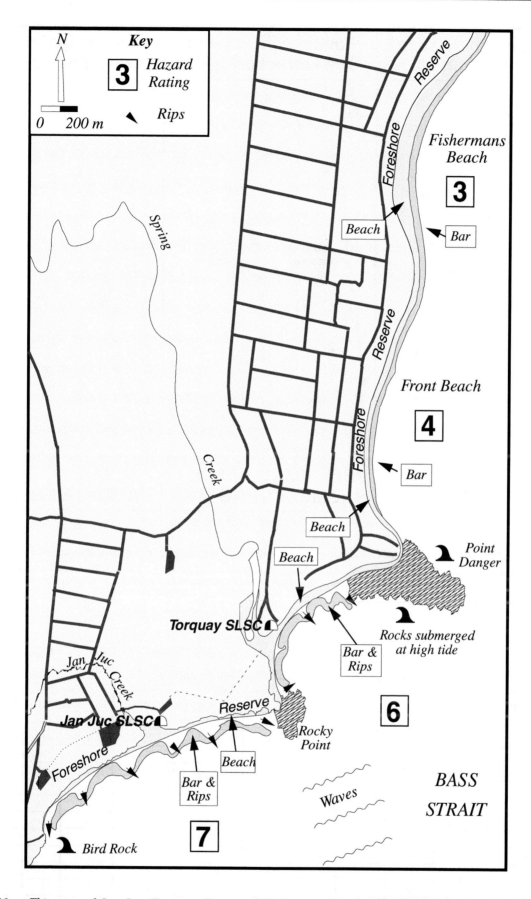

Figure 46.   This map of Jan Juc, Torquay, Front and Fishermans beaches illustrates the impact of the longshore decrease in wave height between Jan Juc (average 1.5 m) and Front Beach (average 0.5 m) on beach type and hazard rating. The higher breakers produce a more hazardous rip-dominated Jan Juc, compared to a less hazardous attached bar at Front Beach.

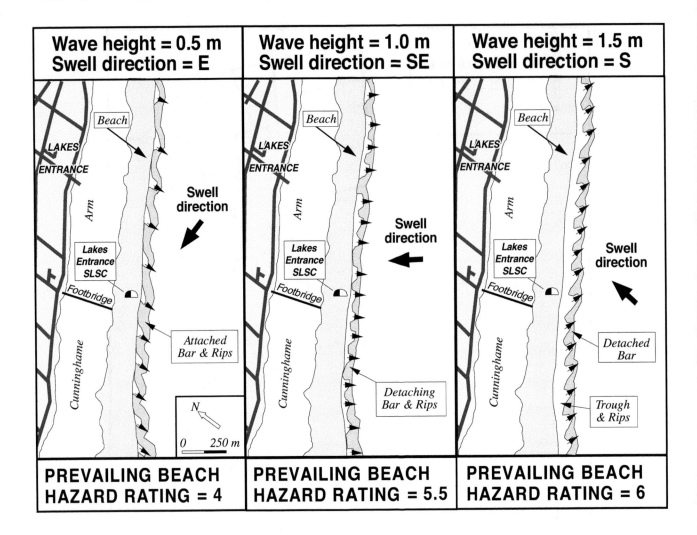

Figure 47.   *Eastern Beach at Lakes Entrance, showing the impact of variable wave direction and height on typical beach, bar and rip topography. Lower waves result in an attached bar with rips, while increasing wave height produces a continuous longshore trough, rips and an offshore bar. Changing wave direction skews both the bars and rip currents.*

Table 14. *Victorian beach hazard rating, by number of beaches and their length (excluding wind hazard)*

Beach Hazard Rating*	No. of beaches		% of total beaches		Mean length (m)		Total length (km)		Proportion of length of beaches (%)		
	Ocean	Bay	Ocean	Bay	Ocean	Bay	Ocean	Bay	Ocean	Bay	Total
1	0	61	0	46	0	1466	0	89	0	50	9
2	9	27	2	21	497	1541	5	42	1	24	5
3	53	37	10	28	741	1102	39	41	5	23	8
4	85	7	15	5	471	781	40	5	5	3	5
5	66	0	12	0	1264	0	83	0	10	0	8
6	109	0	19	0	3477	0	379	0	47	0	38
7	148	0	26	0	1091	0	162	0	20	0	16
8	77	0	14	0	1281	0	99	0	12	0	10
9	11	0	2	0	386	0	4	0	<1	0	<1
10	2	0	0	0	175	0	0.4	0	0	0	0
TOTAL	**560**	**132**	**100**	**100**	**1448**	**1343**	**811**	**177**	**100**	**100**	**100**

* Beach Hazard Rating: 1 = least hazardous; 10 = most hazardous

4.2.3   Rip identification - how to spot rips

To the experienced surfer rips are not only easy to spot, but they are the surfer's friend, providing a quick way (and at times the only way) out the back; as well as carving channels to produce better waves.

To the inexperienced however, rips are not only unknown or 'invisible' to them, but if caught in one it can be a terrifying and even fatal experience.

Most recreational bathers and visitors do not have the time or desire to become experienced swimmers and surfers. In order to assist them, a check list of features that indicate a rip or rips are present on the beach, is noted below:

**Rip Current Spotting
Check List**

Note: any one of these features indicates a rip, but not all will necessarily be present.
    * indicates always present
    + indicates may be present

---

*   A seaward movement of water (Figure 42) either at right angles to or diagonally across the surf. To check on currents, watch the movement of the water or throw a piece of driftwood or seaweed into the surf and follow its movement.

*   Rips only occur when there are **breaking waves** seaward of the beach. If water is moving shoreward, it must return seaward somewhere.

*   Disturbed water surface (ripples or chop) above the rip; caused by the rip current as it pushes against incoming waves and water. May be difficult to spot.

+   Longshore rip feeder channels and/or currents running alongshore, hard against the base of the sloping beach face. Rips are usually supplied by two rip feeder channels, one on either side of the rip.

+   Rhythmic or undulating beach topography, with the rips located in the indented rip embayments.

+   Areas where waves are not breaking, or are breaking less across a surf zone, owing to the deeper rip channel.

+   A deep channel or trough, usually located between two bars or against rocks. The channel may contain inviting clear, calmer water compared to the adjacent turbulent surf on the bars. However, do not be fooled. In the surf, calm water usually means it's deeper and contains currents.

+   A low point in the bar where waves are not breaking, or break less. This is where the rip channel exits the surf zone.

+   Turbid, sandy water moving seaward; either across the surf zone and/or out past the breakers.

+ In the rip feeder and rip channel, the stronger currents produce a **rippled seabed**. These ripples are called megaripples and are sandy undulations up to 30 cm high and 1.5 m long. If you see or can feel large ripples on the seabed, then strong currents are present, so stay clear.

+ If you see one rip on a **long beach,** there will be more if wave height remains the same along the beach. Rip spacing can vary from 150 to 500 m, depending on wave conditions.

+ If there is surf and **rocks, reef or a headland**, a rip will always flow out close to or against the rocks. These rips are often permanent and have local names like the 'express', the 'accelerator', the 'garbage bowl', etc.

Rips - how to escape if caught in one

- If the water is less than chest deep adults should be able to **walk out** of a rip. Conversely, avoid going into deeper water. So if you are in any surf current, become very careful once the water exceeds waist depth. Get out while the water is shallow.

- Most people rescued in rips are **children**. Never let them out of your sight and if they get into difficulties, go to them immediately while the water is still relatively shallow.

- As long as you can swim or **float,** the rip will not drown you. There is no such thing as an undertow associated with rips, or for that matter, with surf zone currents. Only breaking waves can drive you under water. Most swimmers caught in rips drown because they panic. So **stay calm**, tread water, float and conserve your energy.

- If there are people/lifesavers on the beach, raise one arm to **signal** for assistance.

- **Do not** try to swim or wade in deep water directly against the rip, as you are fighting the strongest current; there are easier ways out.

- Where possible, **wade rather than swim**, as your feet act as an anchor and help you fight the current.

- If it is a relatively weak and/or shallow rip, **swim or wade sidewards to the nearest bar.** Once on the bar, walk or let the waves or wave bores return you to shore.

- If it is a strong and/or deep rip, go with it through the breakers. Do not panic. When **beyond the breakers,** slowly and calmly swim alongshore in the direction of the nearest bar, indicated by heavier wave breaking. If you are not a surfer, simply wait for a lull in the waves, swim into the break and allow the waves to wash you to shore. Stay near the surface so the broken wave can wash you shoreward. Do not dive under the waves as they will wash over you.

- To summarise: stay calm, swim sideways toward breakers or the bar and let the broken waves return you to shore. Raise an arm to signal for help if people are on the beach.

- If rescued, thank the rescuer.

4.3   Surf Safety

Surfing, as opposed to bathing, requires the surfer to go out to and beyond the breakers, so he or she can catch and ride the waves; in other words, go surfing. This can be done using just your body (body surfing) or a range of surfboards, bodyboards and skis.

Surfing safely requires a substantially greater knowledge of the surf, compared to bathing on the shore or bar. The following points should be observed before you begin to surf:

1. You must be a strong swimmer.

2. You must also be experienced at swimming in the surf.

3. You must be able to tell if and when a wave will break.

4. You must know the basics of how breaking waves and the surf zone operate. You should be able to **spot rip currents.**

5. Equally, you should know what **hazards** are associated with the surf, including breaking waves, rips, reefs, rocks and so on.

6. You must only use **equipment** that is suitable for you, ie. the right size and level.

7. You must know how to use your equipment; whether it be flippers, bodyboard, surfboard or wave ski.

8. You should use **safety equipment** as appropriate, including a legrope or handrope, wetsuit, flippers and in some cases a helmet.

9. You should ensure your **equipment is in good condition,** with no broken fibreglass, frayed legrope, etc.

Some tips on safe surfing:

- Remember surfing is fun, but it is also hazardous.

- **Never surf alone**. If you get into trouble, who will help you?

- Before you enter the water, always **look at the surf for at least five minutes.** This will enable you to first, gauge the true size of the sets, which may come only every few minutes; and secondly, besides picking out the best spot to surf, you can also check out the breaker pattern, channels, currents and rips; in other words, the circulation pattern in the surf. This is important as you can use this to your advantage.

- On patrolled beaches **observe the flags**, surfboard signs and directions of the lifesavers. Do not surf between the red and yellow flags or near a group of bathers.

- If you are surfing out the back, the safest, quickest and usually the easiest way to get out is via a rip. This is because the water is moving seaward, making it easier to paddle; the rip flows in a deeper channel, resulting in lower waves; and the rip will keep you away from the bar or rocks where waves break more heavily.

- Once out past the breakers, particularly if paddling out in a rip, move sidewards and position yourself behind the break.

- Buy and read the **Surf Survival Guide,** published by Surf Life Saving Australia and available at all newsagents.

- Obtain an **SLSA Surf Survival Certificate** from your school.

Some general tips:

Surfers conduct many rescues around the Australian coast, so be prepared to assist if required. Remember, if you are on a surfboard, bodyboard or wave ski and have a legrope and wetsuit, you are already kitted out to perform rescues. The board is a good flotation device that can be used to support someone in difficulty. The wetsuit will keep you warm and buoyant and thereby give you more energy and flotation to assist someone in distress; and the legrope or board can be used to tow someone in difficulty, while you paddle them toward safety.

The simplest way to get someone back to shore is to lay them on your board while you swim at the side or rear of the board, and let the waves wash the board, patient and you back to shore.

Some surfing hazards to watch out for when paddling out:

* Heavily breaking/plunging waves, particularly the lip of breaking waves.

* Rocks and reefs.

* Strong currents, particularly in big surf.

* Other surfers and their equipment.

...when you are surfing:

* Other surfers - the surfer on the waves has more control and is responsible for avoiding surfers paddling out, or in the way.

* Heavily breaking waves.

* Your own and other surfboards. They can and do hit you, and can knock you out.

* Shallow sand bars.

* Rocks and reefs.

* Close-out sets and big surf

...when returning to shore:

* Heavy shorebreaks.

* Rocks and reefs.

* Strong longshore/rip feeder currents.

Remember: The greatest danger to surfers is to be knocked out and drown. Most surfers are knocked out by their own boards or by hitting shallow sand or rocks. This can be avoided by always covering your head with your arms when wiped out, by wearing a wetsuit for flotation, by wearing a helmet and by surfing with other surfers who can render assistance if required.

4.4   Rock fishing safety

The rocky sections are the most hazardous part of the Victorian coast (Figure 48), with most fatalities due to fishermen being washed off the rocks and drowning.

Rock fishing is hazardous because:

* Deep water lies immediately off the rocks, often containing submerged reefs and rocks, and heavily breaking waves.

* Occasional higher sets of waves can wash unwary fishermen off the rocks.

* Most fishermen are not prepared or dressed for swimming, as they are often wearing heavy waterproof clothing, shoes and tackle.

* Many fishermen are not experienced surf swimmers, and many cannot even swim.

To minimise your chances of joining this distressing statistic, two points must be heeded. Firstly, avoid being washed off and secondly, if you are washed off, make sure you know how to handle yourself in the waves until you can return to the rocks or await rescue.

The biggest problems usually occur when inexperienced fishermen are washed off rock platforms. To compound the problem, they either cannot swim or are not prepared for a swim. You only need to watch experienced board and body surfers surfing rocky point and reef breaks, to realise rocks are not a serious hazard to the experienced and the properly equipped.

So the rules are:-

1. Before you leave home:

* Check the weather forecast. Avoid rock fishing in strong winds and rain.

* Phone the boat or surf forecast and check the wave height. Avoid waves greater than 1 m.

* Check the tide state and time. Avoid high and spring tides.

* Are you suitably attired for rock fishing?

* Are you suitably attired in case of being washed off the rocks?

* A loose fitting wet suit is both comfortable and warm; and it will keep you afloat and protected if washed off the rock platform.

2. Before you start fishing:

* Check the waves for ten minutes, particularly watching for bigger sets.

* Choose a spot where you consider you will be safe.

Figure 48.   Rock fishermen often place themselves in areas of extreme risk. Experienced rock fishermen know how to identify and avoid hazardous areas or situations. The inexperienced or unwary rock fisherman can however, be knocked over by waves and at worst washed off the rocks and into the sea. This is the major cause of drownings on the Victorian coast. So be very wary, particularly if you are not experienced and if you cannot swim.

- When choosing a spot to fish:

if the rocks are wet, then waves are reaching that spot;
if the rocks are dry, waves are not reaching them, but may if the tide is rising or wave height is increasing.

- Ensure you have somewhere to easily and quickly retreat to, if threatened by larger waves.

- Place your tackle box and equipment high and dry.

3. When you are fishing:

- Never turn your back on the sea, unless it is a safe location.

- Watch every wave.

- Be aware of the tide; if it is rising, the rocks will become increasingly awash.

- Watch the waves, to check for:

  - increasing wave height, leading to more hazardous conditions;

- the general pattern of wave sets; it is the sets of higher waves that usually wash people off rocks.

- Remember, 'freak waves' exist only in media reports. No waves are freak; all that happens is that a set of larger waves arrives, as any experienced fisherman or surfer can tell you. These larger sets are likely to arrive every several minutes.

- Do not fish alone; two can watch and assist better than one.

- If you see a larger set of waves approaching - retreat. If you cannot retreat, lie flat and attach all your limbs to the rock. Forget your gear; you are more valuable. As soon as the wave has passed, get up and retreat.

- Wear sensible clothing. A wetsuit provides warmth, protection and safety; particularly if you are washed off or knocked over.

4. If you are washed off, here are some hints:

- If you have sensible clothing; that is, clothing that will keep you buoyant, such as a wetsuit or life jacket; then you should do the following:

- Head out to sea away from the rocks, as they are your greatest danger.

- Abandon your gear, it will not keep you afloat.

- Take off any shoes or boots and you will be able to swim better.

- Tread water and await rescue, assuming there is someone who can raise the alarm.

- If you are alone, or can only be saved by returning to the rocks, try the following:

  - Move seaward of the rocks and watch the waves breaking over the rocks in the general area, then:

  - Choose a spot where there is either:

    a **channel** - this may offer a safer, more protected route;

    a **gradually sloping rock** - if waves are surging up the slope, you can ride one up the slope, feet and bottom first, then grab hold of the rocks as the swash returns;

    or a **steep vertical face** with a flat top reached by the waves - swim in close to the rocks, wait for a high wave that will surge up to the top of the rocks, float up with the wave, then grab the top of the rocks and crawl onto the rock as the wave peaks. As the wave drops, you can stand or crawl to a safer location.

Fishing Information: The *Victorian Recreational Fishing Guide* is published by the Department of Conservation and Natural Resources and is available for free from all DCNR offices.

Local Fishing Guides are published by some councils, for example the *Fishing Guide - Torquay to Lorne*, published by the Surfcoast Shire.

---

5   VICTORIAN BEACHES

5.1   How to find a beach in this book

This chapter contains a description of every beach on the Victorian coast. It is divided into nine coastal regions (Figure 8), including a section on Port Phillip Bay beaches. A regional map is located at the beginning of each section locating all beaches. In addition, individual maps are provided of all beaches patrolled by surf clubs and/or lifeguards. The beaches are arranged in order from east to west.

To find a beach, you can use any of four systems:

1.   By using the alphabetical **BEACH INDEX** at the rear of the book.

2.   By location on the coast; using either the **REGIONAL MAPS** of the particular coastal region, or by following the beaches up or down the coast until you find the beach.

3.   By name of the **SURF LIFESAVING CLUB**, if the beach has one. If it differs from the beach name, then both are listed in the **BEACH INDEX**.

4.   If you are a surfer and only know the name of the surfing break, which may differ from the beach name, then use the **SURF INDEX**.

5.2   What the beach description tells you

The description of each beach contains the following information:

- **Beach number:** 1 to 560 on ocean coast, plus P1 to P132 for Port Phillip Bay beaches

- **Beach name** or names

- **Surf lifesaving club:** name of surf lifesaving club, if present

- **Lifeguards:** lifeguard patrol hours, if lifeguards are present

- **Beach hazard rating:** 1 to 10

- **Wind hazard rating:** 1 to 4

- **Beach type:** (R, LTT, TBR, RBB, LBT, D); see page 37 for explanation
---

- **Inner bar:** refers to inner bar, close to or attached to the beach

- **Outer bar:** refers to the outermost bar on double bar beaches, lying seaward of the inner bar
---

Comments on beach location, access, length, beach and surf conditions, including beach type and presence and location of rips.

Specific comments are made for most beaches on:

- **Swimming** - suitability and safety

- **Surfing** - good surfing spots

- **Fishing** - presence and location of gutters and holes

- **Summary** - general overview of beach

5.3   Surf lifesaving club patrol dates and hours

In Victoria, all surf lifesaving clubs voluntarily patrol the beach on weekends and public holidays over summer. The surf lifesaving clubs are established on 41 of the 692 Victorian beaches.

5.3.1   Patrol dates

Patrols commence in late November and run each weekend until the end of the Easter Holidays.

5.3.2   Patrol days and hours

- Saturday   10 am to 4 pm
- Sunday   9 am to 5 pm
- Public holidays   9 am to 5 pm

Note: Times may vary in regional areas.

5.4   Lifeguards and Contract Lifesavers

Surf Life Saving Victoria lifeguards patrol 29 of Victoria's popular and/or dangerous surf beaches during the summer holidays. The patrols work on a Monday to Friday basis over the peak summer holiday period (Boxing Day to the end of the school holidays in late January).

5.4.1   Lifeguard patrol days and hours

Most lifeguards patrol all weekdays between the hours of 10 am and 6 pm.

5.5   Victorian Coastal Regions

The nine regions of the Victorian coast are described in the following sections. Table 15 lists the names of these regions, as well as the total length of coast in each region and the number of beaches and lifesaving clubs that it contains.

Table 15. Victorian coastal regions - length, beaches and lifesaving clubs.

No	Region	Length of coast (km)	Beach Nos	No of beaches	No of lifesaving clubs
1	Croajingolong National Park	150	1 - 74	74	1
2	Ninety Mile Beach & Corner Inlet	228	75 - 83	9	3
3	Wilsons Promontory	122	84 - 118	35	0
4	Waratah, Venus & Kilcunda Bays	117	119 - 182	64	5
5	Phillip Island & Mornington Peninsula	145	183 - 280	98	5
6	Port Phillip Bay	259	P1 - P132	132	27
7	Point Lonsdale to Cape Otway	139	281 - 410	130	12
8	Cape Otway to Childers Cove	118	411 - 498	88	1
9	Warrnambool to Nelson	211	499 - 560	62	3
	Ocean	**1230**	**1 - 560**	**560**	**30**
	Bay	**259**	**P1 - P132**	**132**	**27**
	Total Victorian Coast	**1489**		**692**	**57**

1. CROAJINGOLONG NATIONAL PARK

> **Croajingolong National Park and**
> **Bemm River To Snowy River**
> **(Cape Howe to Marlo)**
>
> Beaches: 1 to 74
> Coast length: 150 km
>
> *For maps of region see Figures 50, 51 & 52*

The eastern coast of Victoria is dominated by the resilient rocks of the Lachlan Fold Belt. The rocks produce a rugged east-west trending coast, interspersed with both rounded granitic and jagged metamorphic rocks, separated by 74 beaches. The beaches are predominantly composed of quartz sand grains derived from long term erosion of the rocks. The beaches range in size from many small pockets of sand a few tens of metres long, usually hidden amongst bold headlands, rocks and reefs; to the longer beaches (Figure 49) near Cape Howe and west of Little Rame Head.

Of the 74 beaches, 20 can be reached by car, with an additional five accessible by 4WD. Of the remainder, only three have sealed walking track access, while the other 46 are accessible via rough tracks, or not at all, and therefore remain in a near-natural state.

Croajingolong National Park occupies the first 105 km of Victoria's coast, extending from Cape Howe on the NSW border, to the mouth of the Bemm River at Sydenham Inlet. It surrounds the popular fishing and holiday town of Mallacoota.

Beyond Sydenham Inlet, the coast continues almost due west for another 90 km to Red Bluff at Lake Tyers. Several long, straight beaches are interrupted by the prominent, resilient headlands of Pearl Point, Cape Conran and Point Ricardo. The shifting Snowy River mouth at Marlo and the crumbling Tertiary sandstone bluffs at Red Bluff also interrupt these beaches. Beyond the bluffs, the great Ninety Mile Beach begins.

Figure 49.   The Croajingolong coast is typified by this view of the western side of Point Hicks, with a long sandy beach exposed to high energy south to south-west waves and winds. These have produced a rip dominated double bar surf zone and extensive parabolic dunes, that in this location have moved over 3 km inland.

CROAJINGOLONG NATIONAL PARK

Area: 87 500 ha
Length of coast: 105 km
Number of beaches: 52

Beaches: 1 to 4
             13 to 56
             61 to 66

Camping Areas (bookings required in holiday seasons):

     Shipwreck Creek
     West Wingan Inlet
     Mueller River
     Thurra River
     Tamboon Inlet

A coastal walking track, which follows both the beaches and The Old Coast Road, runs from the New South Wales border to Sydenham Inlet.

Information: Department of Conservation, Forests and Lands Offices: Cann River   (051) 958 6351
                        Mallacoota   (051) 958 0219

The only settlements along this 200 km section of coast are at Mallacoota, Bemm River, Marlo and Lake Tyers. The bulk of the coast is given over to the National Park and State Forests. The beaches are backed alternately by dense forests and massive coastal sand dunes, some containing extensive active sand sheets. It remains one of the most natural sections of the Australian coast and provides a wealth of beautiful, natural beaches, dunes, estuaries and lakes; many of which rarely see a visitor.

Victorian Beaches - Region 1: East (Cape Howe to Sandpatch Point)

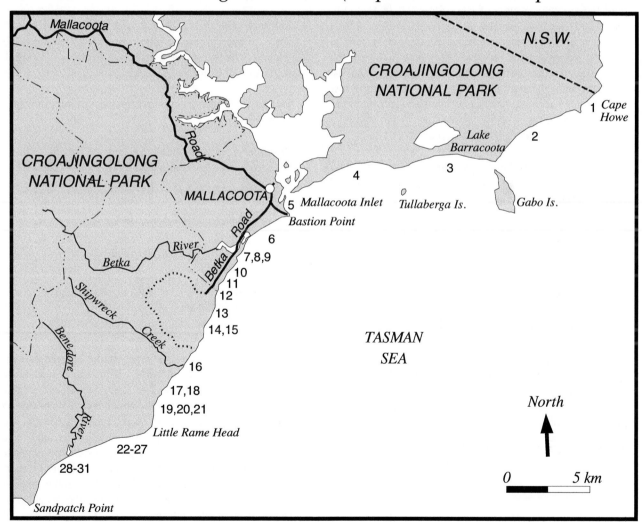

Figure 50.   Region 1 (eastern section): Cape Howe to Sandpatch Point (Beaches 1 - 31)

1   CAPE HOWE

Unpatrolled
Single bar: TBR/RBB
Beach Hazard Rating: **7**
Length: 800 m

Cape Howe Beach is the most eastern beach in Victoria. It begins right on the border of New South Wales and runs south-west for 800 m to the lee of Iron Prince rocks; site of a wreck of the same name in 1923. The beach is backed by a kilometre of active sand dunes that are steadily moving sand northward, to cascade over low rocks into New South Wales waters just over the border. The beach is accessible only on foot from New South Wales (14 km walk), up the beach from Mallacoota (19 km walk), or by boat. As a result, it sees few visitors and even fewer bathers and surfers.

The beach itself is pinned in between the sand sheets and the energetic waters of Bass Strait. It faces south-east and receives waves averaging 1.5 m.   These produce a 100 m wide surf zone, usually cut by four to five strong rips, which increase in strength toward the low rocks at the northern end (Cape Howe). At the southern end, the southerly waves wrap around the low Iron Prince rocks and produce a right hand surf break, as well as offering a little protection.   However, a permanent rip runs out against the rocks.

Swimming: An isolated, rip dominated beach. Use extreme care if bathing. Stay toward the south end and away from the strong rips.

Surfing: A right hand point break runs along the north side of Iron Prince rocks.

Fishing: The beach is dominated by deep rips and often has a trough running its full length.

Summary: A remote, rip dominated beach; a nice part of the coast walk, but hazardous in the surf.

Captain Cook sighted **Cape Howe** on 20 April 1770 and named it after Admiral Earl Howe. It has been the site of a number of shipwrecks, including the Prince of Wales in 1853, Ellen Simpson in 1866 and Harlech Castle in 1870.

2   IRON PRINCE

Unpatrolled
Single bar: TBR/RBB
Beach Hazard Rating: **7**
Length: 5 600 m

Iron Prince Beach begins at Iron Prince Rocks and runs in a broad arc for 5.6 km south-west to the sandy foreland in lee of Gabo Island. A kilometre of active sand dunes back the beach. The beach receives waves averaging 1.6 m and usually has up to 18 rips, spaced every 300 m. Following high seas, the rips are often linked to form a continuous deep trough running the length of the beach. The beach can be reached on foot from Mallacoota Inlet (12 km walk) or by boat.

Swimming: This is a remote and little used beach, with a surf dominated by strong, deep troughs and strong rips. Use extreme caution if bathing.

Surfing: Beach breaks dominate the beach. However, wave shape improves after wrapping around Gabo Island in the south; and a left point break runs along the south side of Iron Prince Rocks.

Fishing: A deep trough and rips occur along the length of the beach.

Summary: An isolated, little used beach, with an energetic surf zone dominated by rips and troughs.

3   GABO - TULLABERGA

	Inner	Outer
Unpatrolled		
Double bar:	TBR	RBB/LBT
Beach Hazard Rating:	**6**	**7**
Length: 5 200 m		

In lee of **Gabo Island**, a large, sandy, cuspate foreland protrudes over 1 km seaward. Until 1892, it extended another 500 m all the way to Gabo Island; but a major storm in that year washed the spit away. This protrusion forms the eastern end of a 5.2 km long beach, that curves essentially due west to a smaller foreland in lee of Tullaberga Island. The southern orientation exposes the beach to the full forces of the dominant southerly waves, which average over 1.5 m. At the same time, the strong westerly winds blow sand from the beach into a series of blowouts and sand sheets. The beach can only be reached on foot from Mallacoota Inlet (7 km along the beach) or by boat. In 1923, the ship Riverina was wrecked toward the eastern end.

The waves interact with the sand to build a 350 m wide surf zone, containing two shore parallel bars. The inner bar is dominated by alternating attached bars and deep rips every 300 m. A deep trough lies beyond this, then a shore parallel outer bar, cut by occasional rips.

Swimming: This is a hazardous beach, dominated by rips. If bathing, stay on the attached sand bars and well clear of rip currents and troughs.

Surfing: Beach breaks occur along the length of the beach, with some cleaner waves near both forelands.

Fishing: Numerous deep rip holes occur along the length of the beach.

Summary: An isolated beach backed by a variety of sand dune types and fronted by a rip dominated surf.

Gabo Island is Commonwealth Territory and contains the 50 m high Gabo Island Lighthouse. The lighthouse was built in 1862, following the tragic wreck of the steamer Monumental City, in 1853. The island has a small harbour on its northern leeward side, which contains a small wharf and the island's only beach. Other wrecks on the island include the Balclutha in 1881, Federal in 1901 and Christina Fraser in 1933.

---

4   MALLACOOTA SPIT (BIG)

Unpatrolled		
	Inner	Outer
Double bar:	TBR	RBB/LBT
Beach Hazard Rating:	6	7
Length: 8 000 m		

The famous Mallacoota Inlet, and infamous bar, form the western end of **Mallacoota Spit**, a 7 to 8 km long south to south-east facing beach, also called the **Big Beach**. The length of the beach depends on the location of the inlet, which can shift a kilometre or more. The beach extends to the sandy foreland in lee of Tullaberga Island. It is backed by a low vegetated foredune in the west, which gives way to a series of blowouts to the east. The exposed surf zone receives waves averaging more than 1.5 m, which produce a 350 m wide double bar system. The inner bar is usually attached to the shore and is cut by rips every 300 m, producing up to 24 rips. In addition, the extensive tidal shoals and channels of the inlet dominate the western kilometre of the beach. The western end can easily be reached by boat from Mallacoota, with a long walk required to reach the rest of the beach. In 1994 the inlet closed, thus linking the beach to Bastion Point.

Swimming: A relatively isolated beach, dominated by rips. Use extreme caution and stay clear of rips and on the attached bars, if present. Also be careful near the inlet, as strong tidal currents and a wide surf zone produce hazardous bathing and boating conditions.

Surfing: Waves break over the inner and outer bars along the length of the beach; plus variable breaks over the shifting inlet sand bars.

Fishing: The inlet mouth and the beach running to the north are the most popular areas for beach fishing, but are hazardous for boats. There are numerous rips and troughs along the beach, if you feel like a good walk.

Summary: A long and little used beach away from the inlet; nice for a walk, but hazardous for bathing.

---

5   MAIN BEACH - BASTION POINT

Mallacoota SLSC
Single bar: LTT + rocks
Beach Hazard Rating: **5**
Length: 500 m

Bastion Point guards the entrance to Mallacoota Inlet and bar. Unfortunately it offers little protection for the boats that still have to negotiate one of the most dangerous entrances on the east coast. Standing on Bastion Point, you can see waves breaking over the bar up to several hundreds of metres out to sea. Local knowledge and experience are required to successfully cross the bar. The inlet and bar also shift from time to time, and occasionally the inlet closes completely. The inlet may be located adjacent to, or up to 1 km east of, the point, at which time it is joined to the point by a stretch of sand called **Main Beach**. When it is present, Main Beach contains rips and numerous bars, as well as shoals and channels associated with the inlet.

Below the 10 m high bluff of the point is a permanent 500 m long beach, that for the most part lies behind eroded lines of low sandstone ridges. The sealed road from Mallacoota ends at the centre of the beach. Here there is the new surf lifesaving club and small car park. The beach is used for boat launching and small boats use this spot to avoid having to cross the bar.

The beach faces north-east and receives waves averaging 0.5 m, which tend to run along the beach and rocks and, where there is enough sand, form a shallow, attached sand bar. The most popular spot is in front of the car park, where the surf is clear of rocks for about 100 m. Elsewhere, tidal pool and rock fossicking are more popular than bathing.

When the inlet is closed or it shifts to the east, the beach can turn and continue northward until the inlet is reached, usually within 1 km. This section of beach and adjacent inlet receives higher waves and is dominated by rips and shifting sand bars and channels, and should not be used for bathing.

Swimming: The car park beach is the safest in the area, so long as you stay on the attached bar and between the flags. Avoid the rocks, and particularly the inlet area.

Surfing: Bastion Point offers the best right hand break in the area. It has two breaks called *Bastion Point* and *Broken Board*. They begin to work in any swell over 0.5 m and hold waves to 2.5 m.

Fishing: A popular spot for fishing into the inlet, and for launching small boats to go outside.

Summary: One of Mallacoota's most popular beaches, with excellent access, usually safe for bathing and protection below the bluffs from westerly winds.

6   SURF or BETKA

Unpatrolled
Single bar: TBR/RBB
Beach Hazard Rating: **6**
Length: 3 000 m

Surf Beach blocks the shallow mouth of the Betka River. The beach and foredune system begins at Bastion Point and runs for 3 km due south-west to the mouth of the river, and terminates at a bluff and rocks a few hundred metres beyond. This is the most popular surfing beach in the Mallacoota area, owing to its usually higher waves and good access. It can be reached on foot around Bastion Point, via the golf course track. This track leads to a small car park, viewing platform and the popular Betka River picnic area that provides access, parking and facilities for both the beach and river mouth area.

Wave height increases north of the river mouth and, for most of the beach's length, waves average 1.5 m. These produce prominent rips every 300 m, with permanent rips running out against Bastion Point and the rocks just south of the golf course track. South of the river mouth, waves decrease slightly and rips tend to diminish in size, however, rocks begin to occupy the surf zone.

Swimming: Safest just south of the river mouth, with higher waves and strong rips to the north.

Surfing: A popular spot that, as the name suggests, picks up all swell and offers both beach and rock

controlled breaks. The eastern end, reached via the golf club, is known as *Tip Beach*.

Fishing: Very popular owing to the good access, persistent deep rip channels and often longshore troughs, especially adjacent to the rocks and point.

Summary: Mallacoota's surfing beach; very popular and moderately safe at the river mouth.

7   SURF SOUTH

Unpatrolled
Single bar: LTT/TBR + rocks
Beach Hazard Rating: **6**
Length: 400 m

Three hundred metres south of the Betka River mouth, a bluff protrudes across the beach with rocks continuing into the surf zone. Around the rocks is **Surf South Beach**; a 400 m long sandy beach littered with numerous rocks and remnants of bluffs, and backed by 20 m high steep bluffs along its southern half. The rocks continue into the surf as rocks and reef. The reefs lower the average waves to 1 m, which produce a more continuous attached bar, only cut by rips following high seas.

Swimming: A relatively safe and quiet beach, however there are numerous rocks in the water and care should be taken, especially near the rocks.

Surfing: Lower beach break, but watch the rocks.

Fishing: The southern point provides access to deeper water.

Summary: Most people walk along this beach from the main Betka Beach. A nice spot to get away from the crowds, but take care if bathing amongst the rocks.

8, 9   BEACH 8, TOWER

Unpatrolled			
	Rating	Single bar	Length
8 Beach 8	5	LTT/TBR	250 m
9 Tower	4	LTT + rocks	150 m
		Total length	400 m

Beach 8 and Tower Beach are two small beaches backed by steep vegetated bluffs. **Beach 8** can be reached on foot around the craggy rocks at the southern end of Surf Beach, and from a vehicle track to the rear of the beach. It is 250 m in length and is bordered by jagged rock headlands, with large rocks outcropping on the beach and in the surf. The sandy beach usually has a continuous attached bar, which is cut by two to three rips following high seas.

Tower Beach is more difficult to find and access. The Betka Road runs 150 m behind the beach. If you know where to look, there is a small track hidden in the bluff-top vegetation that leads to a steep descent to the beach. It can also be accessed with difficulty from Beach 8, around the rocks at low tide. The beach is surrounded by 20 to 30 m high bluffs and has several large rocks lining the beach and in the surf. The sandy beach usually has a narrow attached bar, with only occasional rips present.

Swimming: Both beaches receive waves averaging less than 1 m and are relatively safe away from the rocks. However, both are little used by bathers (particularly Beach 8) and have numerous rocks and reefs. So use caution if bathing.

Surfing: Usually only a low shorebreak.

Fishing: Plenty of rocks and reef, with the headlands giving good access to deeper water. Only occasional rip holes are present, after larger seas.

Summary: Two small attractive beaches; readily accessible if you know where to look, but little used except by locals.

---

10   **QUARRY**

Unpatrolled
Single bar: TBR
Beach Hazard Rating: **6**
Length: 1 100 m

Quarry Beach is named after a disused gravel quarry that now serves as a small car park toward the northern end of the beach. The sandy beach is 1 100 m in length and is bordered by prominent 20 to 30 m high jagged headlands, formed of steeply dipping metamorphic rocks. These rocks also outcrop along the beach and in the surf. Waves averaging just over 1 m arrive at the beach, to produce, along with scour holes next to the larger rocks, a continuously attached bar, that is usually cut by rips every 250 m.

Swimming: Usually best in front of the car park where it is relatively clear of rocks. Waves decrease slightly to the south, but the rocks increase. Stay clear of the rocks; and rips, when present.

Surfing: Beach breaks, which are best on a low to moderate swell following high seas.

Fishing: A popular spot, owing to the excellent access and range of beach, rock, reef and headland fishing available.

Summary: A relatively popular beach for bathing and fishing.

11, 12   **SEACAVE NORTH & MAIN**

Unpatrolled			
	Rating	Single bar	Length
11 Seacave North	4	R/LTT + rocks	80 m
12 Seacave Main	5	LTT + rocks	200 m
		Total length	280 m

Seacave Beach is located at the end of Betka Road, just where it turns inland to become the Betka 4WD track. Short vehicle tracks off the road lead to the northern and central areas of the beach, with short walking tracks from the road to the beach. The beach is bounded by 30 m high jagged rocks and headlands, with rocks toward the northern end dividing the small northern 80 m long beach (11), from the main 200 m long beach (12). A sand filled sea cave through these rocks links the two beaches, which are also joined by sand at low tide.

Both beaches receive waves averaging less than 1 m and usually have a moderately steep sandy beach fronted by a narrow attached bar, with waves breaking close to the base of the beach. Rocks outcrop on the northern beach and toward the southern end of the main beach.

Swimming: Two relatively safe beaches, with only occasional rips during bigger seas. However, the water is deep close inshore, so children and poor swimmers should be closely supervised.

Surfing: Usually a low shorebreak.

Fishing: Relatively deep water off the beach and particularly off the headlands.

Summary: Two attractive, timber-fringed beaches, usually with low waves. The sea cave should only be entered at low tide, as it is awash at high tide.

CROAJINGOLONG NATIONAL PARK

Beaches: 13 to 56

Behind Beach 13, the Coastal Walking Track leaves the Betka vehicle track and winds its way south. There are a number of usually small pocket beaches along the adjacent coast, commonly at the mouths of creeks. Some can be reached off small side tracks, while a few have no formed access and are very difficult to reach. If bathing at any of these beaches, use extreme caution, as they are all very remote, many have rocks and reefs and some have strong rips.

---

13   COBBLE

Unpatrolled

Single bar: R + rocks
Beach Hazard Rating: **4**
Length: 100 m

Cobble Beach is the last beach to be reached off the Betka Road. A narrow vehicle track leads to a small parking area on the slopes above the northern end of the beach. A steep walking track runs down the 40 m high vegetated bluffs to the beach. The beach blocks the mouth of a small creek, that has delivered cobbles and boulders to form the small, 100 m long beach. The beach consists of sand, broken by prominent cobble cusps; and is backed by a cobble berm and fringed by jagged rocky platforms. The sand terminates at the shoreline and rock flats and reefs lie off the beach. Waves average only 0.5 m and usually lap against the beach.

Swimming: Largely rock strewn at low tide, so bathing is best at high tide. The beach is particularly good for snorkelling.

Surfing: None.

Fishing: Plenty of rocks and reef off the beach, with some raised platforms just around the northern rocks.

Summary: An attractive little cobble beach; nice for a quiet picnic, but not really suitable for bathing.

---

14   BOULDER

Unpatrolled

Single bar: R + rocks
Beach Hazard Rating: **4**
Length: 60 m

One kilometre south of Cobble Beach, a short side track leads to the 50 m high slopes overlooking **Boulder Beach**. A moderately steep track leads to the deeply embayed, 60 m long cobble and boulder beach. The beach is partly protected by a low, several hundred metre long southern headland, which is fronted by pockets of cobble and rocks. Waves average about 0.5 m; and the beach is steep and fronted by a shallow, rocky seabed.

Swimming: Okay for getting wet, but watch the rocks.

Surfing: None.

Fishing: Numerous shallow rocky flats off the beach and headlands.

Summary: An attractive pocket beach, with timber running right to the beach.

---

15   SHIPWRECK CREEK NORTH

Unpatrolled

Single bar: TBR + rocks
Beach Hazard Rating: **7**
Length: 250 m

On the eastern side of **Shipwreck Creek** headland is a narrow, 250 m long sandy beach wedged in between steep, 50 m high bluffs and a rock strewn surf zone. Waves averaging 1.5 m carve deep rips amongst the rocks, producing a dangerous surf zone. The beach has no formed access and is rarely seen or used.

Swimming: Not recommended. Very remote and hazardous with waves, rips and rocks.

Surfing: Some beach breaks amongst the rocks.

Fishing: Deep rip holes and rocks dominate the beach.

Summary: If you can find this beach, look, but do not bathe.

---

16   SHIPWRECK CREEK

Unpatrolled

Single bar: TBR
Beach Hazard Rating: **6**
Length: 350 m

Shipwreck Creek is the site of a small but popular National Park camping area. It can be reached by 4WD along the Betka Track from Mallacoota, or on foot along the Coastal Walking Track. Overnight camping is provided for hikers, while longer term campers need to book in holiday periods. The camp is located on the bluffs above the beach. The beach blocks the mouth of Shipwreck Creek, which flows as a shallow stream across the northern end of the wide, 350 m long sandy beach. The beach faces almost due south and picks up most waves, with wave heights averaging 1.5 m. These usually form two strong rips across the beach, with one permanent rip against the eastern headland.

Swimming: Be careful if you are not experienced, as rips dominate. Stay close to shore and on the bar, if it's attached. Avoid the headlands where rips persist.

Surfing: Good beach breaks during low to moderate south swell.

Fishing: Excellent rip holes, particularly against the eastern headland. A good rock platform extends out from the western headland.

Summary: A relatively popular spot for sunbathing, bathing, surfing and fishing for those using the camping area. However, it is little used for most of the year.

17, 18   BEACHES 17 & 18

Unpatrolled			
	Rating	Single bar	Length
17 Beach 17	7	TBR + rocks	200 m
18 Beach 18	7	TBR + rocks	50 m
		Total length	250 m

One and a half kilometres south along the coast from Shipwreck Creek are two small adjacent beaches, neither accessible from the walking track. The first **(Beach 17)** is 200 m long and consists of two pockets of sand, both fronting small creeks and backed by steep, vegetated bluffs. Two hundred metres further south is **Beach 18**, a 50 m long pocket of sand, bordered by 30 m high bluffs. Both beaches face south-east and receive waves averaging 1 to 1.5 m. These produce rock controlled rips on both beaches.

Swimming: Both beaches have waves, rips and rocks and are rarely seen or used. Not recommended for bathing, unless calm conditions are prevailing.

Surfing: Beach breaks amongst the rocks, if you can get your board there.

Fishing: Deep rip holes, rocks and headlands provide plenty of opportunities.

Summary: Two small, rarely seen or used beaches, surrounded by steep bluffs with no formed access. They

are off the coastal walking track and both are fronted by permanent rips.

19, 20, 21   SEAL CREEK NORTH, MID & MAIN

Unpatrolled			
	Rating	Single bar	Length
19 Seal Creek North	7	TBR + rocks	200 m
20 Seal Creek Mid	7	TBR + rocks	80 m
21 Seal Creek Main	7	TBR	200 m
		Total length	480 m

Seal Creek Beach straddles the coast walking track and has a small informal camping area. The beach can only be reached on foot and is therefore only used by passing hikers. The main beach (21) blocks Seal Creek, which ponds behind the beach and flows out across the northern end. The beach is 200 m long and bordered by steeply dipping, 50 m high jagged bluffs. Toward the northern end, these extend into the surf and separate two additional small beaches beyond. The first is 80 m long (20), the second 200 m (19). The two northern beaches have prominent rock headlands and rocks crossing the beach and in the surf. All three beaches face south-east and receive waves averaging 1.5 m. The waves produce persistent rips on each beach, particularly adjacent to the northern rocks.

Swimming: Safest on the main Seal Creek Beach; but even here, one to occasionally two strong rips persist. Do not bathe on the northern beach at all; and not at any of the three, unless you are an experienced surfer. These beaches are remote and little used.

Surfing: Beach breaks at the main beach which, if you have carried a board this far, are a must.

Fishing: Permanent rip holes, plus rocks and a southern rock platform provide plenty of opportunities.

Summary: A relatively popular spot for hikers, particularly as it's the first beach to be reached after the 6 km long track from Benedore River.

22 - 27   LITTLE RAME HEAD 1 to 6

Unpatrolled			
	Rating	Single Bar	Length
22 Little Rame Head 1	7	RBB + rocks	80 m
23 Little Rame Head 2	7	RBB + rocks	40 m
24 Little Rame Head 3	7	RBB + rocks	70 m
25 Little Rame Head 4	7	RBB + rocks	150 m
26 Little Rame Head 5	7	RBB + rocks	1 300 m
27 Little Rame Head 6	7	RBB + rocks	300 m
		Total length	1 940 m

At Little Rame Head the coast turns to face almost due south, exposing it to the full force of the southerly waves and south-westerly winds. Between Little Rame Head and Sandpatch Point 9 km to the south-west, are a series of nine beaches spread amongst the jagged metamorphic rock, headlands and reefs that extend all the way to Sandpatch Head. The first six are here referred to as **Little Rame Head Beaches 1 to 6**. They range in size from 40 to 1 300 m. All receive waves averaging 1.5 m which produce a 150 m wide surf zone, consisting of a continuous outer bar which links the beaches. The bar is cut by strong rips every 300 m on the long beach and against the headlands on the shorter beaches. In addition, numerous rocks and reefs litter the beaches and surf zone. Access to all beaches is by foot along the beach, from the coastal walking track at Benedore River, 1 km west of the sixth beach.

Swimming: These are remote and little used beaches, off the coastal walking track. They all experience strong surf and are dominated by persistent rips, rocks and reefs. Use extreme caution if bathing.

Surfing: Numerous beach and reef breaks along the beaches. Best during low to moderate southerly swell.

Fishing: The whole section is full of rip holes, rocks and reefs, providing numerous opportunities for beach and rock fishing.

Summary: An interesting rocky section of coast with pocket beaches; however, access past the easternmost beach (22) is over rocks and should only be attempted during low tide and calm seas.

28 - 31   BENEDORE RIVER 1 to 4

Unpatrolled			
	Rating	Single Bar	Length
28 Benedore River 1	7	RBB	1 300 m
29 Benedore River 2	7	RBB	1 100 m
30 Benedore River 3	9	RBB + rocks	250 m
31 Benedore River 4	9	RBB + rocks	300 m
		Total length	2 950 m

The central portion of the open embayment between Little Rame Head and Sandpatch Point is occupied by four beaches, which lie either side of the Benedore River entrance. The river mouth is dammed by the beach and forms a 1 km long, narrow, winding lake behind the beach. The four **Benedore River Beaches** all face south-east and receive waves averaging 1.5 m, which interact with the sand to produce a continuous outer bar, cut by rips every 300 m on the longer beaches. The bar continues off the two western beaches (30 & 31), with the beach itself and inner surf zone dominated by steeply dipping jagged rocks.

The only access is via the coastal walking track, which reaches the coast at the Benedore River mouth, after bypassing Little Rame Head. It then follows the beach for 1.5 km, exiting at the end of beach 31; to bypass Sandpatch Point. Water is available at this exit.

Swimming: The main beaches are dominated by rips and occasional rocks, while the western two are rock strewn. It is best to bathe near the river mouth (if closed), however the beach is steep and there is a deep trough and rips between the beach and the bar. So use extreme caution if entering the surf on this remote beach.

Surfing: Beach breaks on the outer bar, however watch the rocks.

Fishing: A deep gutter usually runs the length of the main beaches, with numerous rocks and reefs just off the beach.

Summary: The beach provides a refreshing change for those on the walking track and a chance to cool off. However, unless you are an experienced surfer, it's better to swim in the Benedore River.

Victorian Beaches - Region 1: Central (Sandpatch Point to Tamboon Inlet)

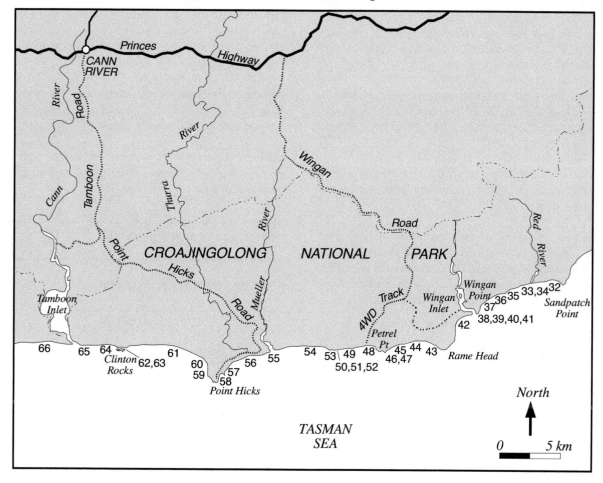

Figure 51.   Region 1 (central section): Sandpatch Point to Tamboon Inlet (Beaches 32 - 66)

32 - 35   RED RIVER 1 to 4

Unpatrolled			
	Rating	Double bar	Length
32 Red River 1	7	TBR/RBB RBB/LBT	3 000 m
33 Red River 2	7	TBR/RBB RBB/LBT	150 m
34 Red River 3	7	TBR/RBB RBB/LBT	100 m
35 Red River 4	6	TBR/RBB RBB/LBT	1 000 m
		Total length	4 250 m

The **Red River** flows out across the beach midway between Sandpatch Point and Easby Head. These four sandy beaches occupy 4.25 km of the 5 km between the two rocky points, both of which are formed of rounded granite. The beaches all face the south-east and receive the full force of the southerly waves. The dominant south-westerly winds blow along the beach and have blown sand diagonally from the beaches east of Red River across Sandpatch Point. The coast track skirts around behind these dunes and reaches the coast at the Red River mouth, from where you can either walk along the beaches to Wingan Inlet, or via the Easby Track, which runs behind the beaches.

The first beach (32) is 3 km long, is backed by a high vegetated foredune and four active parabolic dunes, and has an energetic surf zone. The sandy beach averages 50 m in width and has a moderate slope to the shoreline. The waves average 1.5 m on all four beaches, producing a 150 m wide surf zone with an inner bar cut by rips every 300 m, resulting in up to 20 rips along the beaches; and an outer bar with more widely spaced rips. Additional strong rips run out against the low granite rocks that separate the four beaches. The middle beaches (33 & 34) are short and consequently they are dominated by the rocky boundaries, and have strong permanent rips. The Red River is usually closed, but when open, poses an additional hazard.

Swimming: The river is the safest place, as all four beaches are extremely hazardous with usually high waves, and strong longshore and rip currents; plus the rocks, reefs and their associated currents. If you do bathe, stay on an attached bar and well clear of rip channels and currents.

Surfing: The wide surf, bars and rips produce beach breaks along the length of the beaches, with more consistent banks near the rocks. Best in northerly winds and low to moderate swell.

Fishing: An abundance of good spots including the river, numerous rip holes and gutters, plus the rocks west of the river mouth.

Summary: A series of four interconnected natural beaches, backed by vegetated dunes and the river, and fronted by an often wild surf. West of the river mouth is part of the coastal track. Do not try to walk around Sandpatch Point, as waves wash over the rocks.

36, 37, 38   EASBY 1, 2 & 3

Unpatrolled			
	Rating	Double bar	Length
36 Easby 1	7	TBR/RBB	160 m
		RBB/LBT	
37 Easby 2	7	TBR/RBB	1 600 m
		RBB/LBT	
38 Easby 3	7	TBR/RBB + rocks	30 m
		RBB/LBT	
		Total length	1 790 m

The three **Easby Beaches** extend for 1 800 m from Easby Head, south-west toward the eastern rocks of Wingan Point. The first (36) is located between low granite rocks; the second (37) is more continuous, with a few rocks in the centre of the beach. The third (38) is a pocket of sand surrounded on three sides by rounded granitic rocks, with a permanent rip flowing out between the rocks. All three face the south-east and are exposed to waves averaging 1.5 m, which maintain a continuous 150 m wide surf zone, containing an inner bar cut by strong rips every 250 m; then a trough and an outer bar with more widely spaced rips. Adjacent to all the major rocks and points are strong permanent rips and reefs.

The coast track runs both along and behind the beach (Easby Track), with the western Wingan Point safe to walk around under normal wave conditions. Easby Creek is usually closed and reaches the coast toward the western end of the main beach.

Swimming: Three exposed rip dominated beaches, with both strong longshore and beach rip currents, as well as permanent rips against the rocks. Use extreme care if bathing. Stay close to shore on the attached portions of the inner bar and well clear of the rocks, and river mouth if open.

Surfing: Numerous beach breaks on the inner and outer bars, with the best conditions during low to moderate swell and northerly winds.

Fishing: An excellent location for beach, rock and river fishing. Numerous shifting and permanent rip holes and gutters.

Summary: A natural long beach easily accessible around the rocks from Wingan Inlet.

39   WINGAN POINT NORTH

Unpatrolled
Single bar: LTT
Beach Hazard Rating: **4**
Length: 70 m

Wingan Point is a 500 m long, 20 m high, sloping granitic headland. On its northern side is a moderately protected, 70 m long beach. The beach faces east and receives waves refracting around the point and offshore reefs and rocks, that average 0.5 m in height. These produce a single attached bar, with a small rip running out against the eastern rocks. The beach is backed by a low vegetated foredune, with rounded granite to either side. It is easily reached around the rocks from Wingan Inlet and provides good shelter from westerly winds.

Swimming: A moderately safe, if isolated beach, so long as you stay on the bar and away from the rocks and rip. Seaweed can be a problem, particularly after bigger seas.

Surfing: Usually a low beach break, which tends to close out in higher seas.

Fishing: The surf zone is usually shallow, so the rocks offer the best spots.

Summary: One of the more protected and calmer beaches on this section of coast.

40, 41   WINGAN POINT & INLET

Unpatrolled			
	Rating	Single bar	Length
40 Wingan Point	**5**	R/LTT + inlet	100 m
41 Wingan Inlet	**4**	R + channel	100 m
		Total length	200 m

On the western side of Wingan Point are two small beaches. The first, **Wingan Point Beach**, is a 100 m long section of sand, bordered by rocks and fronted by the ebb tidal shoals of Wingan Inlet. The shoals can extend up to 200 m off the beach, with low waves usually breaking across them.

The second, **Wingan Inlet Beach**, forms the eastern side of the inlet and is a narrow beach with a steep drop off into the several metre deep inlet channel. While waves are low, there are strong tidal currents flowing along the beach.

Swimming: Both beaches have low waves and are a lovely stop to sit and watch the seals on the rocks and reefs of the 'Skerries'. However, if bathing, be very careful of the strong tidal currents which flow across both beaches, particularly on a falling tide.

Surfing: Usually only low waves on Wingan Point Beach, with spilling breakers further out.

Fishing: The inlet is a very popular spot, with the deep channel swept by strong tidal currents between the lagoon and the sea.

Summary: Two lovely small beaches; seemingly tranquil but adjacent to the strong inlet currents.

NOTE: CROSSING WINGAN INLET

Many hikers have to cross Wingan Inlet when walking the coast track. If no boats are available for a lift, it is possible to walk the seaward side of the inlet at low tide. To do this, you must wait until low tide and walk out to where the low waves are breaking and the water is about waist deep. Avoid the main channel as it is several metres deep. It is a good idea to place your backpack in a plastic bag as a precaution.

---

42   WINGAN

Unpatrolled
Inner bar: TBR Outer bar: RBB
Beach Hazard Rating: **6**
Length: 2 000 m

Wingan Beach and camping area is one of the most popular spots in Croajingolong National Park. Camping is so popular that a booking system is required for the summer holidays. The camping area can be reached by car along a 34 km gravel road from the Princes Highway. It is another kilometre walk along the Wingan Nature Walk, which includes a board-walk through the Melaleuca swamp and a trek across the dune to the beach. It is well worth the effort, as the beach is one of the nicest on the coast.

The beach is 2 km long and faces east at the western end, then swings around to face south by the inlet. It is backed by 60 m high scarped dunes, that have blown across Rame Head. The western end, called Fly Cove after the sand flies, is moderately protected by Rame Head and has a continuous attached bar. However as soon as the waves pick up, rips begin to cut the inner bar every 300 m and an outer bar is soon present. The beach has a low gradient and the surf zone is 200 m wide.

The access track from the camping area comes out in the middle of the beach, where waves average 1.4 m and rips dominate. At the eastern end is Wingan Inlet (see Beach 41 for description), while off the inlet are granite rocks and reefs called the Skerries, which are a protected seal colony.

Swimming: Stay close to shore and on the attached portions of the bar. Do not bathe up the beach or near the inlet as waves are larger, rips stronger and a tidal current is present.

Surfing: A popular spot with good, easy beach breaks over the outer bar. Best with a moderate swell and north to north-westerly winds.

Fishing: A very popular spot owing to the good access and choice of lagoon, inlet, beach and rock fishing.

Summary: One of the nicer beaches on the Australian coast, with a good camping area set in eucalyptus forests; a lovely natural river, lagoon and the beach.

Wingan Inlet is also called **Fly Cove**; the name given it by **George Bass** in 1797 when he explored the south coast in an open whaleboat. He sheltered here for ten days from a Bass Strait gale and was no doubt bothered by the sand flies which inhabit the seaweed, usually washed up on the beach in the western corner.

43, 44, 45   RAME HEAD 1, 2 & 3

Unpatrolled			
	Rating	Double bar	Length
43 Rame Head 1	6	TBR + rocks RBB/LBT	200 m
44 Rame Head 2	7	TBR/RBB + rocks RBB/LBT	1 500 m
45 Rame Head 3	7	TBR/RBB + rocks RBB/LBT	1 000 m
		Total length	2 700 m

The coast track from Wingan Inlet crosses behind Rame Head, following one of the ridges of the vegetated parabolic dunes that cover the head. The track reaches the western side of Rame Head at the first of three beaches (Rame Head 1). The coast track follows the beaches from here for 22 km to Point Hicks. The first two beaches (Rame Head 1 & 2) are near continuous and bordered by rounded granitic rocks. A few hundred metres of dune covered rocks separate them from the third beach (Rame Head 3).

The three beaches and rocks cover a total length of 3.5 km and all face the south, receiving waves averaging 1.5 m. The waves have developed a 50 m wide beach fronted by a near continuous, 150 to 200 m wide, double bar surf zone. The inner bar is cut by strong permanent rips against the boundary rocks, with shifting beach rips every 250 m. The outer bar has more widely spaced rips. Numerous smaller rocks are also found along the beach and in the surf.

The waves have also delivered large volumes of sand over the past several thousands of years, to be blown inland and across Rame Head by the strong west-south-west winds. The now largely stabilised dunes parallel these winds.

Swimming: The three beaches are exposed to all southerly waves, resulting in strong and persistent rips and troughs, together with numerous rocks and reefs. Use extreme caution if bathing; stay close inshore on the bars and well clear of the rips and rocks.

Surfing: These beaches pick up all southerly swell and have numerous beach breaks. Best with a low to moderate swell and northerly winds.

Fishing: Excellent permanent rip holes and gutters can be fished from the beach or rocks.

Summary: Three relatively isolated natural beaches; beautiful for a beach walk or fish, but should only be surfed by experienced groups of surfers.

Rame Head was sighted by Captain Cook on the 19th April 1770, and named after a similar hill of this name in Plymouth Sound. He also commented on the sandy nature of the shore and dunes.

The schooner Sarah was wrecked on the head in 1837.

46, 47   PETREL POINT EAST 1 & 2

Unpatrolled			
	Rating	Single bar	Length
46 Petrel Point East 1	7	TBR	50 m
47 Petrel Point East 2	7	RBB/LBT	50 m
		Total length	100 m

In amongst the 2 km of granitic rocks that form Petrel Point are two small pocket beaches on the western side. They both face south-east and are dominated by high waves and rocks, with permanent rips draining their small inner surf zones. Both are difficult to access and should not be used for swimming or surfing.

48 - 53   PETREL POINT WEST 1, 2 (GALE) 3 to 6

Unpatrolled			
	Rating	Single bar	Length
48 Petrel Point West 1	5	LTT/TBR	400 m
49 Gale	6	Double bar TBR/RBB RBB/LBT	1 000 m
50 Petrel Point West 3	3	Single bar R	20 m
51 Petrel Point West 4	3	R	150 m
52 Petrel Point West 5	7	Double bar TBR/RBB RBB/LBT	1 200 m
53 Petrel Point West 6	7	TBR/RBB RBB/LBT	400 m
		Total length	3 170 m

Petrel Point is a 1.5 km wide, 20 m high sloping granite headland, covered by both vegetated and active sand dunes, that have been blown across the point by the strong west-south-west winds. Running for over 3 km due west of the point in a broad open embayment are six near continuous beaches. They all face due south and are exposed to the dominant southerly waves and south-westerly winds. They are backed by continuous vegetated and active parabolic dunes, particularly towards Petrel Point; and are fronted by a high energy and predominantly double bar, 200 m wide surf zone.

Rounded granite rocks and boulders separate the beaches. Beach 48 is partially protected by Petrel Point and begins with a continuous bar and small rips; however, waves and rips are more prevalent adjacent to the rocks separating it from Gale Beach.

Gale Beach is accessible by 4WD, on the 10 km long Gale Hill track. This track provides the best access to these beaches, other than walking the coast track. At beaches 50 and 51, rocks outcrop sufficiently in the surf to dissipate the waves, so that only low 0.5 m high waves reach the shore. At these two shorter beaches, the beach is steep and there is no bar, with waves breaking by surging up the beach face. Either side however, the inner bars are cut by rip channels every 300 m, with strong, permanent rips against the rocks. A deep trough lies seaward, then the outer bar with more widely spaced rips.

Swimming: The safest bathing is in lee of Petrel Point on Beach 48 and at the two central protected beaches, which usually have a low shorebreak and no surf zone. On either side, the waves and rips dominate the beaches.

Surfing: Gale Beach is the most popular because of the 4WD access. It has several beach breaks which are best in low to moderate southerly swell and northerly winds. Few surfers frequent these beaches; as a result there is usually no need to go further afield.

Fishing: This is also a relatively popular fishing spot, again because of the vehicle access; and like much of this section of coast, has excellent rip holes and gutters. Rock fishing is also possible off the rounded granite rocks.

Summary: Six natural beaches with 4WD access; backed by magnificent dunes; some with fresh water draining across the beach; and fronted by surf ranging from low to the more energetic.

54, 55   MUELLER RIVER & MUELLER RIVER SOUTH

	Rating	Double bar	Length
Unpatrolled			
54 Mueller River	**6**	TBR/RBB RBB/LBT	4 500 m
55 Mueller River South	**6**	TBR/RBB RBB/LBT	200 m
		Total length	4 700 m

The Mueller River flows into an elongated lagoon, which crosses the beach toward the western end of **Mueller River Beach**. The beach begins around the granite rocks from Beach 53 and runs in a broad south-facing arc for 4.5 km to the scattered rocks just past the river mouth. Past the rocks, **Mueller River South Beach** continues on for 200 m to another set of rocks.

Mueller River can be reached by car on a short gravel road off the (Point Hicks) Lighthouse or Thurra Road. There are small camping and day visitor areas, both located on the western shores of the inlet. It is a short walk over the dune or along the inlet from the road to the beach.

The beaches face the south and receive waves averaging 1.5 m. These waves produce a double bar system, with rips cutting the usually detached inner bar every 300 m; then a trough and the outer bar 200 m offshore, which is cut by more widely spaced rips. The South Beach has slightly lower waves, but rocks and reef are present in the surf zone. The entire beach is backed by dunes, which toward the eastern end are very active and have large sandy blowouts up to 2 km long. These blowouts are clearly visible from the river mouth.

Swimming: The safest place is in the inlet. Only bathe on the beach when waves are low; otherwise stay on the shallow sections of the bars and avoid the rips and longshore trough, as strong currents abound, even on relatively calm days.

Surfing: Most surfers stay near the river mouth, as it's a long walk to anywhere else. Beach breaks abound, with best conditions occurring during low to moderate swell and northerly winds.

Fishing: A popular spot with the choice of inlet and beach fishing. Deep gutters are guaranteed; with a few rocks to fish from just west of the usually closed inlet.

Summary: An accessible beach with a small, but popular, summer camping area next to the lagoon and beach.

56   THURRA

Unpatrolled
Inner bar: LTT/TBR   Outer bar: TBR/RBB
Beach Hazard Rating: **6**
Length: 3 000 m

The Thurra River reaches the coast 1.5 km west of the Mueller River. It also has a shallow lagoon and flows across the beach as a shallow channel. The dunes have been blown for up to 3 km across Point Hicks and now descend up to 80 m into the river, producing a sand chokcd rivcr mouth. During floods, the sand is washed out onto the beach to continue its movement along the coast. This whole process is known as headland bypassing.

Thurra Beach is 3 km long, beginning at the rocks just east of the river mouth and running west then south-west to the first rocks of Point Hicks. The beach receives waves averaging 1.4 m at the river mouth, which decrease in height toward the western end. The beach is composed of fine dune sand and is low and flat, with a 200 m wide surf zone containing a usually attached inner bar, cut by rips every 300 m; and an outer bar with more widely spaced rips. The outer bar is only active during big seas.

The Thurra or Lighthouse gravel road reaches the coast at the river mouth and runs along the back of the beach to the lighthouse gate. The Thurra River Camping Area begins at the bridge and follows the river mouth around to the beach. This is the largest camping area in Croajingolong National Park and is very popular in the holiday periods. Bookings are essential.

Swimming: Most people swim at the river mouth and in front of the camping sites. These sites are moderately safe under average waves. However, rips dominate and bathers should stay close inshore and on the bars, avoiding the rip channels and river mouth, if open. Further down the beach are lower waves and safer conditions.

Surfing: A very accessible and popular spot with summer visitors, offering a long stretch of beach breaks, which decrease in height toward Point Hicks. Best in a low to moderate swell and northerly winds.

Fishing: A very popular spot with river, river mouth, beach and rock fishing available.

Summary: Probably the most visited and popular spot in the National Park, with good camping sites adjacent to the river beach; and spectacular dunes.

> **Point Hicks Lighthouse:** The lighthouse is open to the public from 10am to 2 pm, Monday to Friday. Cars must be left at the gate and it is a 2.2 km walk to the lighthouse. Phone (051) 58 4208.

57   THURRA SOUTH

> **Unpatrolled**
> Single bar: LTT
> Beach Hazard Rating: **4**
> Length: 200 m

At the western end of the main Thurra Beach are some low rounded granite rocks, just past which is **Thurra South Beach**. This 200 m long beach faces the east which, together with Point Hicks, affords considerable protection from the southerly waves. They average about 0.5 m at the beach and have built a wide, low foredune and beach, fronted by a shallow attached bar, with a rip running out against the eastern rocks.

Swimming: A relatively safe, if remote beach. Best on the bar away from the rocks. Seaweed can be a problem.

Surfing: Usually a low shorebreak which tends to close out in bigger seas.

Fishing: Best off the rocks, as the beach is quite shallow with few holes.

Summary: A nice spot to walk to, however sand flies can be a problem in summer.

58   POINT HICKS EAST

> **Unpatrolled**
> Single bar: TBR
> Beach Hazard Rating: **6**
> Length: 150 m

On the eastern side of **Point Hicks** is a 150 m long, south facing beach, bounded by prominent granite headlands extending a few hundred metres seaward. A lighthouse is located on Point Hicks.

The beach can only be reached on foot via the lighthouse track. It is moderately exposed and receives waves averaging 1.4 m which, contained by the headlands, produce a single bar beach with strong permanent rips against each headland.

Swimming: A remote and hazardous beach with moderate waves and strong permanent rips against the rocks.

Surfing: It's a 2 km walk to see what the waves are like here, so it is little used. There are however, consistent beach breaks running into the two rips. Best in northerly winds and moderate swell.

Fishing: An excellent spot for permanent deep rip holes and gutters which can be fished from the beach or rocks.

Summary: If you walk out to the lighthouse, you should also visit this beach. However, do not bathe unless conditions are calm or you are with experienced surfers.

Point Hicks is one of Australia's best known points, being the first land sighted on Cook's voyage. Lt Hicks saw the point on 19 April 1770. As usual, a 'large Southerly Sea' was running at the time.

The lighthouse was built in 1890.

59, 60   POINT HICKS WEST 1 & 2

Unpatrolled			
	Rating	Single bar	Length
59 Point Hicks West 1	7	TBR	50 m
60 Point Hicks West 2	7	TBR	50 m
		Total length	100 m

On the western side of **Point Hicks** are two adjoining, relatively small, exposed, west facing beaches. Both are 50 m long and are separated by protruding granite rocks. They receive waves averaging 1.4 m which maintain strong permanent rips in each. A dune blowout backs each beach.

They can only be reached on foot along a 1 km long track that runs down from the lighthouse.

Swimming: Remote, rippy and hazardous.

Surfing: Persistent swell, but a small, rock dominated surf zone.

Fishing: Rip holes can be fished from the beach or rocks.

Summary: A nice spot to walk to, however only bathe if conditions are calm.

61   POINT HICKS - CLINTON ROCKS

Unpatrolled
Inner bar: TBR/RBB   Outer bar: RBB/LBT
Beach Hazard Rating: 6
Length: 6 900 m

West of **Point Hicks**, the beach sweeps in a broad south facing arc for 7 km to **Clinton Rocks**. This beach is exposed to southerly waves averaging 1.5 m, and to the strong westerly winds, that have blown sand inland from the beach, as massive longwalled parabolic dunes up to 5 km long. These dunes cross behind Point Hicks and spill into Thurra River down a steep 80 m high slip-off slope.

The beach can only be reached on foot along the coast track from Thurra River, or by 4WD along the 8 km long Clinton Rocks track. The track terminates in thick banksia scrub on the 40 m high dunes behind the beach, with a steep sandy track leading to the beach. The beach is composed of fine to medium sand, which results in a moderately sloping beach face and a 200 m wide, double bar surf zone. The inner bar may be attached to or detached from the beach, with rips crossing it every 300 m, resulting in up to 25 rips along the beach. A deep trough separates it from the straighter outer bar. A large permanent rip flows out against the rocks just west of the track entrance.

Swimming: An attractive though hazardous beach, owing to the persistent rips and deep longshore trough and rip channels. Stay close inshore and well clear of the rips if swimming. The safest location is at the western end inside the rocks.

Surfing: There are numerous beach breaks on the inner and outer bars; the best occurring on the outer bar during low to moderate swell and northerly winds.

Fishing: Excellent rip holes and longshore gutters dominate the beach.

Summary: It is worth the rough drive out along the Clinton Rocks Track (in a 4WD) to see this and the adjoining beaches. The clean sand, waves and dunes provide some brilliant scenery. However, use extreme caution if bathing, as longshore and rip currents dominate the deep channels which run along the beach.

62, 63, 64   CLINTON ROCKS WEST 1, 2 & 3

Unpatrolled			
	Rating	Double bar	Length
62 Clinton Rocks West 1	6	R TBR/RBB	250 m
63 Clinton Rocks West 2	7	TBR RBB	150 m
64 Clinton Rocks West 3	7	TBR/RBB RBB/LBT	800 m
		Total length	1 200 m

A few hundred metres west of where the Clinton Rocks Track reaches the coast, are a series of rounded granite rocks and points, that form the boundaries of the three **Clinton Rocks West Beaches**. All three are only accessible by foot along the beach and around the low rocks, with the closest vehicle access (4WD) being at Clinton Rocks.

All three face due south and are backed by 20 to 30 m high, predominantly vegetated dunes. The first two beaches (62, 63) are relatively small and are dominated by the surrounding rocks and reefs. Both have relatively steep beach faces and permanent rips flowing along the beach and out by the rocks at each end. The third beach (64) is 800 m long and has rips both against the rocks and every 250 m along the beach. All three are connected by a continuous outer bar, lying up to 200 m off the beaches, which has more widely spaced rips.

Swimming: Three attractive natural beaches; however all are dominated by permanent rips against the rocks, and wide energetic surf zones that also contain scattered rocks. Only bathe if conditions are calm; and stay clear of the trough and rip channels.

Surfing: It's a long walk for beach breaks that are similar to those at Clinton Rocks.

Fishing: Excellent permanent and deep rip holes and gutters which can be fished from the beach or rocks.

Summary: Three natural and remote beaches which form a lovely part of the coast track, but which can be very hazardous when waves are running.

65   TAMBOON INLET

Unpatrolled
Inner bar: TBR/RBB   Outer bar: RBB/LBT
Beach Hazard Rating: **6**
Length: 2 500 m

Tamboon Inlet consists of a 3 km long coastal lagoon which connects the Tamboon River with the sea. The inlet mouth forms the western boundary of this beach, which extends for 2.5 km from the western Clinton Rocks beach. A narrow field of active sand dunes backs the beach. The beach can be reached by boat from Tamboon Landing; otherwise it is a long walk along the beach from Clinton Rocks.

The beach faces due south and receives waves averaging 1.5 m. These produce a 200 m wide double bar system, with an inner bar that is often detached and cut by rips every 300 m; then a deep trough and an outer bar. A strong permanent rip runs out against the eastern rocks; and when the inlet is open, strong tidal currents are present.

Swimming: This is an exposed and isolated beach dominated by a deep inshore trough and rips. So use extreme care if bathing here.

Surfing: Beach breaks occur along the length of the beach, with best conditions during low to moderate swell and northerly winds.

Fishing: Tamboon is very popular for bream, while the beach usually has deep rip holes and gutters right off the beach.

Summary: An isolated beach mainly used by fishers from Tamboon Inlet; nice for a walk but hazardous when a sea is running.

66   TAMBOON - SYDENHAM INLET

Unpatrolled
Inner bar: LTT/TBR   Outer bar: RBB/LBT
Beach Hazard Rating: **6**
Length: 11 000 m

Between **Tamboon Inlet** and **Sydenham Inlet** is a straight 11 km long beach. It faces due south and is backed by a narrow field of active dunes, blown along the beach by strong westerly winds. The beach can only be reached by boat across both inlets, or by foot along the coast track.

Waves average 1.5 m along the length of the beach. They produce a 50 m wide sandy beach, fronted by a 200 m wide surf zone, containing an attached inner bar that is usually cut by rips every 300 m; with a deep trough separating them from the straighter outer bar.

Swimming: This is an isolated and exposed beach, prone to strong longshore and rip currents. Only swim if conditions are calm or you are very experienced.

Surfing: The best breaks are on the outer bar, which works in a low to moderate swell and northerly winds.

Fishing: Most fishers go by boat down to the inlet to try beach fishing in the persistent deep rip holes and currents. A deep trough runs the length of the beach.

Summary: A remote and usually empty beach, where great care should be taken if bathing.

Victorian Beaches - Region 1: West (Tamboon Inlet to Snowy River)

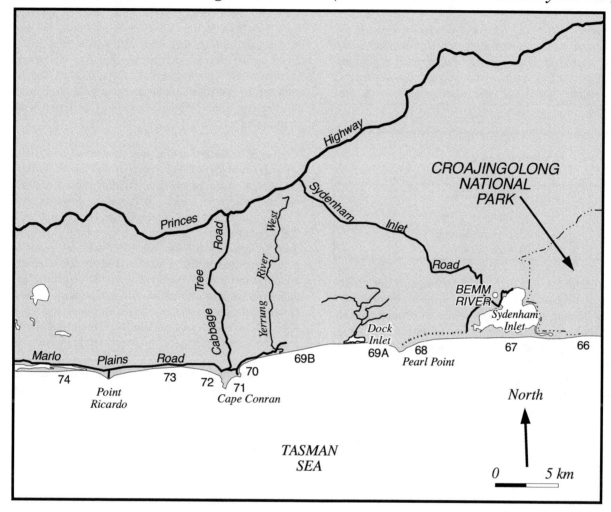

Figure 52.   Region 1 (western section): Tamboon Inlet to Snowy River (Beaches 67 - 74)

67   BEMM RIVER

Unpatrolled
Inner bar: LTT/TBR   Outer bar: RBB/LBT
Beach Hazard Rating: **6**
Length: 11 000 m

The **Bemm River** flows into Sydenham Inlet via a winding delta. The small town of Bemm River is located on the northern side of the inlet, with the mouth located approximately 4 km by boat from the town. The inlet is advertised as the 'bream capital of the world' and is extremely popular in-season. A gravel road also runs out to the coast, joining the beach midway along; where it becomes a 4WD track and runs behind the beach to Pearl Point. There are three parking and beach access points off the track.

The beach runs dead straight east-west for 11 km from the inlet to the rocks east of Pearl Point. It is exposed to all southerly waves which average 1.5 m. It is backed by dunes which narrow to the west; and is fronted by a 200 m wide surf zone with an attached inner bar, which is cut by rips every 300 m; then a long deep trough and the outer bar.

Swimming: This beach is dominated by the inner bar rips and longshore trough, both of which contain strong currents; so bathe only when conditions are calm or if you are very experienced.

Surfing: One of the few beaches accessible by car and hence a moderately popular beach, with numerous beach breaks to choose from. The outer bar works best in low to moderate swell with northerly winds.

Fishing: A popular fishing area with river, lagoon, inlet and beach fishing available. The beach usually has good rip holes, as well as the longshore trough beyond the inner bar. The inlet mouth is most easily accessible by boat.

Summary: Both the inlet and beach are adjacent to the popular holiday and fishing town of Bemm River. There is good car access to the beach, which is popular for fishing, surfing, bathing and walking. However it is a hazardous beach, so take care when waves are breaking.

68   PEARL POINT EAST

Unpatrolled
Single bar: TBR
Beach Hazard Rating: **5**
Length: 500 m

Pearl Point is the first major headland after Point Hicks; a distance of 36 km. The point is composed of steeply dipping sandstone and shales. Low outcrops of these rocks form the boundaries of this 500 m long beach. The point also partially protects the beach, and waves average 1.3 m. The beach can be reached via the road and 4WD track from Bemm River, with an informal parking and camping area behind the low foredune. The beach faces the south and usually has an attached bar cut by strong permanent rips against the boundary rocks, with another one or two rips along the beach.

Swimming: Use care if bathing, as rips and associated currents dominate the inner surf zone.

Surfing: Worth checking out as it offers a little protection from westerly winds, working best in moderate swell and northerly winds.

Fishing: A popular spot for those with 4WDs; with a choice of beach or rock fishing into deep permanent rip holes and gutters.

Summary: A nice spot for a picnic; however rips dominate the surf, so stay on the attached sections of the bar and well clear of the rocks and rips.

69 A, B   DOCK INLET & YERRUNG

		Rating	Double bar	Length
		Unpatrolled		
69A	Dock Inlet	6	TBR RBB/LBT	6 500 m
69B	Yerrung	6	R/LTT RBB/LBT	5 300 m
			Total length	11 800 m

Between Pearl Point and Cape Conran is a 12 km beach interrupted only by the occasional outbreaks of **Dock Inlet** and the **Yerrung River**. The beach is accessible by 4WD from Bemm River at Pearl Point; the track terminating high on the point, with rough steps leading down to the beach. At the western end is the large Banksia Point camping area at Cape Conran, together with a road out to a lookout and access to the beach at the Yerrung River mouth. The bulk of the beach is only accessible by foot.

The beach faces due south, only curving around slightly towards each end. For most of its length it has a moderately steep beach face, which results in either a barless reflective beach (particularly towards Pearl Point) or a continuous attached bar cut by rips every 300 m (particularly towards the river mouth). A broad, deep trough lies off the beach/bar with an outer bar paralleling the entire beach. The beach continues around the base of Pearl Point, with vegetated sand dunes sitting atop the point. High foredunes back the eastern half, with a lower scarped foredune toward the river. Between the river mouth and Cape Conran, a few rocks and reefs outcrop on the beach and in the surf.

Swimming: Moderately safe when waves are low close inshore, however once waves break on the outer bar, strong rip and longshore currents are generated. Deep water parallels the beach when it is reflective. At Yerrung, the wave height decreases to the west, as do the rips and currents.

Surfing: The numerous beach breaks work best with a low to moderate swell and northerly winds. *Lynns Bombora,* just east of the river mouth, can produce some good waves in moderate swell. The far western section is best for learners.

Fishing: Pearl Point is popular with its permanent deep rip holes against the rocks and usually a deep trough along the beach. At Yerrung there is the river, however the bar is shallower, so you need to look for rip holes.

Summary: Both ends of this beach are accessible; but in stark contrast to Pearl Point, it offers an undeveloped cliff-top camp site. Banksia Bluffs at Yerrung has a good range of facilities in a well-organised beach front camping area.

70   EAST CAPE

Unpatrolled
Single bar: R/LTT
Beach Hazard Rating: **4**
Length: 300 m

East Cape Beach is a 300 m long east-south-east facing beach, lying on the eastern side of Cape Conran. A sealed road runs down to the back of the low beach, where there is a picnic area. The Banksia Bluff camping area begins on the bluffs just to the east of the beach.

The beach is shielded by the cape and receives lower waves averaging 0.6 m, which break across a low attached bar, with small rips usually against the eastern rocks. Rocks and reefs are also located along the beach towards the cape.

Swimming: A relatively safe beach under normal low waves; however rips will intensify with higher waves.

Surfing: Usually a small beach break suitable for learners. A reef break just off the beach, known as *The Houses,* works in moderate swell and is protected from westerly winds.

Fishing: The bar is usually shallow, even off the rocks.

Summary: A lovely picnic spot next to the safest beach in the area and the best place to take the family swimming, under normal conditions.

71   COWRIE BAY

Unpatrolled
Single bar: TBR
Beach Hazard Rating: **7**
Length: 250 m

Cowrie Bay is located out on the eastern side of Cape Conran. It can be reached by walking along the East Cape walking track, or from the West Cape parking area. It is a 250 m long south-east facing sandy beach, with dipping rocks and reefs outcropping along half of the beach. Consequently, the surf zone is dominated by rocks and strong rip currents flowing out along the larger reefs.

Swimming: Unsafe unless conditions are calm; owing to the rocks, reefs and permanent rips.

Surfing: There are more rocks than good surf here.

Fishing: A good fishing spot with deep permanent rip holes and gutters that can be fished off the rocks or beach.

Summary: A nice beach to walk to and to fish; however unsuitable for swimming or surfing.

Cape Conran protrudes 1.5 km out to sea and breaks up what is otherwise a sandy coastline. The cape is composed of steeply dipping sandstones and shales, with granitic intrusions particularly prominent on the western cape, where they form large rounded boulders on the shore and in the surf. The Cape Lighthouse was constructed in 1966.

72   SALMON ROCKS

Unpatrolled
Single bar: R
Beach Hazard Rating: **4**
Length: 50 m

Salmon Rocks is on the exposed, west facing side of Cape Conran. It is a small, 50 m long break in the rocks and is the site of the only boat launching ramp in the area. A sealed road runs to the ramp, which is backed by a large car park. The ramp can only be used during low waves and easterly winds, as it is awash under other conditions.

Surfing: During large swell, rideable waves break on a reef located out from the boat ramp, with lefts running in toward the beach. It is best reached from the boat ramp.

Summary: This beach is unsuitable for bathing or surfing and should be left to the boat users.

73   CAPE CONRAN - POINT RICARDO

Unpatrolled
Inner bar: LTT/TBR   Outer bar: TBR/RBB
Beach Hazard Rating: **6**
Length: 11 500 m

Between **Cape Conran** and **Point Ricardo** is an 11.5 km long, south facing beach. The Marlo - Cape Conran road runs behind the beach; however the easiest access is on the western side of Cape Conran and at Point Ricardo, where there are sealed roads and car parks.

The beach receives waves averaging 1.5 m. The eastern end has medium sand which produces a steep beach face, usually fronted by a narrow bar; with a deep trough running between the beach and the outer bar. Toward Point Ricardo, the sand fines, the inner bar widens and rips increase on the inner bar. The point is a sandy foreland in lee of offshore reefs.

Swimming: The exposed location and usually moderate to high waves produce a heavy shorebreak and strong currents in the rips and longshore trough. A strong permanent rip runs out just below the West Cape car park and should be avoided by bathers. Use extreme care if bathing.

Surfing: There are numerous beach breaks on the outer bar along this beach; with Point Ricardo occasionally producing a long right hand break over the reefs.

Fishing: The deep trough can be fished from the beach, with more rip holes at the Point Ricardo end.

Summary: A long exposed beach with access at each end, but no facilities. Nice for a visit, but take care if bathing.

74   POINT RICARDO - MARLO

Unpatrolled
Inner bar: R/LTT   Outer bar: RBB/LBT
Beach Hazard Rating: **6**
(Snowy River Mouth: **8** - deep tidal channels and strong currents when open)
Length: 7 000 m

West of **Point Ricardo**, the beach can extend for anywhere between 2 and 7 km; the actual length at any time depending on the location of the Snowy River mouth. When the river flows straight out to sea at **Marlo**, the beach is 7 km long. However, as the south-westerly waves push the entrance eastward, the beach shortens; until a large flood occurs, that will usually take the entrance back to Marlo.

The Marlo - Cape Conran road parallels the beach, but access can only be obtained along the western few kilometres; east of the water-filled French Narrows. The road runs along the top of 40 m high bluffs, with dune ramps providing access to the beach below.

The beach faces due south and receives waves averaging 1.5 m; which combine with the medium sand to build a relatively steep beach face. This is fronted by a narrow continuous bar on which waves break heavily. A deep trough separates this bar from the outer bar. Towards the river mouth, tidal currents and shoals produce strong variable currents and very hazardous conditions for bathers and boaters.

Swimming: Not recommended on this isolated and hazardous beach. Even during low waves, the water is deep close in and currents can flow along the trough.

Surfing: Beach breaks on the outer bar are best with low to moderate swell and northerly winds. Closer to the Snowy River mouth, the river bars can produce some long rights.

Fishing: The deep longshore trough can be fished from the beach, while the river mouth provides additional attractions.

Summary: A very dynamic beach and river mouth, which can be viewed from the road. Usually unsuitable for swimming.

2. NINETY MILE BEACH

> ### Ninety Mile Beach and
> ### Corner Inlet
> ### (Marlo to Corner Inlet)
>
> Beaches: 75 to 83
> Coast length: 228 km
>
> *For maps of region see Figures 54, 56, 57, 58 & 61*

Ninety Mile Beach officially begins at the Snowy River mouth and runs with a few minor interruptions for 177 km (111 miles) to Shoal Inlet (Table 16). At this point, increasing tidal dominance breaks the beach into a series of low, sandy islands and large, tide dominated inlets; collectively known as Corner Inlet. These islands and inlets terminate at the large Snake Island. Only one major town; Lakes Entrance; lies adjacent to the long beach, with smaller towns at Marlo, Lake Tyers and Seaspray. There are also small settlements at Loch Sport, Paradise Beach, Golden Beach and Woodside. In lee of the islands of Corner Inlet are the two small towns of Port Albert and Welshpool.

The entire system, from the Snowy River to Snake Island, is dominated by sandy beaches. These are shaped by the prevailing westerly winds and waves, with tides increasing in importance in Corner Inlet. While the whole system curves around from facing south to south-east; overall the impression is given of very long and very straight beaches; generally backed by moderately high, vegetated foredunes (Figure 53).

Figure 53. Ninety Mile Beach is the longest beach in Victoria and the second longest in Australia. This view, looking south-west down Bunga Arm, clearly shows the single bar separated from the beach by a deep, rip dominated longshore trough; backed by the relatively narrow barrier (beach and foredune), then the elongated waters of Bunga Arm, with Boole Boole Peninsula behind.

Table 16. How long is Ninety Mile Beach?

No	Name	Miles	Kilometres	Cumulative kilometres
75	Corringle - Lake Tyers	25.5	41	41
76	Shelly Beach - Red Bluff	1.2	2	43
77	Eastern Beach (Lakes Entrance)	5.0	8	51
78	Ninety Mile Beach (Barrier Landing - Shoal Inlet)	79	127	178
79	Shoal Inlet Island	3.7	6	184
80	Shallow Inlet Island	4.3	7	191
81	Clonmel Island	5.0	8	199
82	Clonmel Banks	3.7	6	205
83	Snake Island	10.5	17	222
	Total	**137.9**	**222**	

Victorian Beaches - Region 2 (Marlo to Lakes Entrance)

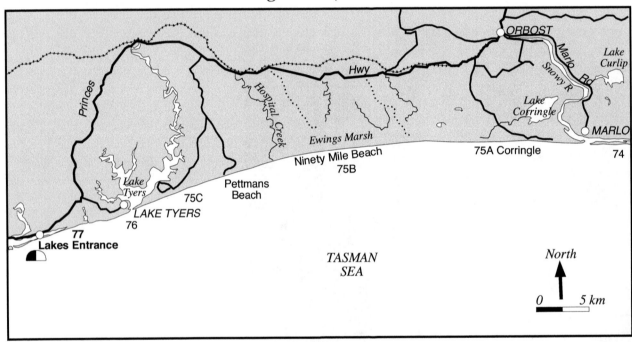

Figure 54.   Region 2 (eastern section): Marlo to Lakes Entrance (Beaches 75 - 77)

75 A, B, C   CORRINGLE - EWINGS MARSH - PETTMAN

Unpatrolled			
	Rating	Single bar	Length
75A Corringle	7	RBB/LBT	10 km
75B Ewings Marsh	6	RBB/LBT	20 km
75C Pettman	6	RBB/LBT	11 km
		Total length	41 km

The first section of Ninety Mile Beach is one of Australia's longest beaches; the 41 km section from the Snowy River mouth to Lake Tyers. It consists of three sections. The easternmost section lies on the western side of the Snowy River mouth, at **Corringle**. There is a camping area and access to the beach over the dune, together with boat access to the river.

The second long section parallels the largely infilled **Ewings Marsh**. The marsh prevents vehicle access to the beach and you have to leave your vehicle at the end of one of the forestry tracks and walk or wade across the marsh, to reach the dune and beach.

In the west, there is road access again at **Pettman Beach**, where there is a car park and picnic area. Finally, the Lake Tyers House road provides access, but no facilities, toward the western end of the beach. The beach is little used outside of Corringle, due to both its length and limited access and facilities.

The beach essentially faces south, curving slightly to the south-south-east toward Lake Tyers. The entire length of the beach receives waves averaging 1.5 m. These interact with the generally medium sand to build a moderately steep beach face, fronted by a 50 m wide and 2-3 m deep trough running the length of the beach. A single bar lies seaward of the trough and may be straight following big seas, or rhythmic and cut by rips every 250 m, under normal conditions. These rips may induce similarly spaced rhythmic topography along the shoreline. A 10 to 20 m high, narrow foredune backs most of the beach, increasing in size east of Corringle.

Swimming: This is a long, isolated and little used beach. The beach is fronted by a deep trough with often strong longshore and rip currents. So use extreme care if bathing and avoid the hazardous Snowy River mouth area.

Surfing: Most popular at Corringle and Pettman where access to the beach is easiest. Corringle has the best potential, with the Snowy River mouth bars capable of producing long lefts. Elsewhere, beach breaks occur on the bar and are best with low to moderate swell and northerly winds.

Fishing: Again Corringle and Pettman are the most popular because of the easy access. The deep longshore trough can be easily reached from the beach.

Summary: A magnificent long beach and foredune system; backed by the pristine Ewings Marsh and bordered by the mighty Snowy River and Lake Tyers. It should be incorporated into the adjoining National Park.

76   SHELLY

Unpatrolled
Single bar: RBB/LBT
Beach Hazard Rating: **7**
Length: 2 000 m

Lake Tyers is a relatively natural estuary containing three drowned valleys. The flooded valleys form the lake and join together to enter the sea at the western end of the long Corringle - Ewings Marsh - Pettman Beach. Tertiary rocks rising to 20 - 40 m high hills and bluffs, form the southern side of the inlet and provide the location for the small town of Lake Tyers. The persistent southerly waves usually build the beach across the entrance, blocking the lake from the sea.

The Lake Tyers road terminates at the inlet where there is a boat ramp, large caravan park and a car park for

Shelly Beach. The beach is also accessible at the western end from the bluff-top car park at Red Bluff.

The beach extends from the usually closed lake mouth 2 km south-west to the 40 m high sandstone and shale bluffs at Red Bluff. This small bluff is the only bedrock outcrop on Ninety Mile Beach between Cape Conran and Wilsons Promontory. The bluff is surrounded by a rock platform, with reefs in the surf zone. The beach is fully exposed to the southerly waves which average 1.5 m. They produce a straight beach, with a moderately steep beach face, usually fronted by a 2-3 m deep, 50 m wide longshore trough, seaward of which the longshore bar is cut by deep rip channels every 250 m.

Swimming: The lake is the safest place to bathe, if the entrance is closed. Be very careful in the surf, because of the deep trough against the beach and its strong longshore and rip currents.

Surfing: Red Bluff is the best break on the entire Ninety Mile Beach and also the most popular. The reefs on either side can produce some excellent right and left handers.

Fishing: This is a very popular destination for fishing, with a range of lake, surf and rock fishing on offer. The beach usually has a deep trough close inshore.

Summary: A nice spot; close to Lakes Entrance, but away from the summer crowds. It offers a good range of lake and beach environments, with basic facilities and accommodation available in the small settlement.

77   EASTERN

Lakes Entrance SLSC
Patrols: late November to Easter holidays
Surf Lifesaving Club: Saturday, Sunday and public holidays
Lifeguard: 10 am to 6 pm weekdays and Saturday morning, during Christmas holidays
Single bar: RBB/LBT
Beach Hazard Rating: **6**
Length: 8 000 m
For map of beach see Figure 55

Eastern Beach is the northern section of the great Ninety Mile Beach. It begins at Red Bluff, a popular surfing spot, and runs straight west-south-westerly for 8 km to the training walls at Lakes Entrance. The beach faces straight into the dominant southerly waves and receives waves averaging 1.5 m. These waves, as for most of Ninety Mile Beach, produce a 100 m wide surf zone characterised by a usually steep beach face, then a

50 m wide, 2-3 m deep trough fronted by a continuous bar. This bar is cut by rip channels and currents every 200 m; resulting in up to 30 or more rips along the beach.

The Lakes Entrance Surf Life Saving Club is located at the end of the footbridge from the town of Lakes Entrance. This is a popular beach for bathing, surfing, walking and fishing; particularly as it is the closest and most accessible beach to Lakes Entrance. The surf lifesaving club was founded in 1956 and averages 17 rescues a year. The beach is only accessible by car at three locations: Red Bluff in the north; at the Eastern Beach car park in the centre; and via the Lakes Entrance footbridge.

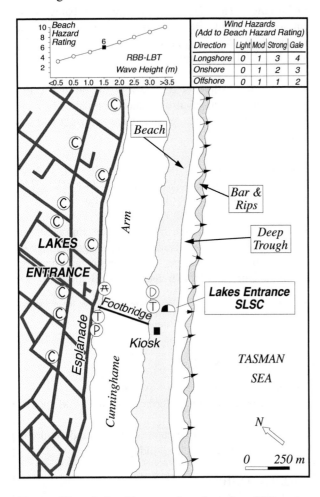

Figure 55.   Lakes Entrance beach and surf lifesaving club are reached via the footbridge from the town. The beach usually has a steep beach face, deep trough and a bar offshore. Stay between the flags and watch for rips in the trough.

Swimming: Only bathe between the flags at the surf club, as this is an exposed beach with deep troughs and persistent rips. Rips increase adjacent to Red Bluff and toward the entrance walls. Strong winds, particularly from the south and west, intensify the longshore and rip currents.

Surfing: A popular spot, more for its accessibility than its waves. Best at Red Bluff, where the reefs provide more consistent breaks, and toward the lake entrance where tidal shoals can produce some longer rides. Otherwise the beach breaks depend on the bars, waves, tide and wind and hence are highly variable. All spots are best in a northerly, with moderate south swell.

Fishing: Like most of Ninety Mile Beach, this is very popular for beach fishing, with a deep trough usually running the length of the beach. Rip channels or holes are also common. Rock and reef fishing on Ninety Mile Beach is only available at Red Bluff and at the north entrance wall.

Summary: This is the most popular spot on Ninety Mile Beach. It offers a safer, patrolled beach, with plenty of room for walking and fishing. However, the deep trough and rips produce hazardous bathing conditions and a reasonable surf. So bathe between the flags and only surf with friends.

78 NINETY MILE BEACH
(Barrier Landing to Shoal Inlet)

The longest continuous section of **Ninety Mile Beach** extends for 125 km from Barrier Landing; on the southern side of Lakes Entrance; down to the next inlet at Shoal Inlet. This magnificent stretch of beach; one of the longest in Australia; begins facing south-south-east, before turning to face south-east, from where it runs as a near straight beach all the way to Shoal Inlet. While the beach and surf zone change little throughout its length, the backing foredune, dunes, barrier and lakes undergo considerable changes. The beach can be divided into 22 identifiable sections.

GIPPSLAND LAKES COASTAL PARK	
Coast Length:	79 km
Number of beaches:	1 (part of Ninety Mile Beach)
Beach:	78

Public Camping Areas
(bookings required in holiday seasons):

- Bunga Arm (lakeside)
- Paradise Beach
- Gippsland Lakes (Golden Beach to Seaspray)
- McGaurans Beach
- Reeves Beach

Park Information:	Bairnsdale:	(051) 952 2656
	Loch Sport:	(051) 946 0278
	Lakes Entrance:	(051) 955 1539

NINETY MILE BEACH

No.	Name	Beach Hazard Rating	Inner bar	Outer bar	Length (km)
Figure 56:					
78A	Barrier Landing	7	RBB/LBT + inlet	-	6
78B	Second Blowhole	6	RBB/LBT	-	5
78C	First Blowhole	6	RBB/LBT	-	5
78D	Bunga Arm	6	RBB/LBT	-	2
78E	Steamer Landing	6	RBB/LBT	-	2
78F	Ocean Grange	6	RBB/LBT	-	7
Figure 57:					
78G	Lake Reeve	6	RBB/LBT	-	6
78H	Stockyard Hill	6	RBB/LBT	-	22
Figure 58:					
78I	Paradise	6	RBB/LBT	-	6
78J	Golden	6	RBB/LBT	-	4
78K	Delray	6	RBB/LBT	-	3
78L	The Wreck	6	RBB/LBT	-	3
78M	Flamingo	6	RBB/LBT	-	3
78N	Glomar	6	RBB/LBT	-	7
78O	The Honeysuckles	6	RBB/LBT	-	6
78P	**Seaspray (SLSC)**	5	LTT/TBR	RBB/LBT	2
78Q	Lake Denison	5	LTT	RBB/LBT	8
78R	McGaurans	5	LTT	RBB/LBT	8
78S	Jack Smith Lake	5	LTT	RBB/LBT	4
78T	**Woodside (SLSC)**	5	LTT	RBB/LBT	8
Figure 61:					
78U	Reeves	5	LTT	RBB	4
78V	Mclaughlins	6	LTT	RBB + inlet	4
				Total length	125 km

Swimming: Ninety Mile Beach is a long, potentially hazardous and largely isolated beach. Approximately 500 rips occur along this beach, which, together with the longshore trough, dominate the beach circulation. A heavy shorebreak also occurs on the steeper section of the beach. In the deep trough that separates the beach from the bar, there are usually strong longshore and rip currents; often with a pronounced easterly drift. Use extreme caution if bathing on Ninety Mile Beach. Do not bathe alone. Stay close inshore on the beach or attached bar, if present. Do not swim out into the trough or to the outer bar, unless you are very experienced.

Surfing: The numerous rips and longshore bar produce literally hundreds of potential beach breaks, particularly on the outer bar. Conditions vary considerably and are best with a low to moderate south swell and north to north-west winds.

Fishing: Ninety Mile Beach and the backing Gippsland Lakes are a fishing paradise. The continuous trough and hundreds of rips produce any number of good spots for beach fishing.

78A   BARRIER LANDING

Barrier Landing refers to the jetty 2 km south of the Lakes Entrance mouth. The landing here has a lakeside picnic area and toilets, with a short walk over the foredune to the beach. A few fishing shacks are located on the beach. This beach has the additional hazard of being close to the lake entrance, where tidal shoals produce additional tidal currents. These improve the fishing and surfing, at the expense of bathing.

Victorian Beaches - Region 2 (Lakes Entrance to Bunga Arm)

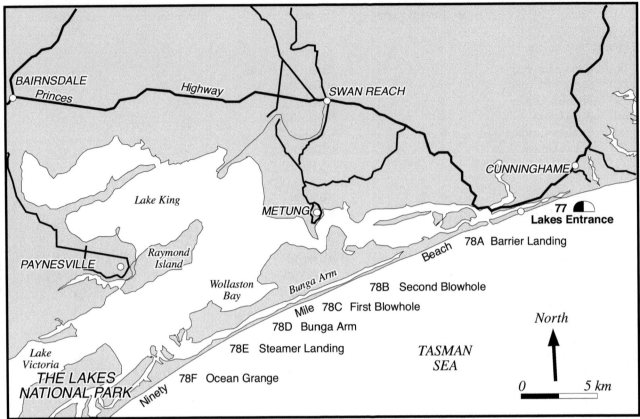

Figure 56.   Region 2: Ninety Mile Beach - Lakes Entrance to Ocean Grange (Beaches 77 - 78F)

78B - F   SECOND & FIRST BLOWHOLE, BUNGA ARM, STEAMER LANDING & OCEAN GRANGE

This 21 km section of Ninety Mile Beach can only be reached by boat, with public jetties at **Ocean Grange** and **Steamer Landing**. There are picnic areas at Ocean Grange and **First Blowhole**, and a camping area at **Bunga Arm**. All are located along Bunga Arm; a narrow 15 km long stretch of water lying between the often narrow dune and beach, and the backing Boole Boole Peninsula.

All these beaches are usually about 50 m wide, with a moderately steep beach face, a longshore trough with rips every 250 m, and an outer bar. When there are crenulations in the beach, the rips are always located in the embayments.

78G   LAKE REEVE

The 6 km long **Lake Reeve** section of Ninety Mile Beach is backed by the shallow Lake Reeve. Most of

the barrier is private property and there is no public entry. The only way to access the public beach is by foot from Ocean Grange or Stockyard Hill.

78H   STOCKYARD HILL

Stockyard Hill refers to a prominent 20 m high foredune on the beach near Loch Sport. There is a road out to the beach and parking at the hill. Private property lies to either side with no public access.

78I, J   PARADISE & GOLDEN

Golden Beach is one of the more popular stops on Ninety Mile Beach. The road from Sale reaches the coast here, and it is the first place south of Lakes Entrance where you can drive to the beach with a shop and some facilities. A road runs north for 3 to 4 km from here to **Paradise Beach**, where there are numerous holiday homes.

Victorian Beaches - Region 2 (Bunga Arm to Paradise Beach)

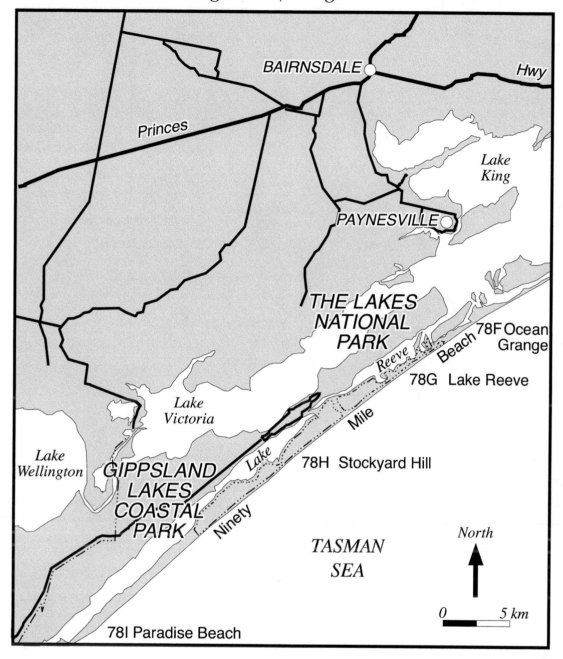

Figure 57.   Region 2: Ninety Mile Beach - Bunga Arm to Paradise Beach (Beaches 78F - 78I)

78K - O   DELRAY, THE WRECK, FLAMINGO, GLOMAR, HONEYSUCKLES

Between Golden Beach and Seaspray, the road follows the beach for 25 km. Along this section are a number of parking areas which provide the best access to the beach. The beaches tend to be named after the parking areas (**Delray, Flamingo, Glomar, Honeysuckles**); except for **The Wreck**, where there is the wreck of a boat on the beach.

The beach remains steep, with a rhythmic longshore trough and bar, and rips every 250 m.

Victorian Beaches - Region 2 (Paradise Beach to Woodside)

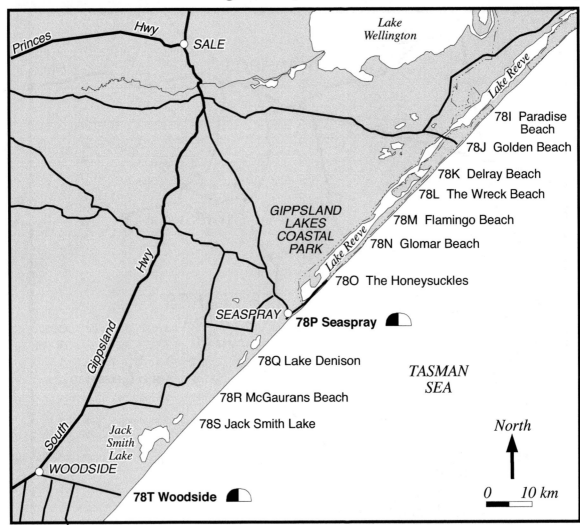

Figure 58.   Region 2: Ninety Mile Beach - Paradise Beach to Woodside (Beaches 78I - 78T)

78P   **SEASPRAY**

Seaspray SLSC

Patrols: late November to Easter holidays
Surf Lifesaving Club: Saturday, Sunday and public holidays
Lifeguard: 10 am to 6 pm weekdays, during Christmas holidays

Inner bar: LTT/TBR   Outer bar: RBB/LBT
Beach Hazard Rating: **5**
Length: 2 000 m

For map of beach see Figure 59

The **Seaspray** section of Ninety Mile Beach is a 2 km long section centred on the small seaside holiday settlement of Seaspray. The beach is straight, faces the south-east, and receives southerly waves averaging 1.4 m. However, higher waves associated with westerly gales and the finer sand along this section produce a wide, low beach; fronted by a 120 m wide surf zone containing two shore parallel bars. The inner bar is usually attached to the shore and is either continuous (LTT) or cut by rip channels every 250 m. A 50 m wide, 2 m deep trough lies between the inner and outer bars. Waves greater than 1 m usually break on the outer bar, which is also cut by rip channels and currents every few hundred metres. There is a tendency for the waves and currents to run up the beach, due to the common south to westerly winds.

The surf club, formed in 1955, is located in the centre of the foreshore reserve and adjacent to the caravan park. Most facilities are available at the club and adjacent park and settlement.

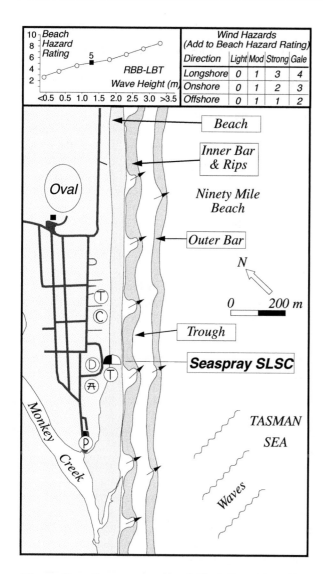

Figure 59. Seaspray township lies immediately behind the beach. A foreshore reserve backs the dune, with the beach usually containing a shallow, attached inner bar, which may have rip holes and currents; then a deeper trough with currents and an outer bar. Stay on the inner bar and between the flags.

Swimming: A moderately safe beach owing to the usually shallow, attached bar. However, watch for the adjoining rips and longshore trough, which can contain strong currents. Do not bathe on the outer bar unless you are experienced. This is a patrolled beach, however 4 rescues occur on average each year, so stay between the flags.

Surfing: Potential beach breaks on both the inner and outer bars; with the outer bar being the best during moderate waves and light north to westerly winds.

Fishing: Excellent beach fishing both in the beach gutters (the rips), together with the longshore trough off the inner bar.

Summary: Seaspray has all the facilities required for a day at the beach with shops and a grassy foreshore reserve providing good parking and access to a moderately safe and patrolled beach.

78Q, R, S　LAKE DENISON, MCGAURANS, JACK SMITH LAKE

South of Seaspray the beach continues to change. The sand remains fine all the way to Reeves Beach. This results in a double bar system, with a usually attached inner bar cut by rips every 250 m, a trough, then the outer bar. In addition, the beach is eroding along parts of this section, which usually results in a scarped dune at the back of the beach.

The first beach access south of Seaspray is **McGaurans Beach**. The McGaurans Beach Road runs straight out to the beach, where there is a small car park but no facilities. Four kilometres to the south is **Jack Smith Lake Beach**; named after the backing lake. Access to this beach is via a 2 km track off the Seaspray Road. The track can be flooded during wet weather.

78T　WOODSIDE

Woodside SLSC
Patrols: late November to Easter holidays **Surf Lifesaving Club:** Saturday, Sunday, and Christmas holidays **Lifeguard:** no lifeguard on duty or weekday patrols
Inner bar: LTT　　Outer bar: RBB/LBT Beach Hazard Rating: **5** Length: 8 000 m
For map of beach see Figure 60

Woodside Beach is the southernmost access with facilities on Ninety Mile Beach. The beach is located in the centre of this straight, 8 km long, south-east facing section. The waves usually arrive from the south-west to south-east and average 1.3 m. These combine with the fine sand to produce a beach with a low slope, fronted by a usually continuous attached inner bar, which is cut by rips during and following higher waves. Seaward of this bar is a continuous 2 m deep trough and an outer bar, which is cut by rips every few hundred metres. Westerly winds both increase the waves and intensify the flow of rip and trough currents to the north.

The beach is patrolled by the Woodside Surf Life Saving Club; the newest in the state, being formed in 1968; and is also backed by parking and picnic areas and a caravan park.

Swimming: A moderately safe beach under average conditions; beware during strong winds when waves and rips intensify. Stay between the flags as 10 people are rescued here each year.

Surfing: Usually low to moderate beach breaks, with shape depending on the bars. Best during moderate swell and light north to westerly winds.

Fishing: A popular fishing spot, with occasional rip holes in the inner bar, and the long trough separating the bars.

Summary: Woodside is a relatively quiet location with just the caravan park and surf lifesaving club. However, it is an easily accessible and moderately safe patrolled beach.

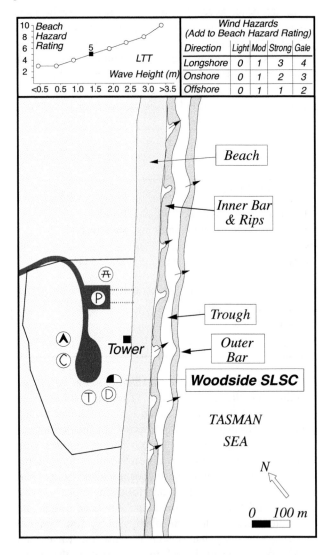

Figure 60. Woodside Beach is the site of a caravan park, picnic area and surf lifesaving club. The beach usually consists of a shallow attached inner bar, often cut by rips; a deeper longshore trough and an outer bar. When bathing, stay on the inner bar and between the flags.

78U, V   REEVES, MCLAUGHLINS

Reeves Beach is the southernmost part of Ninety Mile Beach that is accessible by road, with the Reeves Beach Road terminating at the beach. There is a camping area behind the beach, but no other facilities. A 4WD track runs for a few kilometres down the back of the dunes, where it links up with the foot track from the estuarine **Mclaughlins Beach**. These tracks provide access to Mclaughlins Beach; the southernmost on Ninety Mile Beach. Two kilometres south of the foot track, the beach ends at Shoal Inlet.

The last kilometre of the beach towards Shoal Inlet is increasingly impacted by the inlet's tidal shoals and currents. These produce variable currents and more hazardous bathing conditions. They do however improve the fishing potential, as well as the surfing breaks.

CORNER INLET	
Beaches:	5
Inlets:	5
Islands:	5
Coast Length:	41 km

Corner Inlet is the name given to a series of inlets, low sand islands and the extensive backing estuarine area. The estuary includes the towns of Port Albert and Port Welshpool. The coast consists of a series of five sand islands that lie between the southern end of Ninety Mile Beach and the high granite rocks of Wilsons Promontory.

Unlike Ninety Mile Beach, the beaches of Corner Inlet are dominated as much by the tide as by waves. Wave height decreases toward Snake Island, owing to the protection afforded by Wilsons Promontory. The dominant westerly winds tend to blow offshore, thereby not producing waves at the coast or building high dunes. At the same time, the tide range increases as the tidal waves are amplified over the shallower continental shelf of eastern Bass Strait. The spring tide range reaches 2.8 m at Port Welshpool.

As a result of the decreasing influence of waves and increasing influence of tides, the sandy coast is broken up by a series of five major tidal inlets; the largest in southern Australia. Each inlet contains a deep tidal channel and extensive tidal shoals on the flood (landward) and ebb (seaward) sides of the inlet. Port Albert and Corner Inlet extend their ebb-tide banks out to sea 3 and 5 km, respectively.

Victorian Beaches - Region 2 (Reeves Beach to Snake Island)

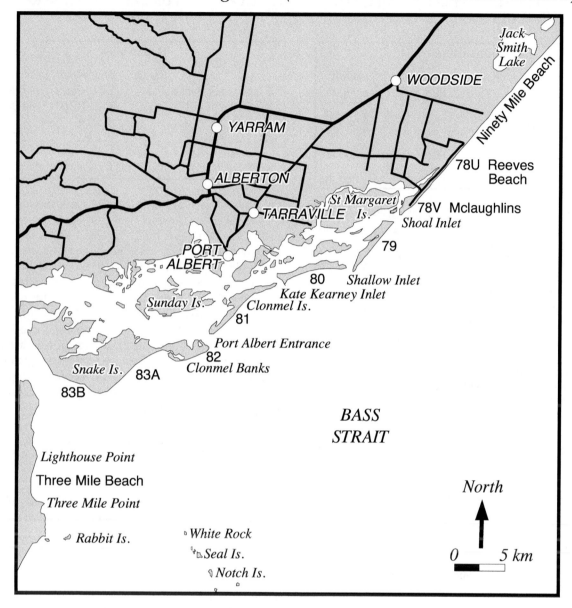

Figure 61.   Region 2: Ninety Mile Beach and Corner Inlet - Reeves Beach to Snake Island (Beaches 79 - 83B)

NOORAMUNGA MARINE AND COASTAL PARK

Open Coast Length:	75 km
Number of beaches:	6
Beaches:	79-83B

Nooramunga Marine and Coastal Park incorporates all the open coast from Reeves Beach to Snake Island, as well as much of the backing estuarine waters, islands and shoals. The park is zoned for amateur and commercial fishing, with amateur spear fishing of flounder permitted.

Information:	Yarram	(051) 982 5995
	Tidal River	(056) 980 8538

CORNER INLET MARINE AND COASTAL PARK

The estuarine waters and shores of western Corner Inlet form part of the Corner Inlet Marine and Coastal Park. All of the park can be used for amateur and commercial fishing and amateur flounder spear fishing.

Information:	Yarram	(051) 982 5995
	Tidal River	(056) 980 8538

79, 80, 81   SHOAL INLET ISLAND, SHALLOW INLET ISLAND, CLONMEL ISLAND

Unpatrolled			
	Rating	Single bar	Length
79 Shoal Inlet Island	5	TBR + inlet	6 000 m
80 Shallow Inlet Island	5	TBR + inlets	7 000 m
81 Clonmel Island	5	TBR + inlets	8 000 m
		Total length	21 000 m

The first three sand islands: Shoal Inlet, Shallow Inlet and Clonmel Island, are nearly a continuation of Ninety Mile Beach, only broken by Shoal, Shallow, Kate Kearney and Port Albert Inlets. The only access to the islands is by boat. All three consist of two parts. The first is the central section dominated by normal waves, which have produced a single, usually attached bar, cut by rips every 200 m. The second part consists of the ends of each island, which border the prominent inlets.

The strong tidal flows through the inlets have produced both deep tidal channels and extensive tidal shoals. The shoals continue to change and modify the shape and size of the ends of the islands. Waves break well seaward of the beach on these shoals, causing the beach in their lee to be steep and barless. However, strong tidal currents flow along the sides of the beaches.

Swimming: These are isolated and little used islands, so use extreme care if bathing. Stay well clear of the inlet and their deep channels; and on the beach, stay clear of the rips.

Surfing: Waves average about 1 m, producing usually low beach breaks.

Fishing: Most fishing is done from boats in the estuary and channels. However, the beaches offer a shore base for fishing the inlet channels and the beach.

Summary: Three totally natural and very dynamic islands and inlets.

82, 83A & B   CLONMEL BANKS, SNAKE ISLAND EAST & WEST

Unpatrolled			
	Rating	Single bar	Length
82 Clonmel Banks	5	LTT + inlet	6 000 m
83A Snake Island East	6	TBR/RBB + inlet	8 000 m
83B Snake Island West	5	R + inlet	9 000 m
		Total length	23 000 m

On the south side of Port Albert Inlet is a very dynamic tidal shoal called Clonmel Banks. It lies a few hundred metres off the normal run of the coast and is low, with little vegetation cover. Even here it is still sheltered by shoals further offshore. Waves average less than 1 m and a narrow continuous bar runs along the island. On adjoining Snake Island, the beach is initially oriented south-east, receives waves averaging 1 m and has a bar cut by rips every 250 m. The orientation of the island changes at Corner Inlet and the western side actually forms the northern side of the inlet. Wave height drops to a few decimetres and the beach becomes steep and barless. The deep inlet and its strong tidal currents lie just off the western beach.

The only access to these islands is by boat and though there are no facilities, camping is permitted on Snake Island.

Swimming: These are remote and little used islands and beaches. Be extremely careful if bathing; even if waves are low, there can be strong tidal currents.

Surfing: Usually only low shorebreaks.

Fishing: Most fishing is done from boats as they provide the only access to the islands. However if it is too rough outside, the beaches can be used to fish the adjoining inlets.

Summary: A highly variable array of low sand islands, tidal shoals and inlets, with low wave energy but potentially dynamic and dangerous beaches.

3. WILSONS PROMONTORY

> **Wilsons Promontory**
> **(Corner Inlet to Shallow Inlet)**
>
> Beaches: 84 to 118
> Coast length: 122 km
>
> *For map of region see Figure 62*

Wilsons Promontory is Australia's southernmost mainland location. Unlike the predominantly sandy coast to either side, it consists of ancient granite hills reaching heights of over 500 m. These hills have been eroded and dissected over many millions of years to produce a series of valleys around the Promontory, many of which now house sandy beaches.

The beaches however, vary considerably between the two sides of the Promontory. Those in the east are in lee of the Promontory and are protected from the direct impact of the strong prevailing westerly winds and waves. Consequently, they are generally low energy beaches composed of pure quartz grains that have been weathered from the surrounding granite (Figure 63).

Victorian Beaches - Region 3 (Wilsons Promontory)

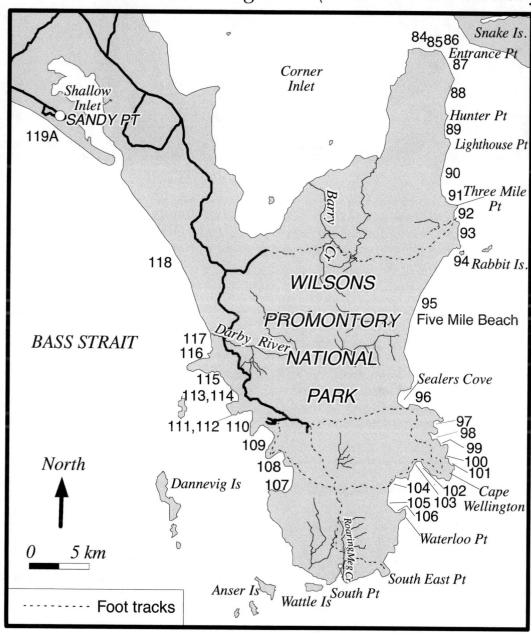

Figure 62.   Region 3: Wilsons Promontory (Beaches 84 to 118)

Figure 63. Sealers Cove, on the protected eastern side of Wilsons Promontory, is backed by the towering granite domes that dominate the entire Promontory. Sediments eroded from this granite supply much of the quartz sand to the eastern beaches. These beaches receive generally lower waves than those on the western side. At Sealers Cove, the additional protection of Horn Point results in very low waves at the southern end and the formation of an extensive intertidal sand flat, seen here at low tide.

The western Promontory beaches face the wind and waves. The waves build wide, low gradient beaches largely composed of shell detritus washed up from the continental shelf. They have also blown the sand up to 10 km inland, to form a series of large transgressive coastal dune systems. The older dunes were deposited hundreds of thousands of years ago during previous high sea levels.

84, 85, 86 SHELTER COVE WEST & EAST, BIDDIES COVE

Unpatrolled			
	Rating	No bar	Length
84 Shelter Cove West	2	R	300 m
85 Shelter Cove East	2	R	350 m
86 Biddies Cove	2	R	600 m
	Total length		1 250 m

The northern end of Wilsons Promontory is dominated by 475 m high Mount Singapore. At its foot are three relatively protected north facing beaches: **Shelter Cove West and East** and **Biddies Cove**. They lie along the southern edge of the 3 km wide Franklin Channel; the main tidal inlet for Corner Inlet. As a result, strong tidal currents flow off the beaches. The beaches can only be reached by boat.

Each beach consists of a steep, reflective high tide beach, fronted by intertidal sand flats that are exposed at low tide. The flats reach a width of 200 m in the middle of the coves, but are narrow off the boundary points. Strong tidal currents run just off the flats.

Swimming: Best at the beach at high tide. The depth over the tidal flats depends on the tide. Do not bathe off the tidal flats, owing to the deep channel and strong currents.

Surfing: None.

Fishing: Best off the points that lie close to the main tidal channel.

Summary: Three similar, protected, sandy beaches facing the tide dominated Franklin Channel.

WILSONS PROMONTORY
National Park, Marine Park & Marine Reserve

NATIONAL PARK
Area:	49 000 ha
Coast length:	122 km
Number of beaches:	34

All of the Promontory south of Kanakie Entrance forms part of the Wilsons Promontory National Park. The park has sealed road access to Tidal River, where there are extensive caravan, cabin and camping facilities. The only other facilities are several overnight camping sites located on the walking trails. The park offers a wide range of activities including bathing, surfing, beach, rock and boat fishing, walking, hiking and sightseeing.

MARINE PARK
Coast length:	east coast:	34 km
	west coast:	32 km
Number of beaches:	east coast:	11
	west coast:	8

The Marine Park borders 66 km of the east and west sides of the Promontory. Commercial fishing is permitted, however there are restrictions on amateur and spear fishing.

MARINE RESERVE
Coast length:	54 km
Number of beaches:	11

The Marine Reserve rings the southern coast of the Promontory. No amateur fishing is permitted, while commercial fishing is being phased out.

Coastal camping areas:
- Tin Mine Cove (Corner Inlet)
- Johnny Souey Cove
- Sealers Cove
- Refuge Cove
- Waterloo Bay
- Roaring Meg
- Oberon Bay
- Tidal River

Information: Tidal River   (056) 980 8538

87, 88, 89   ENTRANCE POINT, ENTRANCE POINT SOUTH, HUNTER POINT SOUTH

Unpatrolled			
	Rating	Single bar	Length
87 Entrance Point	4	R	1 500 m
		Double bar	
88 Entrance Point South	5	LTT LBT	3 600 m
89 Hunter Point South	5	LTT LBT	1 500 m
		Total length	6 600 m

Entrance Point forms the northern tip of a large accumulation of sandy ridges, which have been deposited by waves and tides at the southern entrance to Corner Inlet. In places, the sand ridges are 3 km wide. They parallel the southern beach, but curve around into the inlet behind the northern beach. A mobile, cuspate, sandy foreland marks the boundary between the tide dominated northern beach and wave dominated southern beaches. **Hunter Point**, which rises steeply to 350 m high Mount Hunter, separates the two southern beaches; with the large Lighthouse Point forming the southern boundary of Hunter Beach. All beaches can only be reached by boat or on foot. The nearest walking track reaches the southern side of Lighthouse Point.

The northern beach lies in lee of the extensive ebb tidal shoals and faces north-east. It consequently receives very low waves and has a steep beach face in lee of variable tidal channel and shoal deposits. Once around the point and down the straight east facing southern beaches, the tidal shoals while still present, begin to move offshore and dissipate. This allows wave height to increase down the beach, reaching 1 m at the southern end. These waves, plus occasional higher waves, have built a steep beach face, fronted by a deep trough and longshore bar paralleling the beach. Low waves surge against the beach, while only larger waves break on the bar.

Swimming: Use caution here, even if waves are low. The water is usually deep against the beach face and strong tidal and/or wave generated currents occur in the troughs and tidal channels.

Surfing: Usually none, with at best a shorebreak on the southern beach. The very southern end, toward Hunter and Lighthouse Points, can provide a better beach break at mid tide.

Fishing: You can fish straight off the beach into the deep trough.

Summary: An interesting, relatively long, straight and steep beach.

90, 91   THREE MILE, ROUNDBACK

Unpatrolled			
	Rating	Single bar	Length
90 Three Mile	4	LTT	4 000 m
91 Roundback	3	LTT	400 m
		Total length	4 400 m

South of Lighthouse Point is the straight, east facing, 4 km long **Three Mile Beach**; with a low granite point separating it from the southern 400 m long, north-east facing **Roundback Beach**. The more extensive 120 m high Three Mile Point forms the southern boundary of the whole system. The only access to these beaches is by boat or along the Mount Margaret Track, which reaches the northern end of Three Mile Beach.

The beaches receive waves averaging 1 m and 0.5 m respectively, which produce a single attached and continuous bar. This is only cut by rips during and following high seas.

Swimming: Relatively safe under low waves and best at high tide. A strong shorebreak can develop at low tide.

Surfing: A low shorebreak is the norm along these beaches; best at mid to low tide and around the points.

Fishing: Usually a shallow beach, so best at high tide; otherwise off the shallow bar at low tide.

Summary: Three Mile Beach is a long, straight beach, with dense vegetation coming right down to the back of the beach.

92   JOHNNY SOUEY COVE

Unpatrolled
Single bar: LTT
Beach Hazard Rating: **4**
Length: 800 m

Johnny Souey Cove is a straight, 800 m long, east facing beach, bounded by the prominent granite headlands of Three Mile and Johnny Souey Points. The Johnny Souey walking track reaches the southern end of the beach, where there is a camping area. The beach receives waves averaging 1 m and usually has a continuous attached bar, only cut by rips during and following larger waves.

Swimming: A relatively safe beach under low waves; best at high tide, with a shorebreak at low tide.

Surfing: A low shorebreak that improves as the tide falls.

Fishing: The bar is shallow, so fishing is best off the rocks, or off the end of the bar at low tide.

Summary: There is a lovely camping area next to a small lagoon at the southern end, fronted by the natural beach. Well worth the long walk.

93   MONKEY POINT NORTH

Unpatrolled
Single bar: LTT
Beach Hazard Rating: **3**
Length: 200 m

On the north side of 60 m high **Monkey Point** is a 200 m long, east facing beach; backed by the steep vegetated slopes of the point. It receives waves averaging 0.5 m and has a wide, low gradient beach fronted by a continuous attached bar. The only access to the beach is by boat.

Swimming: Best at high tide, very shallow at low tide.

Surfing: Usually a low shorebreak.

Fishing: Better off the rocks than the shallow beach.

Summary: An isolated beach surrounded by steep slopes.

94   MIRANDA BAY

Unpatrolled
Single bar: LTT
Beach Hazard Rating: **4**
Length: 500 m

Miranda Bay lies on the southern side of Monkey Point. It is 500 m long and faces south-east, thereby picking up waves that average 1 m. They have built a steep high tide beach, with a continuous bar exposed at low tide. Rips may develop toward the more exposed western end, particularly against the rocks.

The beach is surrounded by steep vegetated slopes that rise to 100 m on the western headland. The only access is by boat.

Swimming: A moderately safe beach, especially toward the eastern end. Watch for rips at the western end, wherever waves are breaking.

Surfing: Chance of a reasonable beach break toward the western point.

Fishing: Best off the rocks and beach at the western end, where there is a rip under moderate to high waves.

Summary: An attractive beach surrounded by steep vegetated slopes.

95   FIVE MILE

Unpatrolled
Single bar: TBR
Beach Hazard Rating: **5**
Length: 7 900 m

Five Mile Beach is 7.9 km or 5 miles long. It runs straight throughout its length and faces the south-south-east. The northern end receives waves averaging 1.3 m, which produce occasional rips along the attached bar. Toward the south, the beach is partially protected by Cape Wellington and Horn Point and wave height decreases to less than 0.5 m. As this occurs, the beach steepens and a wider, shallow bar fronts the beach.

The beach can be reached by a walking track that comes out 1 km south of the northern point. A camping area is provided in the northern corner of the beach next to Miranda Creek and is called Five Mile Camp.

Swimming: Safest at high tide and towards the south, as the northern rips will intensify at low tide.

Surfing: Best in the north when a moderate swell is running.

Fishing: Also best in the north, where there are rocks and the possibility of rip holes along the beach.

Summary: A long, steep, sandy beach backed by low dune ridges. The northern camping area has the beach, creek and rocks.

96   SEALERS COVE

Unpatrolled
Single bar: TBR/LTT
Beach Hazard Rating: **5**
Length: 2 000 m

Sealers Cove is a picturesque, circular cove bordered by 300 to 500 m high forested headlands. The sandy beach is backed by a vegetated foredune, then the 1 to 2 km wide Sealers Swamp. The Sealers Cove walking track crosses the swamp to reach the southern end of the beach, where there is a camping area next to Sealers Creek.

The beach sweeps around for 2 km, running south, then east in the southern corner. The southern end is sheltered by Horn Point and the low waves here allow extensive tidal flats to fill the corner. Up the beach, wave height increases to average almost 1 m. During higher wave events, a series of several equally spaced rips cut the northern bar.

Swimming: Safest at the camping area at high tide. Watch the creek mouth on a falling tide; and the chance of rips up the beach when waves are breaking.

Surfing: None in the south, unless there is a large swell. Best chance is up the beach, where at best there will be a beach break.

Fishing: The creek mouth and southern rocks are the best sites.

Summary: An attractive cove with a protected camping area at the southern end.

97   SMITH COVE

Unpatrolled
No bar: R + rocks
Beach Hazard Rating: **3**
Length: 40 m

Deep inside **Smith Cove** is a small, protected, 40 m long beach backed by dense forests and ringed by large, rounded, granite boulders. You can reach the sandy beach by boat at high tide; however at low tide, the boulders block boat access.

Swimming: Better for snorkelling than swimming.

Surfing: None.

Fishing: Best from a boat.

Summary: A small but picturesque beach.

98, 99   REFUGE COVE NORTH & SOUTH

Unpatrolled			
	Rating	No bar	Length
98 Refuge Cove North	3	R	350 m
99 Refuge Cove South	2	R	150 m
		Total length	500 m

Refuge Cove, as the name implies, offers protection and refuge from the ocean. For 200 years, sailors and boaters have used this cove to escape the hazards of Bass Strait and as a safe and attractive anchorage. The entrance to the cove is 400 m wide and opens up inside to reach 1 km across. The cove is surrounded by steep forested slopes, rising to 200 - 300 m. A walking track connects the cove with Sealers Cove to the north and Waterloo Bay to the south. A camping area is located behind the southern beach.

The cove houses two sandy beaches; both composed of pure, rounded, yellow quartz grains. The coarse quartz and low waves produce steep beach faces and deep water close to shore. The cove is therefore ideal for boat anchorage. The northern beach faces south-east toward the entrance, while the smaller southern beach faces due north, and is the preferred anchorage. Dense vegetation grows right to the back of the beach, with the camping area located under the trees. A creek cuts through the vegetation and across the eastern end of the beach.

Swimming: A relatively safe location so long as you can swim, as the water is deep right off the beach. When a swell is running, there is a strong surge on the beach.

Surfing: None.

Fishing: There is always deep water right off the beach.

Summary: Probably the most attractive bay on the whole Promontory, rarely without a few boats at anchor. Worth visiting by boat or foot.

100, 101   LARKIN & BARE BACK COVES

Unpatrolled			
	Rating	No bar	Length
100 Larkin Cove	3	R	80 m
101 Bare Back Cove	3	R	50 m
		Total length	130 m

Larkin Cove lies a few hundred metres inside Brown Head; while **Bare Back Cove**, 1 km to the south, is almost a replica. Both face north and receive only low waves at the shore. The Refuge - Waterloo Track runs on the slopes 200 m west of Larkin and 400 m west of Bare Back. Steep, forested slopes surround both coves. Both beaches are composed of coarse, white quartz sand that results in a steep, barless beach, with seagrass growing in the relatively deep water just off the shoreline. Larkin is 80 m long, with Bare Back just 50 m in length.

Swimming: Relatively safe, so long as you can swim in the deep water of the coves.

Surfing: None.

Fishing: Deep water, rocks and seagrass just off the beach.

Summary: Two attractive, protected and relatively safe coves; best accessed by boat.

102, 103   NORTH WATERLOO BAY EAST & WEST

Unpatrolled			
	Rating	Single bar	Length
102 North Waterloo Bay East	3	R	400 m
103 North Waterloo Bay West	4	LTT	300 m
		Total length	700 m

Waterloo Bay contains four beaches. The northern two are located in the open, south-east facing **North Waterloo Bay**. The two beaches are separated by a small granite headland at high tide and are joined at low tide. Steep, forested slopes rise to 200 - 300 m behind the beaches. The walking track from Refuge Cove comes out at the eastern end of the east beach and runs along both beaches, before exiting at the southern end of the west beach. The beaches receive waves averaging 0.5 m, which produce a steep, barless east beach; and a steep beach fronted by a continuous bar to the west.

Swimming: Both beaches are relatively safe, with usually low waves but deep water off the east beach; and a shallow bar off the west beach. Rips only occur against the rocks when waves exceed 1 m.

Surfing: Usually a very low shorebreak.

Fishing: Best off the rocks and the deeper eastern beach.

Summary: Two pristine beaches with near white quartz sand.

104, 105   LITTLE WATERLOO & WATERLOO BAYS

Unpatrolled			
	Rating	Single bar	Length
104 Little Waterloo Bay	4	LTT	300 m
105 Waterloo Bay	5	TBR	1 500 m
		Total length	1 800 m

The southern half of Waterloo Bay contains two white quartz sand beaches separated by a small granite headland. The northern **Little Waterloo Bay** beach faces east, is 300 m long and is backed by steep, forested slopes with large protruding granite boulders. Small creeks drain across the northern end of each beach. The southern **Waterloo Bay** beach is 1 500 m long, faces east and is backed by a low vegetated foredune and a swamp up to 1 km wide. The Refuge Cove walking track reaches the camping area at the northern end of Little Waterloo Bay, then turns inland through the swamp behind Waterloo Bay.

The beaches receive waves averaging 1 m, which build steep reflective beach faces, fronted by continuous bars, with often perfectly spaced rips (200 m apart) cutting across the bars. The rips are more likely to occur on the main beach.

Swimming: Moderately safe when waves are low; however any swell will intensify the rip currents, so stay close inshore and on the bars.

Surfing: Usually a low shorebreak. A moderate swell can produce some reasonable waves at high tide.

Fishing: Rip holes are common along the main beach after high waves.

Summary: Two lovely white sand beaches usually fronted by crystal blue water, which offer the best chance of a surf on the eastern side of the Promontory.

106   HOME COVE

Unpatrolled
Single bar: R + boulders
Beach Hazard Rating: **3**
Length: 50 m

Home Cove is the most southern mainland beach in Australia and is located just 5 km from the southern tip at South Point. It is a protected beach lying in lee of Waterloo Point; it faces the north and is 50 m in length, with a few large granite boulders on the beach and in the water. It can only be reached by boat and its deep, usually calm water provides a good anchorage. The beach is composed of white quartz sand; and steep, forested slopes rise immediately behind the beach.

Swimming: Relatively safe, apart from the deep water and boulders.

Surfing: None.

Fishing: This is part of the Marine Reserve and amateur fishing is prohibited.

Summary: A nice safe anchorage, however, watch the boulders.

SOUTH POINT

Australia's Southernmost Point

The southern tip of Australia's mainland is not South East Point, where the Wilsons Promontory Lighthouse is located; rather it is a smaller, low protrusion of granite rock and large wave-washed boulders called South Point.

It is located 4 km west of the lighthouse and 1 km further south. While a walking track leads to the lighthouse, there is no track to the southernmost point. The best way to reach it is by a small boat on a calm day. Even then, this requires a leap onto the rocks.

The lighthouse, which stands 117 m above sea level, was constructed in 1859. It was supplied from a small rocky landing below the northern slopes, with a steep rail link to the ridge.

107, 108   OBERON & LITTLE OBERON BAYS

Unpatrolled			
	Rating	Single bar	Length
107 Oberon Bay	6	LTT/TBR	1 900 m
108 Little Oberon Bay	4	LTT	250 m
		Total length	2 150 m

Oberon Bay contains two beaches: the main 2 km long, west facing Oberon Bay and the smaller, 250 m long, south-west facing Little Oberon Bay. Both beaches can be reached on the walking track from Tidal River. A popular camping area is located next to a small creek that crosses the southern half of the beach.

Oberon Bay Beach is exposed to strong westerly winds, which have built sand dunes extending up to 3 km inland. Two large sand blowouts are presently active, while Growler Creek flows out across the northern end of the beach. Waves are only moderate, averaging less than 1.5 m, being reduced somewhat by the Glennie Group of islands 10 km offshore. The beach is wide and low, with a usually wide, shallow, continuous surf zone, but only occasional rips.

Little Oberon Bay Beach is backed by a high, unstable foredune, with steep slopes rising 250 m to Little Oberon Mountain. It is protected by the islands and Norman Point and receives waves averaging 1 m, which produce a shallow continuous bar.

Swimming: These beaches can be hazardous when a surf is running. Stay in the shallow parts of the inner surf zone of these wide beaches and away from the rocks. Watch for rips when waves exceed 1 m.

Surfing: The wide, shallow surf is best at mid to high tide and a low to moderate swell, with easterly winds.

Fishing: This is part of the Marine Reserve and amateur fishing is prohibited.

Summary: An energetic beach backed by massive dunes, with a camping area toward the centre of the beach.

109   TIDAL RIVER/NORMAN

Unpatrolled
Single bar: LTT/TBR
Beach Hazard Rating: **6**
Length: 1 600 m

Tidal River is the centre of all activities taking place on Wilsons Promontory. It is the site of a very large cabin, caravan and camping area, and provides most essential facilities. In summer it assumes the size of a small town. The settlement lies on low dunes just behind the beach; officially known as **Norman Beach**.

The 1.6 km long beach is bounded by the prominent granite headlands of Norman and Pillar Points; both extending over 1 km seaward. The beach faces south-west, with Pillar Point and the offshore Glennie Group of islands providing some protection from westerly waves. The waves average 1.4 m at the beach and interact with the fine carbonate sand to build a wide, low gradient beach and surf zone. Rips are prevalent when waves exceed 1 m.

Swimming: Best at high tide in the inner surf zone. Stay clear of the rocks and watch for rips during higher waves.

Surfing: Usually a wide beach break, which is best at mid to high tide and during easterly winds.

Fishing: The beach and creek at high tide, and the rocks, are the most popular. The southern rocks are part of the Marine Reserve and fishing is prohibited.

Summary: A very popular summer destination with a range of activities in the area, including the wide beach and usually wide surf.

110   SQUEAKY

Unpatrolled
Single bar: TBR
Beach Hazard Rating: **6**
Length: 700 m

Squeaky Beach is named after the noise the dry sand can make as you walk across it on a hot day. The beach is located in Leonard Bay and is bordered by the large granite headlands of Pillar and Leonard Points. There is sealed access to a car park located 200 m behind the beach and high foredune. Dunes from the beach have been blown over 1 km inland.

The beach is 700 m long and faces south-west. It receives waves averaging 1.5 m, which usually produce four rips across the wide shallow bar; two located against the rocks and boulders at each end.

Swimming: This is one of the Promontory's more hazardous beaches, with usually high waves and persistent rips. Use care if bathing here. Stay inshore on the bar and clear of the rips.

Surfing: A popular surfing spot owing to the higher waves that can move in between Norman and Great Glennie Islands.

Fishing: One of the better spots, owing to the more persistent rips and rocks.

Summary: This is one of the main surfing beaches for the Promontory.

111, 112   LEONARD BAY 1 & 2

Unpatrolled			
	Rating	Single bar	Length
111 Leonard Bay 1	5	LTT	50 m
112 Leonard Bay 2	4	LTT	50 m
		Total length	100 m

Leonard Bay, beside Squeaky Beach, has two small pocket beaches set amongst the rocks on the southern side of Leonard Point. Both beaches are 50 m in length, are backed by steep vegetated slopes, face south and receive waves averaging less than 1 m. The best access is by boat or around the rocks at low tide. Both beaches have a continuous attached bar and one rip against the rocks. The rips intensify when waves exceed 1 m.

Swimming: Moderately safe during low waves, on the bar and clear of the rip. However, these beaches are isolated and should be avoided when waves exceed 1 m.

Surfing: Usually a low shorebreak.

Fishing: Better off the surrounding rocks than the beach.

Summary: Two small isolated beaches.

113, 114   PICNIC BAY, WHISKY BAY

Unpatrolled			
	Rating	Single bar	Length
113 Picnic Bay	7	LTT	500 m
114 Whisky Bay	7	TBR	200 m
		Total length	700 m

Picnic and Whisky Bays are two adjoining beaches on the northern side of Leonard Point. They are separated by the small Picnic Point. Car parks are located to either side of Picnic Point, with walking tracks leading across the foredunes to each beach.

Picnic Bay Beach is 500 m long, faces west and is sheltered toward its southern end by Leonard Point. Waves average 1.2 m and produce a usually continuously attached bar, with a permanent rip against Picnic Point.

Whisky Bay Beach is shorter and receives waves averaging 1.4 m. This results in a rip dominated surf zone, with the bar often detached from the beach, and strong permanent rips against the boundary rocks.

Swimming: Safest at the southern end of Picnic Bay Beach. Elsewhere strong, permanent rips dominate, particularly on Whisky Bay Beach. Use extreme care if bathing here, stay on the bar and well clear of the rips and rocks.

Surfing: Both beaches are popular surfing spots, with beach breaks over the bars and either side of Picnic Point.

Fishing: Good permanent rip holes are located either side of Picnic Point and along Whisky Bay Beach.

Summary: Two accessible, high energy beaches; nice for a picnic as the name implies, or a surf or swim during low waves.

115   TONGUE POINT

Unpatrolled
Single bar: LTT
Beach Hazard Rating: **4**
Length: 250 m

On the inner southern side of 2 km long **Tongue Point** is a protected 250 m long, south-west facing beach. The walking track to Tongue Point runs along the ridge 300 m behind the beach. It is bordered by high granite headlands and backed by steep vegetated slopes. The beach only receives refracted waves averaging less than 1 m, which produce a usually continuous narrow bar. There is a chance of rips against the rocks when waves exceed 1 m.

Swimming: A moderately safe but isolated beach. Watch for the boundary rips when waves exceed 1 m.

Surfing: Usually a low shorebreak.

Fishing: Better off the rocks, however deep water can be reached over the narrow bar at low tide.

Summary: A small, isolated beach surrounded by steep slopes.

116   FAIRY COVE

Unpatrolled
Single bar: LTT
Beach Hazard Rating: **4**
Length: 250 m

Fairy Cove is located on the inner portion of the north side of Tongue Point. It is surrounded by granite boulders and steep slopes. The boulders break the high tide beach into two parts, which are joined at low tide into a 250 m long, low gradient beach and a usually attached and continuous bar. The beach is accessible on foot from a side track off the Tongue Point track.

Swimming: A moderately safe beach under low waves, with rips developing when waves exceed 1 m; and granite boulders in the surf.

Surfing: Normally a long walk for a low shorebreak.

Fishing: The rocks at the western end extend out into deep water.

Summary: A lovely spot for a day walk. It is sheltered from the prevailing westerlies.

117   DARBY BAY

Unpatrolled
Single bar: LTT/TBR
Beach Hazard Rating: **6**
Length: 250 m

Darby Bay is a small beach located around the rocks from Darby Beach and 1 km north of Fairy Cove. The beach is backed by steep bluffs, which are capped by cliff-top dunes. The Tongue Point track runs along the ridge 100 m above the beach, however there is no formed track to the beach.

The beach is 250 m long and faces south-west. Owing to some protection from Tongue Point, it receives waves averaging 1.3 m. These produce a low beach fronted by a continuous bar, which has two permanent rips; one against a large boulder in the surf and the second against the northern point.

Swimming: An isolated, potentially hazardous beach with permanent rips, rocks and reefs to contend with.

Surfing: If you are keen, you can paddle around the southern Darby Beach point to reach this hidden beach; where you will usually find a wide shorebreak.

Fishing: Rip holes are usually found against the northern boulder and point.

Summary: This is a difficult to access beach, hemmed in between steep bluffs and the surf.

118   DARBY/COTTERS

Unpatrolled
Inner bar: TBR   Outer bar: RBB/LBT
Beach Hazard Rating: **7**
Length: 15 000 m

Darby/Cotters Beach is the longest on Wilsons Promontory; it is also the most exposed and energetic. The beach is accessible at the southern end, where there is a car park and short track to the beach, along the side of the shallow Darby Creek.

The beach runs for 15 km, from the steep granite headland that forms the southern boundary of Darby Beach, to the mouth of Shallow Inlet. It faces the west-south-west, receives waves averaging 1.5 m, and is exposed to the full force of the westerly winds. The waves interact with the fine, calcareous sand, to break over a 250 m wide, low gradient surf zone. More than 30 large rips, spaced about every 300 m, dominate the inner surf zone; with waves breaking further offshore on the outer bar.

The beach is backed by extensive Pleistocene calcarenite, best seen at the creek mouth. These cliffs are capped by active sand dunes that extend up to several kilometres inland. Some of the calcarenite also outcrops on the beach as rocks, and in the surf as reefs, as at Buckleys Reef. Beyond Buckleys Reef the beach is known as Cotters Beach. At the northern end, the extensive ebb tidal shoals and channel of Shallow Inlet dominate the last kilometre of beach.

Swimming: This is a hazardous, high energy beach, so use extreme care if bathing here. Stay close inshore on the shallow bar and well clear of the rips.

Surfing: A popular spot to stop on arriving at the Promontory, to check out the size of the waves. It works best in a low to moderate swell and north-east winds, with *Buckleys Reef,* 2.5 km up the beach, producing good rights that are best at high tide.

Fishing: The beach is quite flat, but cut by rip holes.

Summary: Worth a visit to see a high energy beach and the backing new and old dunes.

4. WARATAH, VENUS AND KILCUNDA BAYS

Waratah, Venus and Kilcunda Bays
(Shallow Inlet to San Remo)

Beaches: 119 to 182
Coast length: 117 km

For maps of region see Figures 65 and 68

The coast between western Wilsons Promontory and Phillip Island contains three large open embayments. The first, **Waratah Bay**, is bordered by Wilsons Promontory on the east and Cape Liptrap on the west. In between is a 23 km wide, open, south-west facing bay, containing 26 km of near continuous, sandy, exposed beaches between Darby River and Walkerville. There are no towns in this bay, only the small settlements of Sandy Point and still smaller Waratah Bay. There is one surf lifesaving club at Sandy Point, called Waratah Beach.

The second bay is **Venus Bay**, which lies between Cape Liptrap and the rocky coast around Cape Paterson (Figure 64). This bay is very similar to Waratah Bay. It is 37 km wide, faces west-south-west, and has nearly 30 km of exposed, high energy beaches between Cape Liptrap and Anderson Inlet. Past the inlet, rocky coast dominates and numerous smaller sandy beaches occupy the coast out to Cape Paterson. There are two towns: at Inverloch on Anderson Inlet, and at Cape Paterson, a holiday settlement at Venus Bay. Inverloch, Venus Bay and Cape Paterson all have surf lifesaving clubs.

The third bay has no official name and is here referred to as **Kilcunda Bay,** after the only town in the area. It lies between Cape Paterson and Cape Woolamai on Phillip Island; a direct distance of 25 km. It has 18 km of discontinuous sandy beaches between Cape Paterson and Kilcunda, then a rocky coast with smaller beaches between Kilcunda and San Remo. Kilcunda is the only town in the bay, with San Remo lying at the western end. There is a Royal lifesaving club near Wonthaggi.

Figure 64.   This view east from Coal Point to Cape Paterson typifies the rocky sections of the Waratah-Venus-Kilcunda coast. Sandstone rocks outcrop at the coast as low headlands, fronted by extensive shore platforms and rock reefs. In between is a continuous strip of sand that has formed a number of crenulate beaches and backing dunes, resulting in some of the beaches being fronted by rocks at low tide.

**SHALLOW INLET
MARINE AND COASTAL PARK**

Open coast length: 4 km
Number of beaches: 1

Shallow Inlet Marine Park covers the inter- and sub-tidal estuarine area of Shallow Inlet.   The Coastal Park covers the western 4 km of Sandy Point.
Information:
    Yarram   (051) 982 5995
    Tidal River   (056) 980 8538

	Width	Orientation	Coast length	Beach length	No. beaches
Waratah Bay (Shallow Inlet - Cape Liptrap)	23 km	220°	33 km	26 km	17
Venus Bay (Cape Liptrap - Cape Paterson)	37 km	235°	53 km	38 km	21
Kilcunda Bay (Cape Paterson - Cape Woolamai)	25 km	220°	31 km	21 km	26
Total			117 km	85 km	64

Victorian Beaches - Region 4: East (Waratah & Venus Bays)

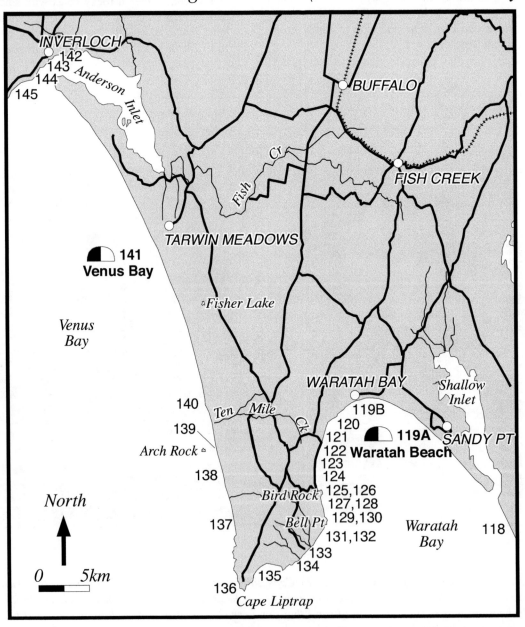

Figure 65.   Region 4 (eastern section): Waratah and Venus Bays (Beaches 119A - 145)

119A,B   WARATAH BAY

Waratah Beach SLSC

Patrols: late November to Easter holidays
Surf Lifesaving Club: Saturday, Sunday and Christmas public holidays
Lifeguard: no lifeguard on duty or weekday patrols

Inner bar: TBR/RBB   Outer bar: RBB/LBT
Beach hazard rating: **6**
Length: 15 000 m

For map of beach see Figure 66

Waratah Bay consists of an open, southerly facing bay that contains a 15 km long beach. The beach runs from the Shallow Inlet entrance to the small settlement of Waratah Bay, at the western end of the beach. The beach is exposed, with high waves averaging 1.6 m at Shallow Inlet. Waves remain high as far as the settlement of Sandy Point in the centre of the beach, where Waratah Beach SLSC is located. From this point, wave height gradually decreases to 1.4 m, toward the western end where the small settlement of Waratah Bay is located.

The high waves and fine sand combine to produce a wide, low gradient beach, fronted by a 300 m wide surf zone consisting of two shore parallel bars. The inner bar is cut by large rips every 300 m and is detached from the shore following high waves. A 100 m wide trough separates this from the outer bar, which is cut by rips every few hundred metres. Toward the eastern end of Waratah Bay, the inner bar becomes more continuous and the rips less active. The entire beach is affected by westerly winds which, when blowing, intensify the waves and rip currents.

The Waratah Beach Surf Life Saving Club, formed in 1963, is located at the Sandy Point settlement. It provides the only good beach access and parking in this area, and the benefit of a patrolled section on this potentially hazardous beach. Access and a beach boat ramp are also provided at the Waratah Bay settlement.

Swimming: Relatively safe in the swash zone, but be very wary of the large rips and the outer surf zone. It's best to stay on the bar between the flags as there are 5 rescues on average each year.

Surfing: The wide surf zone can offer a range of breaks, with more experienced surfers making it to the outer bar, and the less experienced using the inner bar. The beach and bar breaks depend on wind, wave and tide conditions, with the best occurring in a moderate swell and northerly winds. The surfing breaks also go by the name of *Sandy Point.*

Fishing: Usually good rip holes along the beach.

Summary: Sandy Point is an attractive, if quiet, coastal settlement, nestled in amongst the densely vegetated sand dunes, with the added attraction of a patrolled beach. Waratah Bay has a caravan park, and boats can be launched from the beach when waves are low.

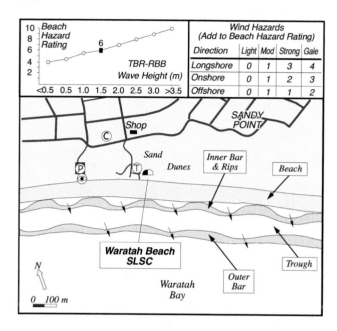

	Wind Hazards (Add to Beach Hazard Rating)			
Direction	Light	Mod	Strong	Gale
Longshore	0	1	3	4
Onshore	0	1	2	3
Offshore	0	1	1	2

Figure 66.   Waratah Beach Surf Life Saving Club is located in the dunes at Sandy Point. It patrols a wide, energetic beach with rip dominated inner and outer bars. Be very careful if bathing here and stay between the flags.

120, 121   COOKS CREEK, PONY CREEK

Unpatrolled			
	Rating	Single bar	Length
120 Cooks Creek	6	LTT/TBR + rock flats	350 m
121 Pony Creek	5	LTT + rock flats	1 500 m
		Total length	1 850 m

The western side of Waratah Bay borders a 10 km wide peninsula of ancient 400 to 500 million year old rocks that terminate at Cape Liptrap. Most of the beaches east of Cape Liptrap are partially protected by the cape. They are also influenced by the steep slopes on this side of the bay and the numerous rocks and reefs that lie on the beaches and in the surf.

The first beach begins immediately west of Waratah Bay Beach. Extensive intertidal rock flats and reefs form the boundary. The beach is 350 m long and faces south-east. It is also bordered by **Cooks Creek** and steep, forested slopes. It is fronted by similar rock flats at its western end, after which the **Pony Creek Beach** continues in the same direction, with rock flats occurring along much of the beach. It terminates at the mouth of Pony Creek, from where rocks, more than sand, dominate the shoreline. The beaches are only accessible by foot from Waratah Bay in the north, or Walkerville North to the south.

Swimming: While waves are moderate here, averaging 1.0 m, these are potentially hazardous beaches owing to the prevalence of rips across the wide shallow bars and the rocks and reefs. If bathing, stay close inshore and avoid the rocks and rips.

Surfing: Some good right hand breaks form over the reefs at high tide. The best known are *Cooks Creek* and *Chicken Rock*. Best in a moderate to high swell and north-west winds.

Fishing: There are plenty of rocks and reefs to fish from off the beach, together with numerous rip holes.

Summary: Two beaches that can only be reached on foot; most suitable for a coastal walk.

122   WALKERVILLE NORTH

Unpatrolled
Single bar: LTT + rock flats & reef
Beach Hazard Rating: **5**
Length: 4 000 m

Walkerville North Beach is located along the base of steep, 100 m high slopes, which widen sufficiently at Walkerville North to fit in a road and caravan park. The road parallels the southern half of this 4 km long beach. The beach faces east-south-east and is fronted by near continuous inter- and sub-tidal rock flats. There is a sandy beach at high tide, however at low tide there is as much rock as sand.

Swimming: Moderately safe close inshore at high tide. However as the tide falls, rocks and reefs dominate the surf. When the waves exceed 1 m, numerous rips form in amongst the rocks.

Surfing: Best at high tide on a moderate to high swell, when some good right hand breaks can form over the reefs. The best known is called *Dunnybrook*.

Fishing: Numerous easily accessible rocks, reefs and rip holes occur along this beach.

Summary: This is an out of the way, but very accessible, strip of beach; most popular with holidaying families and fishers.

123, 124   THE BLUFF, WALKERVILLE

Unpatrolled			
	Rating	Single bar	Length
123 The Bluff	3	LTT + rocks	250 m
124 Walkerville	3	R/LTT + rocks	200 m
		Total length	450 m

Walkerville is the site of historic lime kilns, where the cliffs were mined for limestone to burn for lime, between 1878 and 1926. Sailing ships then transported the lime to the Melbourne building market. The ruins of the kilns stand above these two small beaches. The Walkerville Road terminates at **Walkerville Beach**, with **The Bluff Beach** lying just past a small headland. A walking track connects Walkerville with Walkerville North via The Bluff Beach.

Both beaches are about 200 m in length and face east. They are backed by steeply rising bluffs and slopes, with houses on the slopes above Walkerville Beach. Eroding bluffs fringe the beaches, and rocks and reefs outcrop on the beaches and in the surf. There is a small car park and a boat ramp at Walkerville.

The beaches receive waves averaging 0.5 m, which produce a narrow but low gradient beach and an attached shallow bar.

Swimming: Relatively safe beaches under normal low waves. Just watch the rocks and reefs.

Surfing: None, unless there is a big swell running.

Fishing: Better off the beach in a boat, or off the numerous rocks.

Summary: A small collection of houses can now be found at the former mine site. It still remains an attractive part of the coast, with narrow beaches, forested slopes and craggy headlands.

125, 126   BIRD ROCK NORTH & SOUTH

Unpatrolled			
	Rating	Single bar	Length
125 Bird Rock North	4	R/LTT + rocks	60 m
126 Bird Rock South	6	LTT/TBR + rocks	100 m
		Total length	160 m

Bird Rock is a series of jagged sandstone and shale bluffs, sea stacks, rocks and reefs that extend 1 km seaward from Walkerville. Tucked in on either side of Bird Rock are two small beaches. The northern beach faces north and is well protected from large swell, with waves averaging less than 0.5 m. The narrow, 60 m long beach has a shallow bar with large rocks and boulders on the beach and in the surf.

The beach on the southern side faces south-east and, being more exposed, receives waves averaging 1 m. These waves weave in amongst the rocks and sand to scour both rips and scour holes.

Both beaches are accessible on foot from Walkerville, via the rocks at low tide, or along a bluff-top walking track.

Swimming: Bathing is safest at the northern beach, however even there, rocks dominate the low tide area. The southern beach is very hazardous even with low waves, owing to all the rocks and reef, and should be avoided.

Surfing: None.

Fishing: Best on the southern beach, where there are many vantage points to fish the numerous holes and gutters.

Summary: Two small scenic beaches dominated by rocks and reefs.

127, 128   **DIGGERS ISLAND**
             NORTH & SOUTH

Unpatrolled			
	Rating	Single bar	Length
127 Diggers Island North	4	LTT + rock flats	350 m
128 Diggers Island South	5	LTT + rock flats	50 m
		Total length	400 m

Diggers Island is a small sandstone island, surrounded by rock platforms, capped by scrubby vegetation and tied to the backing bluffs by the beaches on either side. A lime kiln ruin is located behind the south beach. The easiest access is along the beach from Bird Rock, as steep, 60 m high bluffs back the beaches.

The northern beach is a 350 m long, narrow, east facing beach. It has vegetated bluffs behind, Diggers Island at the southern end, and numerous rocks and reefs scattered along the beach and surf. An old jetty also runs out next to the island. Waves average only 0.5 m and the bar is usually narrow and continuous.

The southern beach is smaller but more exposed, receiving waves almost 1 m high. These are sufficient to cut a permanent rip against the rocks and reefs that dominate the surf zone.

Swimming: Moderately safe when waves are low. However when waves exceed 1 m, rip currents are generated which, coupled with the rocks and reefs, produce hazardous conditions.

Surfing: Usually nothing but a low shorebreak, however when the swell is very big outside, a small right hander wraps around the island.

Fishing: The island has a good rock platform, while there is a deep rip hole off the south beach.

Summary: Two scenic beaches, together with the old jetty, lime kiln and island, make this an interesting area for a walk or fish, and a swim when waves are low.

---

129 - 132   **BELL POINT NORTH, MAIN,**
              SOUTH & BOULDER

Unpatrolled			
	Rating	Single bar	Length
129 Bell Point North	4	LTT + rocks	150 m
130 Bell Point	4	R/LTT + rocks	250 m
131 Bell Point South	5	LTT/TBR	250 m
132 Bell Point Boulder	5	R + boulders	100 m
		Total length	750 m

The coast along Bell Point consists of 20 to 40 m high limestone bluffs. They face east and are fronted by narrow sand and cobble beaches, scattered with numerous rocks and reefs, including one rock called the Mushroom Rock. At the point, the rocks are more prominent and the coast swings to face south-east. On the north side are two beaches (129, 130), with two more to the south (131, 132). The point is surrounded by private property and the only access is along the beaches and rocks, from Bird Rock in the north or Maitland Bay to the south.

Bell Point North and **Bell Point Beaches** (129, 130) are, respectively, 150 m and 250 m long stretches of sand lying below the bluffs. They usually receive low waves, which produce a moderately steep and narrow beach, and a continuous bar. Rips only occur when waves exceed 1 m in height.

South of Bell Point, the beaches face south-east and receive waves averaging 1 m. The first beach (131) is sandy and usually has a rip against the eastern rocks. The second beach (132) is a steep cobble and boulder beach, fringed by low tide rock flats.

Swimming: These beaches are relatively isolated and, while moderately safe in low waves, they all contain numerous rocks and reefs, and have strong rips when waves exceed 1 m.

Surfing: Usually low shorebreaks, with the beach just south of the point offering the best chance in a moderate swell. Some of the reefs can produce right hand breaks when there is a big swell outside.

Fishing: There are numerous holes and gutters along this sandy and rocky section of coast.

Summary: These beaches combine to make a pleasant coastal walk when waves are low. However, use care if bathing.

133   MAITLAND

Unpatrolled
Single bar: LTT + rock flats
Beach Hazard Rating: **6**
Length: 1 200 m

Maitland Beach is a 1 200 m long beach consisting of a narrow, crenulate, high tide sand and cobble beach; fronted for much of its length by rocks and low tide rock flats. The Bear Gully Road provides car access to the southern end of the beach, where there is a camping area, but no facilities.

Waves average 1 m; however the surf is dominated by rocks, which generate numerous currents.

Swimming: Only suitable at high tide and even then, rocks dominate the surf. Best to find a quieter break in the rocks to swim.

Surfing: Usually no rideable surf.

Fishing: A popular fishing beach with sand, rocks, reefs and numerous gutters, as well as good access.

Summary: This is a popular fishing beach and camping area.

134, 135   GRINDER POINT EAST & WEST

Unpatrolled			
	Rating	Single bar	Length
134 Grinder Point East	6	LTT + rocks	500 m
135 Grinder Point West	6	LTT/TBR	1 300 m
		Total length	1 800 m

Grinder Point is a low sandstone headland located 3 km north-east of Cape Liptrap. A small dune covers the back of the point. On either side of the point are two south facing beaches. The cape provides some protection for the beaches, which receive waves averaging between 1 and 1.5 m. The east beach is 500 m long, with extensive rocks and reefs in the inter- and sub-tidal zone. The west beach is 1 300 m long and receives slightly higher waves. It has fewer rocks, with permanent rips forming against the rocks at either end and the reef in the centre. The beaches are backed by private property and public access is either along the coast from Maitland Beach, or with permission through the farm.

Swimming: Both beaches are potentially hazardous, particularly at low tide when rocks are exposed and the rips are at their strongest. It is safer to swim at high tide, if you stay close inshore and clear of the rocks and reefs.

Surfing: The reef on the west beach has a right hand break called *Porcupines,* which works in a moderate to large swell.

Fishing: There are numerous good spots along this section of coast with rocks, reefs and, increasingly, rip holes and gutters on the west beach.

Summary: Two natural beaches backed by a scrubby foreshore; good for fishing and surfing on a good swell.

136   CAPE LIPTRAP

Unpatrolled
No bar: R + rocks & reef
Beach Hazard Rating: **10**
Length: 200 m

Cape Liptrap is the oldest part of Victoria, consisting of rocks over 500 million years old. Today these resilient rocks form the prominent 100 m high cape and the jagged rocky shores to either side. A road runs right to the cape where there is a car park, lookout and lighthouse. The lighthouse was built in 1913 and stands 93 m above sea level.

At the western foot of the cape is a 200 m long, steep, cobble and boulder beach, fronted by jagged rocks that are exposed at low tide. The beach is unsuitable for bathing or surfing, however the rocks around the cape are more popular for adventurous rock fishers. In all cases, beware of the high seas that dominate this cape. Just off the eastern side of the cape is a reef which, in a moderate to large swell, produces some rideable, if fullish, waves.

CAPE LIPTRAP & VENUS BAY

Once around Cape Liptrap, the coast swings to the north into the large, open Venus Bay. The bay contains nearly 30 km of open, west facing, exposed beaches with high waves and wide surf zones. These extend all the way to Anderson Inlet, where a large tidal channel and shoals dominate the entrance, with the town of Inverloch on its western shore. Past Inverloch, the coast trends south-west and consists of rocky coast and headlands, with smaller sandy beaches all the way to Cape Paterson and beyond.

---

137   MORGAN

Unpatrolled
Inner bar: TBR   Outer bar: RBB/LBT
Beach Hazard Rating: **8**
Length: 2 500 m

Morgan Beach is a 2.5 km long, south-west facing beach that extends from the western rocks of Cape Liptrap to a series of 40 m high calcarenite bluffs around Arch Rock. The beach is backed by farms and is only accessible via the Munbilla farm.

The beach has both modern and older dune systems extending up to 2 km inland, rising over the bedrock to heights of 100 m. The beach is exposed to the full forces of Bass Strait and receives waves averaging 1.8 m. These interact with the fine to medium beach sand to build a low gradient beach with two bars. The inner bar is usually attached and cut by large rips every 450 m, while the outer bar 250 m offshore has more widely spaced rips. In addition there are several reefs, particularly toward Cape Liptrap, that induce additional currents.

Swimming: Use extreme care as this is an isolated, exposed, high energy beach with large, strong rips and a wide surf zone. Stay inshore on the attached bars and well clear of the reefs and rips.

Surfing: The beach provides some good beach/reef breaks during low to moderate swell. The breaks work best in easterly winds.

Fishing: There are large rip holes and gutters along the beach, together with the rocks and reefs at either end.

Summary: An isolated, high energy beach occasionally used by experienced surfers and fishers.

138, 139   ARCH ROCK 1 & 2

Unpatrolled			
	Rating	Single bar	Length
138 Arch Rock 1	5	R + reefs	300 m
139 Arch Rock 2	7	TBR	400 m
		Total length	700 m

Arch Rock is the most prominent of a series of calcarenite bluffs, cliffs and sea stacks that extend for 2 km along the shore, north of Morgan Beach Rock (the actual arch collapsed about 1920). These rocks are remnants of coastal dune systems deposited during previous high stands of the sea at least 120 000 years ago; and are possibly many hundreds of thousands of years older at their base. The rocks also form extensive reefs off the two beaches.

The first beach (138) is 300 m long and fronted by continuous, submerged reefs which lower waves at the beach to 1 m. These produce a usually steep beach face fronted by the reefs. The water is deep off the beach and strong currents are generated within the reefs.

The second beach (139) is 400 m long and, while fronted by reefs, is more open and receives waves averaging 1.5 m. These combine with the reefs and sand to develop three strong permanent rips across the wide surf zone.

Swimming: Be very careful bathing here as rocks and reefs dominate the surf and currents. Stay close inshore and well clear of any currents.

Surfing: There are a number of reef breaks out from both beaches, which work best in a moderate to low swell and north-east winds.

Fishing: Plenty of good spots along this reef dominated beach, with numerous holes.

Summary: An attractive sand and rocky coast, with narrow beaches and wide reef dominated surf. Suitable for experienced surfers and fishers, but not for bathing.

140   TEN MILE CREEK

Unpatrolled

Single bar: TBR/RBB

Beach Hazard Rating: **7**

Length: 2 500 m

Ten Mile Creek flows through a deep gully cut in the dune calcarenite, to reach the coast in the centre of the 2.5 km long, south-west facing, exposed beach. The beach is backed by calcarenite bluffs and some active dunes, and is fronted by broken calcarenite reefs. Waves average 1.6 m and combine with the reefs and sand to produce a 250 m wide surf zone, dominated by ten permanent rips.

Swimming: A potentially hazardous beach owing to the usually moderate to high waves and strong permanent rips and reefs. Use extreme care if bathing.

Surfing: Plenty of reef breaks along this section, which work best on a low to moderate swell and north-east winds.

Fishing: Several good, deep, permanent rip holes and gutters occur along the beach.

Summary: This beach is backed by dunes and farms, with the best access on foot along the gully. Consequently, it is rarely seen or used.

141   VENUS BAY

Venus Bay SLSC

Patrols: late November to Easter holidays
Surf Lifesaving Club: Saturday, Sunday and Christmas public holidays
Lifeguard: 10 am to 6 pm weekdays, during Christmas holiday period

Inner bar: TBR/RBB   Outer bar: RBB/LBT
Beach Hazard Rating: **7**
Length: 23 700 m

For map of beach see Figure 67

Venus Bay is a very open, exposed bay, facing into the south-west winds and waves. The beach is nearly 24 km long and has public access only toward the northern end at the Venus Bay settlement. This is also the site of the Venus Bay Surf Life Saving Club, founded in 1961. High vegetated dunes up to 2 km wide and 50 m high back the beach. Public access and parking are provided at the surf lifesaving club and four other points adjacent to the 5 km long Venus Bay settlement.

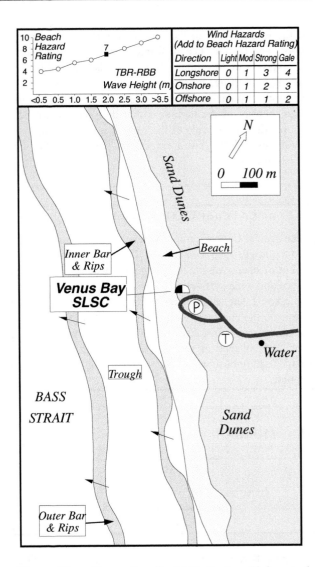

Figure 67.   Venus Bay Surf Life Saving Club patrols the accessible section of the 24 km long beach at the settlement of Venus Bay. The beach faces south-west and receives high waves, which produce a wide, rip dominated, double bar surf zone. Only bathe between the flags on this potentially hazardous beach.

The beach receives high south-west waves averaging 1.8 m which, with the fine sand, produce a wide, low gradient beach, fronted by a 400 m wide surf zone. This consists of an inner bar cut by strong rips every 350 m, then a deep trough, with an outer bar cut by rips every few hundred metres. Due to its orientation, the beach is very prone to westerly winds that blow out the surf and intensify the currents.

Swimming: A potentially hazardous beach usually with high waves and strong rips, so stay on the inner bar and in the patrol area between the flags. There is an average of 23 rescues each year, indicating the level of hazard.

Surfing: Best with a low to moderate swell and north to north-east winds, however the shape depends on the bar configuration.

Fishing: Usually good rip holes along the beach.

Summary: Venus Bay is a holiday settlement nestled in amongst the vegetated sand dunes, which back an exposed and high energy beach. Be very careful if bathing here and stay in the patrolled section.

142, 143, 144   INVERLOCH, POINT HUGHES, MAIN

Inverloch SLSC			
	Rating	Single bar	Length
142 Inverloch	3	R + tidal channel	900 m
143 Point Hughes	3	R + tidal channel	800 m
144 Main	4	R + tidal shoals	1 800 m
		Total length	3 500 m

The town of Inverloch is located on the northern side of Anderson Inlet. A low energy, sandy shore runs along the front of the town, with the beach facing south-east into the mouth of the inlet, which is 1 km wide at its mouth opposite Point Norman. Three crenulate beaches lie along the town foreshore, all backed by a foreshore reserve, with car access and numerous facilities, including a new surf lifesaving club on Main Beach.

The first **Inverloch Beach** is located nearly 2 km inside the wide mouth and only receives very low ocean waves or none at all. A boat ramp crosses the narrow, steep beach, with a seawall built behind much of the beach. The second beach runs west of **Point Hughes** and is also steep, narrow and backed by seawalls along much of its length. Both these beaches face the deep tidal channel with a strong tidal current running in close to the beaches. **Main Beach** is a wider beach that terminates at sandy Point Norman; on the western side of the inlet mouth. It is narrow but fronted by extensive intertidal sand flats and shoals, including some areas of rock flats.

Swimming: These three beaches usually have calm water conditions; however beware of the deep tidal channel, particularly off the inner two beaches, and the strong tidal flows in the channel.

Surfing: None.

Fishing: Very popular for fishing off the beaches into the tidal channel.

Summary: Three quiet beaches backed by a shady foreshore reserve. They are popular for family picnics and for fishing, however all are fronted by deep tidal channels with strong currents.

BUNURONG MARINE AND COASTAL PARK

Coast length: 16 km
Number of beaches: 15 (144 to 159)

This park encompasses all the coastline lying south of the Inverloch - Cape Paterson Road, together with the William Hovell Flora and Fauna reserve in the dunes between Cape Paterson and Coal Point. The main road follows the coast between the two towns and provides some excellent viewing spots and access points to the rocky headlands and usually small pocket beaches.

Information: Inverloch   (056) 974 1236
               Wonthaggi   (056) 972 1066

145   PETREL ROCK

Unpatrolled
Single bar: LTT/TBR
Beach Hazard Rating: **5**
Length: 1 800 m

Petrel Rock is a sandstone reef lying off the western end of a 1.8 km long, south to south-east facing beach. The beach extends from the sand and inlet at Point Norman to the rocky bluffs in lee of the reef. It consists of a narrow, crenulate high tide beach fronted by a wide, continuous bar. This becomes increasingly replaced by rock flats to the south, to become a platform beach. The Cape Paterson Road runs along the back of the beach, with a low foredune right behind the beach. There is access, parking and a boat ramp at Flat Rock.

Swimming: A moderately safe beach when waves are low, however higher waves will produce a strong sweep along the beach, together with rips. Waves and currents intensify close to the inlet mouth at Point Norman.

Surfing: Usually a low, wide shorebreak, with a left hand break off the tidal shoals seaward of Point Norman.

Fishing: Most of this beach is shallow and exposed at low tide.

Summary: A lower energy, protected beach right next to the main road.

Victorian Beaches - Region 4: West (Inverloch to San Remo)

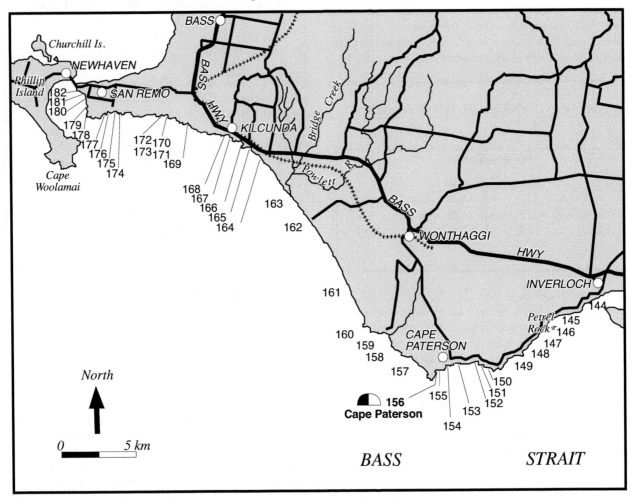

Figure 68.   Region 4 (western section): Inverloch to San Remo (Beaches 145 to 182).

146, 147   **THE CAVES 1 & 2**

Unpatrolled			
	Rating	Single bar	Length
146 The Caves 1	6	R + rock flats	500 m
147 The Caves 2	7	R + rock flats	500 m
		Total length	1 000 m

South of Petrel Rock, 20 to 40 m high sandstone bluffs and cliffs form the coastline, with extensive intertidal rock flats at their base. Along the base of the cliff, sand has collected to form a usually 20 to 30 m wide platform beach. This beach is awash only at high tide and is fronted by exposed rock flats at low tide. The Cape Paterson Road runs along the top of the bluffs, however beach access is restricted by the steep, high bluffs. Both beaches are 500 m long and face south-east, receiving waves averaging 1 to 1.5 m. These break over the

extensive reefs at high tide, or on the edge of the rocks at low tide.

Swimming: Only possible at high tide on the first beach and even then, care must be taken owing to the dominant reefs and rocks. The reefs are more discontinuous along the second beach and strong rips occupy the deeper water.

Surfing: Usually none at the beach, however there is a reef break off the beach called *Suicide,* which breaks both left and right.

Fishing: Like the bathing, this can only be achieved at high tide, unless you walk out on the rock flats to find deep water.

Summary: Two long, narrow, rock fronted beaches.

148   EAGLES NEST

Unpatrolled
Single bar: R/LTT + rock flats
Beach Hazard Rating: **7**
Length: 250 m

Eagles Nest is a sandstone sea stack that sits on a wide rock platform and is awash at high tide. On the eastern side of the Nest is a 250 m long platform beach lying at the base of 40 m high, vegetated bluffs. It is fronted by continuous, low tide rock flats and several rock reefs. The Nest and reefs reduce waves at the beach to about 1 m and produce a narrow, steep beach. The surf is dominated by the rocks and reefs.

Swimming: Only possible at high tide when the rocks are covered, even then be very careful as rip currents sweep over the rocks and reefs.

Surfing: On the east side of the point, a good right hander breaks over the reef in big swells.

Fishing: The rocks around the Nest are the most popular spot, however use caution as large waves can wash over the rocks.

Summary: Eagles Nest is a popular photo stop. The beach is accessible from the roadside car park, however be very careful if swimming at the beach.

149   SHACK BAY

Unpatrolled
Single bar: LTT + rock flats
Beach Hazard Rating: **7**
Length: 150 m

Shack Bay is named after the fishers shacks that used to occupy the back of this small beach. The 150 m long beach lies in a 200 m wide gully, and is fronted by near continuous rock flats. A small gap in the middle is used to launch fishing boats. The main road runs on the bluffs above the beach, and there is a walking track from the road down to the beach.

Swimming: Bathing is safest at high tide, on the sand and close inshore, when the rocks are covered. At low tide, bare rock and deep holes are exposed.

Surfing: There are two breaks here: an outer break off the western point, and an inner break over the reefs. However, watch the rocks.

Fishing: A popular spot, not only for the shack owners, but for anyone looking for a relatively protected bay to fish off the rocks.

Summary: A small, readily accessible bay, suitable for a picnic or fishing, but usually unsuitable for swimming or surfing.

150   TWIN REEFS

Unpatrolled
No bar: R + rock flats
Beach Hazard Rating: **9**
Length: 500 m

The **Twin Reefs** are two protruding, exposed reefs, in lee of which is a narrow, 500 m long, crenulate platform beach, lying between the base of the vegetated bluffs and the continuous intertidal rock flats. The Cape Paterson Road runs along the top of the bluffs and there is a car park and track down to the beach and rocks.

Swimming: Hazardous at high and low tide, as rocks dominate the energetic surf.

Surfing: The outermost reef can produce some good waves in a moderate to large swell.

Fishing: There are good rock platforms that are awash at high tide. They should only be fished in low waves and by experienced rock fishers.

Summary: The car park provides a nice view of this beach, which is unsuitable for bathing or surfing.

151   OAKS EAST

Unpatrolled
No bar: R + rock flats
Beach Hazard Rating: **8**
Length: 80 m

This is a small, 80 m long, narrow platform beach, located immediately east of Oaks Beach. It is backed by steep, crumbling, 30 m high sandstone bluffs and fronted by boulders and rock flats. It is unsuitable for swimming or surfing, but has some good rock platforms for rock fishing.

152   THE OAKS

Unpatrolled
Single bar: TBR
Beach Hazard Rating: 7
Length: 250 m

The Oaks is one of the few beaches with a sandy surf zone along this rocky section of coast. The beach lies in a 250 m wide opening in the rocks. It is backed by a small, vegetated dune, then 30 m high, vegetated bluffs and the main road. A foot track enters the middle of the beach. The beach faces south and receives waves averaging 1.6 m, which produce a 200 m wide surf zone dominated by two permanent rips, both against the rocks. In addition, reefs parallel the rocks at each end of the beach.

Swimming: Usually a high energy beach dominated by strong rip currents. So use extreme care if any surf is running.

Surfing: This beach provides some of the few beach breaks in the area.

Fishing: Good, deep rip gutters run out along both headlands and can be fished from the rock platforms.

Summary: A nice, sandy beach next to the main road, however be careful if swimming here as rips dominate.

153   UNDERTOW BAY

Unpatrolled
Single bar: LTT/TBR
Beach Hazard Rating: **6**
Length: 300 m

Undertow Bay, as the name suggests, can be subject to strong currents in the surf. The beach is located immediately east of the town of Cape Paterson and next to the major bend in the main road. It is 300 m long, faces south-east, has a headland with a broad rock platform at the east end, a rock reef tied to the beach by a sandy foreland at the west end, and a foredune running the length of the beach. In addition, some rocks outcrop at either end of the beach. Waves average 1.3 m, usually producing a continuous attached bar, cut by permanent rips at either end and often one toward the centre. A heavy shorebreak occurs at high tide and when the bar narrows.

Swimming: A moderately hazardous beach owing to the shorebreak and rips.

Surfing: Usually a dumping beach break.

Fishing: Good, high rock platforms border the beach and provide access to deep water off the rocks, with the best gutter off the headland.

Summary: An accessible, sandy beach next to the main road. Only suitable for swimming during low waves, and most suitable for rock fishing.

154   THE BAY

Unpatrolled
No bar: R + rock flats & reefs
Beach Hazard Rating: **4**
Length: 400 m

The Bay is a crenulate, sheltered beach tucked in behind extensive rock platforms, reefs and a headland. The 400 m long, curving beach is located adjacent to the centre of Cape Paterson township, with a foreshore camping reserve and degraded foredune backing the beach. A boat ramp and car park are located at the western end. Waves average about 0.5 m and the beach is narrow and steep at high tide. Most of the rocks are covered by water, however at low tide more rocks than sand occupy the low tide area.

Swimming: A relatively safe beach with usually low waves. Just watch the rocks and reefs.

Surfing: Usually none at the beach, though there are some reefs offshore.

Fishing: Used for boat launching for Cape Paterson. There are also numerous usually lower energy rock platforms to fish from.

Summary: Cape Paterson's most popular protected beach, located at the foot of the caravan and camping area.

155   THE CHANNEL

Unpatrolled
No bar: R + rocks
Beach Hazard Rating: **6**
Length: 50 m

The Channel is a small, 50 m long beach fronted by a near continuous intertidal rock platform, with just a narrow, 5 m wide, straight channel through the centre of the platform. It is located immediately east of the main Cape Paterson Surf Beach and can easily be accessed along the beach. Waves average about 1 m and the beach face is steep and narrow at high tide, and fronted by exposed rock at low tide, apart from the surging channel.

Swimming: Only at high tide, but still be careful of the rocks.

Surfing: There is a right hand reef break off the platform that is very popular with the locals.

Fishing: A popular spot with excellent rock platforms and, of course, the channel itself.

Summary: A more secluded beach near the main surfing beach, most suitable for sunbathing, surfing on the reef and fishing, but not for bathing.

156   CAPE PATERSON (FIRST SURF)

Cape Paterson SLSC

Patrols: late November to Easter holidays
Surf Lifesaving Club: Saturday, Sunday and Christmas public holidays
Lifeguard: no lifeguard on duty or weekday patrols

Single bar: LTT/TBR
Beach Hazard Rating: **5**
Length: 500 m

For map of beach see Figure 69

Cape Paterson Beach (also called First Surf) is the bathing and surfing beach for the holiday town of Cape Paterson, which lies immediately behind the beach. The beach is in the lee of the dune covered Cape Paterson and its extensive sandstone rock reefs. It faces the south-east and extends for 500 m to the rocky platforms and smaller beaches that front much of the town. The Cape Paterson Surf Life Saving Club, founded in 1960, is located at the western end of the beach, and provides basic facilities, parking and access.

The beach is partially sheltered from high waves by the cape and reefs. It receives waves averaging 1.3 m, which decrease toward the cape. These waves combine with the fine sand to produce a beach containing a low beach face, fronted by a single attached, 50 m wide, bar. The bar is usually continuous and is more likely to be cut by one or two rips toward the surf club and the eastern rocks.

Swimming: A moderately safe beach fronted by a wide shallow bar. However, stay clear of the rocks and reefs at each end; watch for rips, especially toward the eastern rocks; and beware of the shorebreak at low tide. Best to bathe on the bar and between the flags, as 9 people are rescued on average each year.

Surfing: Usually a shorebreak on the bar, blown out by westerlies, best in northerlies. However, on the reef at the eastern end is a good quality right hander called *Insides* that works in a large outside swell.

Fishing: Best off the rocks as holes are rare on the beach.

Summary: A relatively safe and popular patrolled bathing beach.

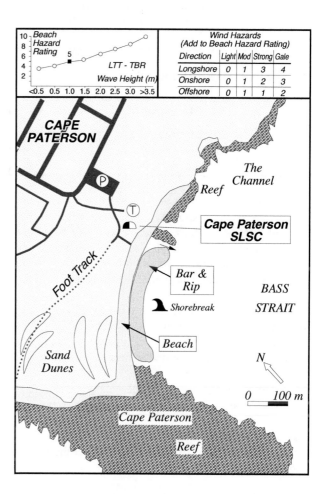

Figure 69.   Cape Paterson is a low sandstone reef that partially protects the main bathing beach, that contains a wide attached bar. The Cape Paterson Surf Life Saving Club is located at the eastern end of the beach. To the west of the cape is the more hazardous surfing beach.

William Hovell Flora and Fauna Reserve

Coast Length: 5 km
Number of beaches: 3 (156 - 158)

The reserve incorporates the scrub covered dune system that backs the three beaches between Cape Paterson and Coal Point. An historic marker at Coal Point marks the discovery of coal here in 1826 by William Hovell.

Information: Wonthaggi (056) 72 1066
             Inverloch   (056) 74 1236

157   SECOND SURF BEACH

Unpatrolled

Single bar: TBR
Beach Hazard Rating: 7
Length: 2 000 m

On the western side of Cape Paterson is a 2 km long, south-west facing, exposed, high energy beach, called **Second Surf** or Seconds. It is backed by up to several hundred metres of dunes that have blown across the cape toward the town. A walking track leads to the beach from the Cape Paterson SLSC car park.

The beach receives waves that are reduced slightly by the offshore reef, and which average 1.4 m. They produce an attached bar cut by rips every 250 m, with some rips permanently located against inshore reefs.

Swimming: Take care on this beach as rips, rocks and reefs dominate.

Surfing: A popular spot that works with a moderate to large outside swell, with best conditions over the reefs at high tide.

Fishing: Cape Paterson's best beach for beach fishing, with persistent rip holes and gutters, together with reefs close inshore.

Summary: This is the surfing beach for Cape Paterson.

158, 159 WRECK BAY, COAL POINT EAST

Unpatrolled

	Rating	No bar	Length
158 Wreck Bay	7	R + rock flats	2 400 m
159 Coal Point East	7	R + rock flats	700 m
		Total length	3 100 m

Between **Wreck Bay** and **Coal Point** is 3 km of crenulate, south-west facing, sandy shore, backed by dunes extending up to 500 m inland and fronted by near continuous intertidal rock flats. At high tide, the 1 to 1.5 m high waves break over the submerged reefs and produce a narrow, steep beach. At low tide, the rocks are exposed (and the beaches high and dry). There is road access to the fishing shacks at Harmers Haven, with a short walk over the dune to the beach.

Swimming: Only at high tide, however beware of the rocks, reefs and strong currents in the surf.

Surfing: There are several reef breaks along this section.

Fishing: The beach can be fished at high tide, while at low tide there are numerous gullies in amongst the rock flats.

Summary: A long beach that changes with the tide, from sandy to rocky.

160 - 165   COAL POINT WEST, CUTLERS, WILLIAMSONS, POWLETT RIVER EAST & WEST, KILCUNDA EAST

Unpatrolled

	Rating	Single bar	Length
160 Coal Point West	7	TBR/RBB	1 800 m
161 Cutlers	7	TBR/RBB	4 700 m
162 Williamsons	7	TBR/RBB	1 300 m
163 Powlett River East	7	TBR/RBB	500 m
164 Powlett River West	7	TBR/RBB	3 300 m
165 Kilcunda East	7	TBR/RBB	750 m
	Total length		12 350 m

At **Coal Point** the trend of the coast swings to face due south-west. Between Coal Point and Kilcunda is a relatively straight 12 km of coast, dominated by sandy beaches but broken by numerous rocks and reefs, particularly toward Coal Point, and the mouth of the Powlett River, toward Kilcunda. Car access is restricted to the **Powlett River** mouth, **Kilcunda**, and the back of **Williamsons Beach**. A rough track off the Chisholm Road reaches the rocks between Coal Point and **Cutlers** beaches.

All the beaches face south-west and receive the full force of the south-westerly winds and waves. Active dunes extend up to 2 km inland, while the beaches throughout are dominated by a high energy, 200 m wide surf zone. There is one bar along the beach, which is often detached and cut by strong rips every 250 m and/or against the major rocks and reefs. There are usually at least 40 rips along the six beaches.

The most extensive rock reefs occur between Coal Point and Cutlers Beach. These extend for a few hundred metres, with sand and dunes backing the rocks. Small rock outcrops separate the other beaches, and also occur at the Powlett River mouth. There are some very strong permanent rips against the rocks, particularly at the more popular Powlett River and Kilcunda, and either side of Cemetery Point, which separates these two beaches.

Swimming: These are very hazardous beaches with usually moderate to high waves and strong transient and permanent rips. Only swim in calm conditions, and avoid the rocks, reefs and rips.

Surfing: There are numerous beach and reef breaks along this section; the biggest problem is getting access to Williamsons and Cutlers beaches. The most popular spots are either side of Cemetery Point at Kilcunda. All these breaks work best in low to moderate swell with north-east winds. The whole section between Kilcunda and Coal Point is worth exploring, if conditions are right and you want your own break.

Fishing: The numerous rips and rocks make for excellent beach fishing for the entire length of the beaches.

Summary: A relatively isolated, energetic section of coast offering the chance of good surf and fishing, but hazardous for bathing.

166, 167, 168　　**KILCUNDA WEST 1 & 2, SHELLY**

Unpatrolled			
	Rating	No bar	Length
166 Kilcunda West 1	5	R + rock flats	300 m
167 Kilcunda West 2	5	R + rock flats	200 m
168 Shelly	6	R + rock flats	150 m
		Total length	650 m

Kilcunda is a small town located where the Bass Highway first reaches the coast. It sits on cleared slopes, with views east along the Kilcunda to Powlett River beaches. There is a caravan park on the bare bluffs between the town and the coast, and three small platform beaches located either side of the caravan park. The first two are accessible from the caravan park, while the third, **Shelly**, has a separate car track leading to it from the Bass Highway.

The three beaches all face south-south-west. They are backed by low foredunes and 20 m high grassy bluffs. They consist of narrow, high tide sand beaches, bordered by more prominent bluffs and are fronted by

near continuous intertidal rock flats. At low tide, the sand is high and dry and only a few channels in the rocks provide access to the water.

Swimming: Bathing is only recommended at high tide in calm conditions. When a swell is running, waves break over the rocks and strong rips are generated.

Surfing: There is a good right hand reef break off Black Head, on the west side of Shelly Beach.

Fishing: All three beaches offer excellent rock fishing, with fishing off the beach only possible at high tide.

Summary: Three small beaches dominated by the surrounding rock, which should be treated with caution if swimming, particularly if there is any surf.

---

169 - 173　　**BEACHES 169 to 173**

Unpatrolled			
	Rating	Single bar	Length
169 Beach 169	5	LTT/TBR	300 m
170 Beach 170	6	TBR	50 m
171 Beach 171	6	TBR	300 m
172 Beach 172	5	LTT	50 m
173 Beach 173	5	R/LTT	200 m
		Total length	900 m

Between Shelly Beach and Punchbowl is six kilometres of sloping, rocky coast backed by farms with no public access. In the central portion of this section are five small unnamed beaches, here called Beaches 169 to 173. Four are backed by cleared slopes and grassy sandstone bluffs, which rise steeply for 20 to 70 m. **Beach 171** is in a gully and has a small creek crossing the beach. All except Beach 171 are backed by small foredunes. The beaches are separated by a series of small sandstone headlands, each fronted by rock platforms.

The beaches all receive waves averaging just over 1 m, which produce a single attached bar on each beach. The bar is usually cut by two rips on the longer beaches and one on the smaller beaches; the rips run out against the small headlands.

Swimming: These beaches are moderately safe, owing to the usual moderate wave height. However, they are all isolated and dominated by rips, and extreme care should be taken if bathing here. Stay on the bars and well clear of the rips and rocks.

Surfing: There are beach breaks on all the beaches which, owing to their location, are rarely surfed.

Fishing: There are good rip holes on each beach, particularly against the headlands.

Summary: Five isolated, pocket sand beaches, offering moderate waves and empty surf.

174   PUNCHBOWL

Unpatrolled
Single bar: R/LTT + rocks
Beach Hazard Rating: **6**
Length: 50 m

The Punchbowl is a steep amphitheatre carved into the 70 m high bluffs. On the eastern side of the Punchbowl is a small, 50 m long beach, almost surrounded by the high bluffs, reefs and rocks. This beach is very difficult to access and hazardous to bathe in, owing to the rocks and permanent rip draining the small beach. It is unsuitable for swimming.

175, 176, 177   BEACH 175, BACK, GRIFFITH'S POINT

Unpatrolled			
	Rating	Single bar	Length
175 Beach 175	5	R/LTT + rocks	300 m
176 Back	4	R/LTT	200 m
177 Griffith's Point	4	R + rocks	50 m
		Total length	550 m

The Potters Hill Road leads straight to the bluffs above **Back Beach** and provides foot access to **Beach 175** and **Griffith's Point Beach**. The three beaches are all backed by steep, grassy, 20 to 30 m high bluffs. They all face south and receive waves averaging less than 1 m; the height of these is reduced by refraction around Cape Woolamai on Phillip Island.

The first beach (175) is 300 m long, with extensive rocks along the western end. Back Beach is backed by a low, grassy foredune and minute swamp, and is bordered by small headlands. Griffith's Point Beach is tucked in lee of the eastern side of the point and receives the lowest waves. Beach 175 and Back Beach have attached bars cut by rips against the headlands and rocks, while Griffith's Point Beach is steep and barless, but with reefs offshore.

Swimming: These beaches are moderately safe, owing to the usually lower waves. However rips still prevail, so use caution. A heavy shorebreak and stronger rips accompany waves above 1 m.

Surfing: Usually a low shorebreak.

Fishing: Best off the rocks.

Summary: Three little used beaches, with Back Beach being the more accessible and most suitable for safe bathing.

178, 179   BONSWICK, BEACH 179

Unpatrolled			
	Rating	Single bar	Length
178 Bonswick	3	R/LTT	150 m
179 Beach 179	3	R/LTT	100 m
		Total length	250 m

Bonswick Beach and **Beach 179** are located 1 km south of San Remo. The road due south of the post office terminates at Bonswick. Both beaches are backed by 20 to 30 m high bluffs, with a foreshore reserve between the bluffs and the road. The beaches face west, receive low waves averaging less than 0.5 m, and are both narrow with a narrow, continuous bar and some rocks in the surf.

Swimming: Relatively safe beaches owing to the usually low waves and absence of rips. Higher waves produce a stronger shorebreak.

Surfing: Usually none. However when big swells reach into the bay, a left hand break, called *Foots*, runs over the rocks off the southern bluffs.

Fishing: The water is relatively deep off both beaches, with rocks to fish from below the bluffs.

Summary: Two small and accessible though little used beaches.

180, 181   BACK, CHILDRENS

Unpatrolled			
	Rating	Single bar	Length
180 Back	4	R + tidal currents	500 m
181 Childrens	4	R + tidal currents	500 m
		Total length	1 000 m

Back and Childrens Beaches used to be part of a continuous strip of sand along the southern shores of San Remo. However in response to shoreline erosion, seawalls have been built between the two beaches and the sand replaced by rock. **Back Beach** is 500 m long and faces south-west. A foreshore reserve and low vegetated dune back the beach with the seawall at the western end. **Childrens Beach** is also 500 m long and faces west into The Narrows, that separate San Remo from Phillip Island. A caravan park backs the beach. The seawall runs along the entire length of the beach and separates it from Back Beach. As a result, the beach is very narrow and in parts is completely awash at high tide.

Both these beaches receive low waves that produce a steep beach face, fronted by extensive tidal shoals that may be exposed at low tide. However, at high tide they are covered and relatively deep water lies off the beach. In addition, Childrens Beach is also influenced by the strong tidal currents that flow through The Narrows.

Swimming: Two relatively safe beaches with low waves and no rips. However, water is deep close to shore and tidal currents run past Childrens Beach, so use caution, particularly with children.

Surfing: Usually none, except when a huge outside swell is running which can push 1 m waves this far into the bay, to produce a right hander over the tidal shoals off Childrens Beach, called *Speedies*.

Fishing: Relatively good access to deep water, off the beach and adjoining seawalls.

Summary: Although these two beaches have been modified by seawall construction, their proximity to San Remo ensures their summer popularity.

182   **SAN REMO**

Unpatrolled
No bar: R
Beach Hazard Rating: **4**
Length: 200 m

San Remo is the gateway to Phillip Island, and below the bridge that links it with the island is a small, 200 m long, north facing beach. The beach is backed by a grassy foreshore reserve and good picnic facilities and is fronted by the bridge and The Narrows. Ocean swell does not usually reach the beach, with only small wind waves lapping at the narrow, steep beach face.

Swimming: This beach is usually calm, however watch the deep water off the beach and particularly the strong tidal currents that flow under the bridge.

Surfing: None.

Fishing: A popular and very accessible spot to fish the tidal channel.

Summary: This beach is a popular picnic area and suitable for wading, however be wary of the tidal currents and boat traffic if swimming.

5. PHILLIP ISLAND AND MORNINGTON PENINSULA

Phillip Island and Mornington Peninsula (Sandy Point to Point Nepean)
Beaches: 183 to 280 Coast length: 145 km
For map of region see Figures 70 & 73

Phillip Island
Coast length: 68 km (Newhaven to Cowes) Beaches: 183 - 226 (44 beaches)
For map of Phillip Island see Figure 70

Phillip Island is a 25 km long island that forms the southern boundary of Western Port. Its southern shores face Bass Strait with its strong winds and waves. Its western and northern shores are exposed to low waves (Figure 71) but strong tidal currents, while in the northeast, the coast is protected between Rhyll and Newhaven and is dominated by tidal flats. The island is primarily composed of volcanic rocks deposited some 50 million years ago. They formed as lava covered 100 million year old sandstone rocks that only outcrop at Watt Point and Helen Head, and form the higher San Remo Peninsula. At Cape Woolamai, the rocks are ancient, 400 million year old granites. The columnar basalt is best displayed along the south coast, particularly at The Colonnades, Pyramid Rocks and The Nobbies at Grant Point. Along the north coast, the weathered basalt forms low red bluffs, as at Red Rocks.

Victorian Beaches - Region 5: East (Phillip Island)

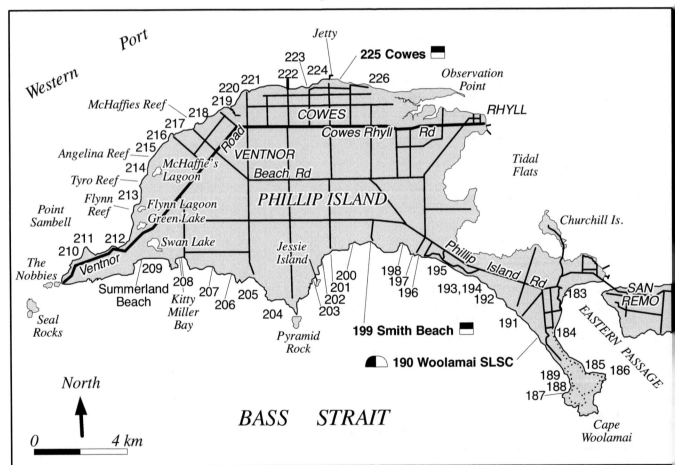

Figure 70.   *Region 5: (Phillip Island) Newhaven to Observation Point (Beaches 183 to 226)*

Figure 71.   A view of the protected north-west coast of Phillip Island, showing McHaffie Point in the foreground, with low energy reflective beaches lying between the sandstone bluffs and reef flats.

On this old landscape, the more recent intrusion of the sea has brought successive deposits of wave and wind borne sand.   The most recent deposits arrived about 6 000 years ago as sea level reached its present level. These deposits are primarily derived from the seabed, worked onshore by the waves, deposited by the swash and in places blown inland as dunes. Today they form the 44 ocean beaches of Phillip Island, from Newhaven to Cowes.

When visiting Phillip Island you should make your first stop at the excellent Phillip Island Information Centre, located on main Phillip Island, 1.5 km west of the bridge.

Phillip Island has three sides:

- **Cleeland Bight**, facing the Eastern Passage, with four low energy beaches;
- The 40 km long **south coast**, with its 23 surfing beaches;
- The **Western Port Bay coast**, with 17 increasingly lower wave beaches between Cowrie Bay and Observation Point.

183, 184   **NEWHAVEN/HOMESTEAD, WOOLAMAI SAFETY**

Unpatrolled			
	Rating	Single bar	Length
183 Newhaven/ Homestead	4	R + tidal currents	2 300 m
184 Woolamai Safety	4	R + tidal currents	2 400 m
		Total length	4 700 m

Cleeland Bight forms the Phillip Island boundary to the Eastern Passage to Western Port. It is a 3 km wide, east facing bight, containing 4.7 km of curving sandy beaches, running from the bridge at Newhaven to the rocks of Cape Woolamai.   Two low wave beaches occupy most of the bight. The first is **Newhaven**, which begins at the bridge, faces south-east and extends to the sandy Manuka Point. It is backed by a low dune and the Newhaven Swamp. Road access is available at each end and a caravan park backs the eastern end. **Woolamai Safety Beach** is, as the name implies, another low wave beach that curves its way south from Manuka Point, then swings south-east toward Cape Woolamai. Esplanade Drive backs the first 1 km of beach after which there is only foot access along the back of the Woolamai sand dunes.

Both beaches receive waves averaging less than 0.5 m, and are exposed to tides reaching over 2 m. Both have relatively steep, narrow beaches with no surf. However at low tide, extensive tidal flats are exposed in front of Newhaven Beach.

Swimming: Two relatively safe low wave beaches, with deep water off the beaches at high tide and tidal flats at low tide. However, be aware of strong tidal currents running parallel to the beaches.

Surfing: None.

Fishing: Best at high tide and over the rocks and reefs along the southern end of the beach.

Summary: Two popular summer beaches for people and families looking for a quieter, usually calm beach. They are also partially protected from westerly winds.

185, 186   CAPE WOOLAMAI NORTH 1 & 2

Unpatrolled			
	Rating	No bar	Length
185 Cape Woolamai North 1	3	R	600 m
186 Cape Woolamai North 2	3	R	200 m
		Total length	800 m

A low, rocky reef separates Woolamai Safety Beach from the **Cape Woolamai North Beach**. Sand dunes spilling over from Woolamai Surf Beach back the western half of this 600 m long beach, with the scarped bluffs of northern Cape Woolamai backing the eastern half. On the eastern side of the bluffs is a similar, but smaller, 200 m long beach, bordered by 30 m high scarped bluffs and backed by grassy slopes. Both the beaches face north-east and lie inside Cape Woolamai. They are sheltered from Bass Strait, with waves averaging less than 0.5 m, that produce two relatively steep, narrow beaches.

Swimming: Two relatively safe though isolated and little used beaches. Water is deep off the beach at high tide and shallow at low tide. There is usually no surf, however strong tidal currents run parallel to the beach.

Surfing: None.

Fishing: Best at high tide and near the boundary rocks and reefs.

Summary: Relatively isolated and sheltered beaches; accessible by boat or along the beach from Woolamai Waters.

187, 188, 189   CAPE WOOLAMAI SOUTH, MAGIC LANDS, MAGIC LANDS WEST

Unpatrolled			
	Rating	Single bar	Length
187 Cape Woolamai South	5	R	60 m
188 Magic Lands	7	TBR/RBB	800 m
189 Magic Lands West	8	TBR/RBB	50 m
		Total length	910 m

On the south-western tip of **Cape Woolamai** is a small, 60 m long pocket sand beach (187), backed by steep, 40 m high bluffs and bordered by similar headlands. Around the western headland the sand continues below the bluffs, that gradually decrease in height toward Woolamai Beach. Rock reefs lie off the beach, resulting in a crenulate shoreline with reef breaks, called *Magic Lands*. At the western end are two low headlands that border a 50 m long pocket beach, and form the boundary with the main Woolamai Surf Beach. All three beaches are only accessible by foot from Woolamai Surf Life Saving Club.

Reefs, rocks and headlands protect the first beach (187), resulting in low waves and a steep, barless beach. The **Magic Lands Beaches** (188, 189) receive waves averaging 1.8 m which, while breaking on the reefs, induce strong, permanent rips in between and against the headlands.

Swimming: These are three relatively isolated and potentially hazardous beaches. Two have high waves and strong rips, and all have rocks and reefs off the beaches. They are more suitable for surfing than swimming.

Surfing: *Magic Lands* has both lefts and rights over the sand and reefs. It is the place to check during north-easterly winds, as they blow offshore in this corner of Cape Woolamai.

Fishing: The reefs that produce the Magic Lands surf also form deep, permanent rip channels, that can be fished from the beach or rocks.

Summary: Three beaches that require a walk down the beach to get there, and are most suited for surfing and fishing.

190   WOOLAMAI

Woolamai SLSC

Patrols: late November to Easter holidays
Surf Lifesaving Club: Saturday, Sunday and
Christmas public holidays
Lifeguard: 10 am to 6 pm weekdays, during
Christmas holiday period

Inner bar: TBR/RBB   Outer bar: TBR/RBB
Beach Hazard Rating: **8**
Length: 4 200 m

For map of beach see Figure 72

Woolamai (Surf) Beach is the longest and most
exposed beach on Phillip Island, and is the site of the
island's only surf lifesaving club, founded in 1959. The
beach is 4.2 km long and faces south-west into the high
waves and westerly winds. The persistently moderate to
high waves average 1.7 m. Together with the fine to
medium beach sand, they produce a moderately steep
beach face. The 250 m wide surf zone contains an inner
bar cut by strong, deep rips every 250 m, with an outer
bar cut by more widely spaced rips. The rips usually
scour deep holes in the beach, making them highly
visible.

The Woolamai Beach Road runs out to the beach and
provides parking in the large car park at the surf club.

Swimming: This is a surf for experienced bathers
only. The waves are usually large, and the rips strong
and close inshore. Stay between the flags at all times
and avoid the rips, as an average of 55 people are
rescued annually. This hazardous beach has the third
highest number of rescues per year in Victoria, lower
only than Portsea and Gunnamatta.

Surfing: A popular spot for more experienced surfers.
It offers some of the best beach breaks in Victoria, often
with long lefts and rights peeling over the wide banks
and into the deep rips. *The Carpark* break is located to
the east of the surf club, while *Anzac Alley* is to the west
of the club and offers good right handers. All Woolamai
breaks are best with a low to moderate swell and
northerly winds.

Fishing: Excellent rip holes are a characteristic of this
beach.

Summary: An exposed, high energy beach for more
experienced bathers and surfers. Best during summer
northerly conditions, as westerly winds blow out the
surf. The dunes are the site of Cape Woolamai State
Faunal Reserve.

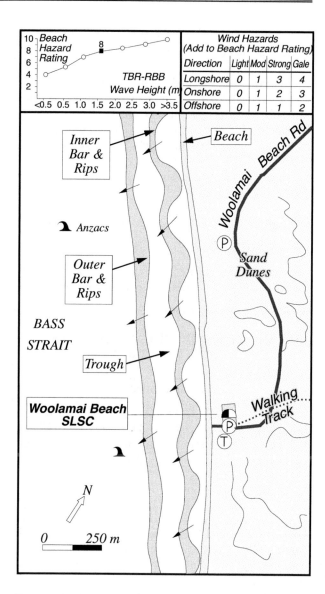

10 Beach Hazard Rating		Wind Hazards (Add to Beach Hazard Rating)			
		Direction	Light	Mod	Strong	Gale
TBR-RBB Wave Height (m)		Longshore	0	1	3	4
		Onshore	0	1	2	3
<0.5 0.5 1.0 1.5 2.0 2.5 3.0 >3.5		Offshore	0	1	1	2

*Figure 72.   Woolamai Beach has the only surf
lifesaving club on Phillip Island. It is also the most
energetic beach on the island, and is dominated by high
waves and rips every 250 m.   Be very careful if
swimming here and stay between the flags.*

191   THE COLONNADES

Unpatrolled

Single bar: TBR
Beach Hazard Rating: **8**
Length: 900 m

The Colonnades is the western extension of Woolamai
Surf Beach, with rock reefs in the surf forming the
boundary between the two beaches. This section is
900 m long, faces south-west, and is bordered in the
west by the red basalt cliffs of Forrest Caves. Columnar
basalt in the cliffs gives this location its name. A few
rocks and reefs also occur along the beach. The beach is

accessible in the east by the road beside the airfield, and in the west via Forrest Caves.

Swimming: An exposed, high energy beach where waves average 1.5 m. Strong rips dominate the wide surf zone, with permanent rips against the reefs. Use extreme care if bathing here, stay close inshore and on the bar, and clear of the rips, rocks and reefs.

Surfing: The reefs and high waves ensure potentially good breaks along this section.

Fishing: There are permanent rip holes against the rocks and reefs, that can be fished from the beach.

Summary: An exposed, high energy beach used by surfers and fishers, but unsuitable for safe bathing.

192, 193, 194   FORREST CAVES, FORREST BLUFF EAST & WEST

Unpatrolled			
	Rating	Single bar	Length
192 Forrest Caves	7	TBR + reefs	100 m
193 Forrest Bluff East	8	TBR + rocks & reefs	200 m
194 Forrest Bluff West	8	TBR + reefs	950 m
		Total length	1 250 m

Forrest Caves and Bluffs are a section of eroding, red, basalt bluffs. The caves are eroded into the basalt rock platform and are most exposed at low tide. Three beaches occupy this south-west facing, 1.3 km section of coast; all three accessible via the Forrest Bluffs Road. They receive waves averaging 1.6 m that produce a 200 m wide surf zone, containing strong rips, particularly against the reefs, rocks and bluffs.

Swimming: Use extreme care if bathing at these beaches, as high waves and rips usually dominate. Even in the calmer inner waters of Forrest Caves, strong currents sweep the reefs.

Surfing: There are a number of breaks over the reefs, which tend to be better at high tide. Wave height begins to decrease along here and waves are usually a little smaller than at Woolamai.

Fishing: This is a popular area off the beaches and headlands at high tide and off the rocks and reefs at low tide.

Summary: These are three exposed beaches with extensive rocks, reefs and rips. Best suited to surfers and fishers.

195, 196, 197   SURF, SURFIES POINT EAST & WEST

Unpatrolled			
	Rating	Single bar	Length
195 Surf	7	TBR + reefs	400 m
196 Surfies Point East	6	TBR	1 000 m
197 Surfies Point West	6	LTT + reefs	200 m
		Total length	1 600 m

Surf Beach and the two beaches either side of **Surfies Point** are located between basalt and calcarenite rock platforms and reefs. The beaches are accessible via the Surf Beach and Sunderland Bay housing estates, with parking either end of Surf Beach and on Surfies Point. The three beaches face south-south-west and all receive waves averaging between 1 and 1.6 m. The waves interact with the sand, rocks and reefs to produce both good reef and point surf, and strong beach and reef rips. Surf Beach usually has permanent rips against each end, while Surfies Point East has similar rips, together with one to two strong central rips. West of the point the waves break over wide reefs and although low at the shore, there are strong currents over the reefs.

Swimming: Beware of the strong rips on all three beaches; stay close inshore and clear of the rips and reefs.

Surfing: *Surfies Point* is, as the name suggests, a very popular and accessible spot. The right hand wave breaks over reefs off the point and can hold to up to 2.5 m. On the west side of the point is a more critical reef break called the *Express*.

Fishing: These three beaches offer small headlands, rock platforms, rocks and reefs, together with rips on the beach running out against the reefs.

Summary: A popular surfing and fishing area, not suitable for safe bathing.

198   SUNDERLAND BAY

Unpatrolled
No bar: R + rocks & reefs
Beach Hazard Rating: 4
Length: 300 m

Sunderland Bay is a 200 m wide gap in the basalt bluffs, into which waves break over extensive reefs and rocks and around a sea stack. This has produced a 300 m long, south facing beach composed of sand and basalt cobbles. Waves are lowered to about 0.5 m by the time they hit the beach, which is steep and narrow, with no bar, but reefs off the beach. The Sunderland Bay Estate borders the eastern side, with parking available on the eastern bluffs and at the back of the beach, via a gravel road. The western side of the bay is bordered by the 40 m high Sunderland Bluffs.

Swimming: A relatively safe beach usually with low waves; however, the beach is steep and composed of cobbles and some boulders, and a heavy shorebreak occurs during high waves. Bathing is best in calm conditions and at high tide, while at low tide, the rock pools at the western end are a popular spot to sit.

Surfing: None at the beach, however at high tide there is a left hander off the eastern reef that breaks toward the sea stack.

Fishing: A popular location, with ample spots for relatively safe rock fishing.

Summary: A small semi-enclosed bay that is most suitable for a picnic and fishing, with surf out off the reefs.

199   SMITH

Summer lifeguard
Single bar: LTT/TBR + reefs
Beach Hazard Rating: **5**
Length: 1 000 m
Lifeguard: 10 am to 6 pm all week, during Christmas holiday period

Smith Beach is a 1 km long, south-west facing beach that fronts the Smith Beach and Beachcomber Estates. Smith Beach Road runs to a large car park on the 20 m bluffs behind the centre of the beach. The beach receives wave averaging 1 m, which combine with the relatively fine sand to form a wide, low gradient beach, with a wide, shallow attached bar. Rips run out against the eastern reef and the western Smith Point.

Swimming: Moderately safe on the inner portion of the bar and clear of the rocks and rips. Stay in the patrolled area between the flags.

Surfing: The shallow bars produce a wide beach break more suitable for learners.

Fishing: The platforms and reefs at each end offer the best spots.

Summary: This beach offers a usually gentle surf with strong rips, more suitable for families and younger surfers.

200   YCW

Unpatrolled
Single bar: LTT
Beach Hazard Rating: **5**
Length: 400 m

YCW is the acronym of the Young Catholic Workers Camp, which lies on the grassy bluffs immediately behind the beach. A road leads to the camp and a car park on the eastern Smith Point. The beach faces south-east, with basalt headlands and rock platforms at each end, and a 400 m long sand beach in between them. The beach receives waves averaging 1 m and usually has a beach that is awash at high tide. The waves then break over a narrow cobble beach between the sand and the backing 10 to 30 m high bluffs. In contrast, at low tide the beach is wide, low and shallow, with rips running out against each head.

Swimming: Moderately safe on the inner portion of the bar and away from the side rips.

Surfing: Usually a wide beach break that is most suitable for younger and less experienced surfers. There is a left off the eastern point.

Fishing: Best off the rock platforms at each end, as both front deeper rip channels.

Summary: Similar to Smith Beach, however not as popular, possibly due to the longer and steeper walk from the car park to the beach.

201, 202, 203   THE GAP, RACETRACK, JESSIE

Unpatrolled			
	Rating	Single bar	Length
201 The Gap	5	R/LTT	200 m
202 Racetrack	5	R/LTT	150 m
203 Jessie	4	R	70 m
		Total length	420 m

At the southern end of the Phillip Island Bike Racetrack is a section of 40 m high basalt cliffs and bluffs, fronted by near continuous rock platforms and reefs. Tucked in at the base of the bluffs are two narrow sand platform beaches in Cunningham Bay, called **The Gap** and **Racetrack**. A similar beach, **Jessie Beach**, lies 200 m to the south-west, in lee of Jessie Island. The Gap Road terminates on the bluffs above the Cunningham Bay beaches, while a walk around the bluffs or rocks is required to reach Jessie Beach.

All three beaches are somewhat sheltered by the reefs, and waves average 0.5 m at the shore. However they only reach the narrow beaches at high tide, with exposed rock flats fronting the beach at low tide.

Swimming: At high tide you can swim off the beach, however at low tide the only suitable water is in the tidal pools. Stay inshore if swimming here as there are rocks, reefs and stronger currents off the beach.

Surfing: There is no surf at the beaches, however on the reefs there are three spots. *The Gap* has a left and right hand reef break that works best at low tide; *Racetrack* has a right that is better at high tide; while *Jessie Island* has a right that works best at low tide.

Fishing: This section of coast is dominated by rocks and reefs and is only suitable for rock fishing.

Summary: These three beaches are relatively isolated and are unsuitable for safe bathing. They are primarily used by surfers and rock fishers.

204   STORM BAY

Unpatrolled
No bar: R + rocks & reefs
Beach Hazard Rating: **8**
Length: 100 m

Storm Bay is a small, 100 m long platform beach located 500 m due west of the Pyramid Rock car park. The beach is surrounded by 20 to 30 m high grassy bluffs, with a boulder beach underlying and backing the narrow sand beach. Basalt boulder flats front the beach. It is only awash at high tide, with surf running over the boulders and reefs at low tide.

Swimming: This beach is unsuitable for bathing, unless conditions are calm and it is high tide.

Surfing: A right hander breaks over the boulders at high tide. However, watch the abundant rocks and boulders.

Fishing: There are a number of exposed rock platforms on either side of Pyramid Rock, however take care, as waves wash over the rocks, particularly at high tide.

Summary: It is well worth the drive and walk to Pyramid Rock; however, Storm Bay is only for experienced surfers and rock fishers.

205   BERRY

Unpatrolled
Single bar: TBR + rocks & reefs
Beach Hazard Rating: **7**
Length: 1 000 m

Berry Beach lies at the end of Berry Beach Road, where there is a car park on the 40 m high bluffs behind the beach, but no other facilities. The beach is 1 km long and faces south-west. It is backed by grassy bluffs, with Redcliff Head and extensive reefs bordering the eastern end, and Berry Reef and Wild Dog Bluff at the western end. Native Dog Creek also drains out at the western end.

The beach receives waves averaging 1.5 m and has a wide, low beach fronted by extensive intertidal rocks and reefs at each end. Only the central 200 m provides a sandy surf zone, although it is bordered by strong, permanent rips against the reefs.

Swimming: Only moderately safe close inshore at high tide, toward the sandy centre, and clear of the rips.

Surfing: There is a wide beach break in the centre, with more experienced surfers going out to Wild Dog Bluff, where the reef can be surfed around high tide.

Fishing: There are wide basalt platforms around both headlands, and extensive reefs off the beach.

Summary: An accessible beach suitable for a picnic, but take care if bathing. More popular with surfers and fishers.

206, 207   HUTCHINSON, THORNY

Unpatrolled			
	Rating	Single bar	Length
206 Hutchinson	6	TBR	250 m
207 Thorny	5	LTT/TBR	400 m
		Total length	650 m

Hutchinson and Thorny Beaches are two similar beaches, separated by 50 m high Helens Head and backed by private farmland, with no public access. Both beaches are backed by steep, grassy bluffs, with prominent grassy headlands and basalt rock platforms at each end. **Thorny Beach** faces south and is protected by the platforms off the western Watts Point, with waves averaging 1 m. **Hutchinson Beach** faces south-west and receives slightly higher waves, averaging 1 to 1.5 m.

Both beaches have a wide, low gradient beach fronted by a wide, shallow, attached bar. On Hutchinson Beach the bar is bordered by two strong, permanent rips against each end, while on Thorny Beach the rips are present, but not as intense.

Swimming: These are two little used beaches, dominated by rips. If bathing, stay on the inner portion of the bar and clear of the rocks and rips.

Surfing: Both have wide beach breaks.

Fishing: Wide rock platforms border each beach, with permanent rip channels along the rocks.

Summary: Two difficult to access beaches.

208   KITTY MILLER BAY

Unpatrolled
Single bar: R/LTT
Beach Hazard Rating: **4**
Length: 500 m

Kitty Miller Bay lies 500 m inside the bordering Watts Point and Kennon Head, with wide rock platforms at the base of the headlands, and reefs in between them. Consequently, the 500 m long, semi-circular, south facing beach is reasonably sheltered and waves average only 0.7 m at the beach, and even lower at low tide. Kitty Miller Road runs straight out to the beach, with a small car park behind the low dune that backs the beach. The beach is narrow at high tide, while at low tide it expands to more than 50 m wide, with deeper water off the centre and extensive intertidal rock flats to either side. Rips are only present during higher waves.

Swimming: One of the safer south coast beaches, owing to the generally low waves and gently sloping central beach section.

Surfing: There are a number of breaks over the reefs that work best at high tide.

Fishing: The bay offers some more protected rock platforms to fish the bay and the reefs.

Summary: A picturesque, circular bay, accessible by car and offering relatively safe bathing and often good surf further out on the reefs, together with fishing off the rocks.

209   SUMMERLAND BAY

Unpatrolled
Single bar: LTT/TBR
Beach Hazard Rating: **6**
Length: 1 200 m
See Figure 37

Summerland Bay is one of Australia's most popular tourist destinations, as every evening hundreds of tourists are bused out to see the sunset penguin parade. A large car park and tourist facility are located at the western end of the beach. Local parking is also provided on top of Phelans Bluff, which forms the western headland.

The beach is 1.2 km long and faces due south. Kennon Head and its platform and reefs form the eastern boundary, while the western end comprises Phelans Bluff, together with its platform, reefs and boulders. A dune field up to 1 km wide backs the beach that, together with backing Swan Lake, is part of Swan Lake Reserve. The reserve is home to both the penguins and a mutton bird rookery.

The beach receives waves averaging just over 1 m which, together with the relatively fine beach sand, produces a wide, low gradient beach fronted by a wide, continuous bar. The bar is usually cut by two or three rips toward the central-eastern end.

Swimming: Bathing is permitted during the day when the penguins are out at sea. The beach is moderately safe, with the safest section being at the low energy western end, as waves and rips increase to the east. However, watch the rip against the western rocks.

Surfing: A popular right hander breaks over the reef and boulders below Phelans Bluff, together with *Centre Crack*, a right hand reef break.

Fishing: Best off the rock platforms at high tide, if the waves are low.

Summary: Victoria's number one tourist destination, together with a moderately safe beach and some good point and reef surf.

210   COWRIE BAY

> **Unpatrolled**
> No bar: R + rocks & reef
> Beach Hazard Rating: **6**
> Length: 400 m

Cowrie Bay is a west facing, rocky bay lying 1 km north of Point Grant. Ventnor Road, that runs to Point Grant, goes past the southern corner of the bay, where there is a small car park on the grassy bluffs overlooking the bay. The bay is backed by low dunes and the bluffs. It is bordered by basalt platforms and reefs, which at low tide almost fill the bay. A steep, narrow, 400 m long sandy beach backs the rocks. Waves average 1 m at the beach at high tide, however at low tide, they break over the rocks and reefs seaward of the beach.

Swimming: Only suitable at high tide, or in the rock pools at low tide.

Surfing: The southern point has a good left hand break known as *Cowries*.

Fishing: There are extensive basalt platforms to fish from. The bay is usually more sheltered than the south coast beaches.

Summary: An open bay exposed to the westerlies, primarily used by surfers and fishers.

211   SHELLY

> **Unpatrolled**
> No bar: R + rock flats
> Beach Hazard Rating: **4**
> Length: 750 m

Shelly Beach lies in the southern corner of Cat Bay. Ventnor Road runs along the top of the grassy bluffs that back the 750 m long beach, with three car parks located behind the beach.   The beach faces north and is sheltered from direct westerly waves by Point Sambell. Generally, low waves averaging 0.5 m arrive at the beach. The beach is narrow and steep at high tide, while at low tide it widens a little, but is fronted by near continuous, low tide rock flats.

Swimming: Relatively safe owing to the usually low waves. Best at high tide when the rocks are covered, however then the water is deep inshore.

Surfing: No surf at the beach, however the well-known *Cat Bay* has two point breaks. *Left Point* is a left hand break over the reefs off Point Sambell, while *Right*

Point is a right hander that breaks off the northern or right point. In addition, spilling waves break over the rock flats when the waves are up and the tide is right.

Fishing: You can fish the reefs off the beach at high tide, or the rock gullies at low tide from the beach or platforms.

Summary: This beach is accessible and usually quiet; suitable for bathing at high tide. A popular surfing spot when the swell is up.

212   FLYNN

> **Unpatrolled**
> No bar: R
> Beach Hazard Rating: **4**
> Length: 1 300 m

Flynn Beach is the northern beach in Cat Bay. Ventnor Road runs behind the low dune field that covers the bluffs backing the beach, with car parks at the northern and southern ends. The 1.3 km long beach is bounded by Flynn Reef in the north and similar reefs to the south. The beach faces west-south-west and receives waves averaging 0.6 m. These produce a steep, narrow beach, which widens as the tide falls. A narrow cobble beach lies between the sand and the bluffs.

Swimming: Relatively safe when waves are low, with deep water usually right off the beach. Larger waves will produce a heavy shorebreak.

Surfing: Usually none at the beach, however the reefs at both ends can produce good waves when the swell is up.   The southern reef has a right hander called *Sunliners*, while the northern *Flynn Reef* is a very popular right hander.

Fishing: You can fish deep water off the beach at high tide, however the rocks and reefs at each end are the more popular spots.

Summary: An exposed beach with usually low waves, suitable for bathing, but most popular with surfers.

213, 214, 215   FARM, WOOLSHED, ANGELINA

Unpatrolled			
	Rating	No bar	Length
213 Farm	4	R + rock flats	500 m
214 Woolshed	4	R + rock flats	600 m
215 Angelina	4	R + rock flats	1 000 m
		Total length	2 100 m

North of Flynn Reef the coast trends north-east. It consists of four prominent, 20 to 30 m high, grassy bluffs at Flynn, Tyro, Angelina, and Hen and Chicken reefs. Three long, narrow, crenulate beaches lie in between the bluffs. The beaches continue around the base of the bluffs and are fronted by extensive intertidal rock flats on either side. Low dunes back the beaches, with the small Flynn and McHaffie lagoons behind **Farm** and **Woolshed Beaches**. The three beaches are backed by private farmland and public access is restricted to walking along the beaches. The nearest car access and parking are located on the south side of Flynn Reef, at the northern end of **Angelina Beach**.

The beaches all face west and receive waves usually less than 0.5 m. This produces the steep, narrow beaches, which have no bar and deep water close inshore, where there is no reef flat. At low tide, the reefs are exposed and much of the beach is high and dry. The beaches themselves are crenulate in between the reefs and rocks.

Swimming: These are all relatively safe beaches, with low waves and usually no surf or rips.

Surfing: The southern point of Farm Beach, known as *The Farm*, is the last recognised surfing spot on the bay coast. It only works in large outside swell and produces moderate sized left handers.

Fishing: There are numerous rocks and reefs to fish along here, but few holes or currents.

Summary: Three relatively isolated beaches, best suited for a coastal walk.

216   HEN AND CHICKEN

Unpatrolled
No bar: R + rock flats
Beach Hazard Rating: **4**
Length: 150 m

Hen and Chicken Beach is named after the offshore reef of the same name. It is a short, 150 m long beach, backed by steep, 20 m high bluffs with bright red bases.

McHaffie or Grossard Point forms the northern boundary, with Grossard Point Road terminating at a car park on the point. A second car park is located on the bluffs behind the beach. The beach itself faces north-west and usually receives low swell, less than 0.5 m; and wind waves, which build the narrow, relatively steep beach. Reef flats are exposed along the length of the beach at low tide.

Swimming: Relatively safe and best at high tide when the rock flats are covered.

Surfing: None.

Fishing: Best off the rocks at low tide, or off the beach at high tide.

Summary: A small, accessible beach, best suited for sunbathing and fishing.

217, 218   ELIZABETH COVE, VENTOR

Unpatrolled			
	Rating	No bar	Length
217 Elizabeth Cove	3	R + reef flats	1 100 m
218 Ventor	3	R + reef flats	1 100 m
		Total length	2 200 m

East of Grossard Point the coast trends east-north-east, fronting the holiday settlement of Ventnor. The beaches are highly crenulate, owing to the predominance of low bluffs and extensive reef flats. Both beaches receive low waves and are relatively narrow and steep, with no bar or surf. Rather, reef flats are exposed at low tide. Low dunes back much of both beaches, particularly away from the bluffs. Boat Creek drains across the northern end of **Elizabeth Cove**, and a few tinnies are stored on the dune behind the beach.

There are car parks at both ends of Elizabeth Cove and at the northern end of **Ventor Beach**.

Swimming: Relatively safe and best at high tide when the rocky reef flats are covered.

Surfing: None.

Fishing: You can fish the reef flats off the beach at high tide, or wander out onto the reef to fish deeper water at low tide.

Summary: Two long, narrow, crenulate beaches, most suitable for sunbathing and swimming at high tide.

219   SALTWATER CREEK

Unpatrolled
No bar: R
Beach Hazard Rating: **2**
Length: 1 000 m

Saltwater Creek Beach blocks the entrance to Saltwater Creek and its small meandering lagoon. Low bluffs and reef flats form the southern boundary, with a narrow, red, basalt dyke called Penguin on the northern border. Scattered holiday houses and car parks lie at each end of the 1 km long beach, with cleared farmland surrounding the central creek. Most of the beach is backed by a low, well-vegetated foredune. It receives low waves and has a relatively steep beach dropping into deep water, particularly at high tide. Reef flats only border the southern bluffs.

Swimming: One of the better bathing beaches along this section of coast, owing to the general absence of reef flats.

Surfing: None.

Fishing: There is usually deep water off the beach.

Summary: An attractive beach, dune and creek area, mainly used by the surrounding holiday makers.

220, 221, 222   RED ROCK WEST & EAST, SHEOAK

Unpatrolled			
	Rating	No bar	Length
220 Red Rock West	3	R + tidal currents	250 m
221 Red Rock East	3	R + reef flats & tidal currents	600 m
222 Sheoak	3	R + reef flats & tidal currents	1 600 m
		Total length	2 450 m

Between Penguin Point and Richardson Point is a near continuous strip of narrow sand beaches, which initially faces north-west, then turns at Red Rock Point to run easterly toward Cowes. The three beaches are all relatively narrow and steep, with low rocky points and reef flats forming the boundaries. The flats are exposed at low tide almost continuously east of **Red Rock**. Waves are usually low, with wind waves being the dominant process. A series of roads runs behind the beaches with car parks at Red Rock Point, **Sheoak**

Beach and toward Richardson Point, where there is also a boat ramp.

Swimming: Relatively safe beaches with deep, calm water against the beach at high tide, and extensive reef flats exposed at low tide. Tidal currents in the Western Passage parallel the beaches.

Surfing: None.

Fishing: Best close inshore at high tide, otherwise off the reef flats at low tide. However, beware of tidal currents flowing parallel to the beach.

Summary: A nice strip of usually quiet sandy beaches.

223   YACHT CLUB

Unpatrolled
No bar: R
Beach Hazard Rating: **2**
Length: 1 200 m

The Cowes Yacht Club is located toward the eastern end of 1.2 km long **Yacht Club Beach**. The beach is relatively narrow and steep, with low bluffs and reef flats bordering it at Richardson Point and Mussel Rocks, and a low dune behind. There is a car park at the yacht club and a caravan park behind the beach. Waves are usually low wind waves.

Swimming: A relatively safe beach with low waves. Better at high tide as reef flats and sand shoals are exposed at low tide.

Surfing: None.

Fishing: Better at high tide or off the reef flats at low tide.

Summary: A popular beach for sailors, boaters and holiday makers from Cowes and the caravan park.

224, 225   MUSSEL ROCKS, COWES

Summer lifeguard			
	Rating	No bar	Length
224 Mussel Rocks	3	R + sand flats & tidal currents	250 m
225 Cowes	3	R + sand flats & tidal currents	300 m
		Total length	550 m

Lifeguard: 10 am to 6 pm all week, during Christmas holiday period

Cowes is the focus of all activities on Phillip Island and where most of the residents and visitors stay. It is located on the middle of the north shore of the island, far enough east for its jetties to be clear of ocean swell, and far enough west to be clear of the tidal flats off Observation Point. The town has all the facilities of a thriving summer tourist and holiday industry.

A well-maintained foreshore reserve runs along the northern shore of the town, with beaches lying below the 10 m high bluffs. There are two beaches adjacent to the centre of Cowes: Mussel Rocks; and the main Cowes Beach, next to the ferry jetty. Both beaches are relatively short, face due north and receive low wind waves. **Mussel Rocks Beach** lies between Mussel Rocks and the ferry jetty; while the main **Cowes Beach** is immediately east of the jetty, with Erehwon Point forming the eastern boundary. At high tide the beaches are relatively narrow and steep; while at low tide, they become exposed with shallow rock and tidal flats extending out toward the jetty.

Swimming: Two relatively safe beaches close to shore. The only problems to watch are the vessels using the jetty and tidal currents, which run off and parallel to the beach.

Surfing: None.

Fishing: The jetty and the rocks are the most popular locations.

Summary: The main beach for Cowes is very popular for picnics in the foreshore reserve, and swimming.

226   OBSERVATION

Unpatrolled
No bar: R + tidal flats & tidal currents
Beach Hazard Rating: **3**
Length: 5 200 m

Observation Beach is a 5 km long beach that runs due east from Erehwon Point at Cowes, out along a low, sandy, vegetated spit that terminates at the narrow, sandy Observation Point. Over the past few thousand years, waves and tidal currents sweeping into the bay have transported sand along the spit to build the point. In the process, the spit has enclosed Rhyll Inlet. A foreshore reserve and caravan park back the western end, while the eastern half is occupied by Observation Hill Reserve.

The beach varies in width and character depending on the amount of sand moving alongshore. Numerous low, wooden groynes have been built near Cowes to try and prevent the inevitable sand movement. Waves are usually low, though strong westerlies will send small wind waves running along the beach. At high tide the water is relatively deep against the beach, while tidal flats are exposed at low tide. Strong tidal currents run along the outside of the tidal flats.

Swimming: Best at mid to high tide when the wide tidal flats are covered. Do not swim far offshore, as tidal currents run along the beach.

Surfing: None.

Fishing: Best at high tide or off the beach in a boat.

Summary: This is Phillip Island's longest beach. The Cowes end is very popular in summer, while the eastern point is a more quiescent nature reserve.

Mornington Peninsula

Coast length: 77 km (Sandy Point to Point Nepean)
Beaches: 227 to 280 (54 beaches)

For map of Mornington Peninsula see Figure 73

The southern shores of Mornington Peninsula are Melbourne's ocean playground. The 77 km of coastline between Sandy Point, in Western Port Bay, and Point Nepean, at the eastern head of Port Phillip entrance, contains 54 beaches. They begin in the east as low wave beaches, exposed to the 2 - 3 m tides of Western Port (Figure 74). However, once around West Head (at Flinders) and particularly Cape Schanck, the beaches are exposed to strong westerly winds and waves, resulting in some of Victoria's most energetic and dangerous beaches.

The peninsula was traditionally farmland, with summer holiday makers heading for a few small coastal settlements along the Western Port shores, and behind the ocean beaches between Rye and Sorrento. However in recent years, these coastal settlements have been rapidly expanding. Today a narrow, but near continuous, strip of houses dots the coastal bluffs from Somers to Merricks Beach and Point Leo to Shoreham, as well as the coastal dunes of the Nepean Peninsula. Elsewhere the coast is too rugged or wind-swept for habitation and belongs to the Mornington Peninsula National Park.

Victorian Beaches - Region 5: West (Mornington Peninsula)

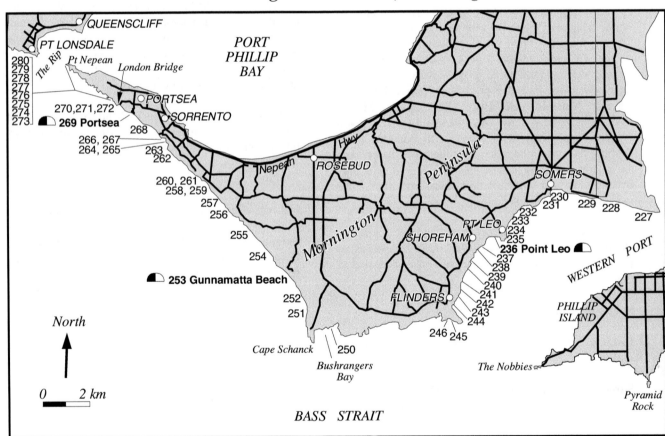

Figure 73.   Region 5: (Mornington Peninsula) Somers to Point Nepean (Beaches 227 to 280)

Figure 74.   Ocean Beach at Flinders typifies the rock dominated, lower energy beaches along the Phillip Island and Mornington Peninsula coast. While the sand beach runs continuously along the southern side of West Head, extensive intertidal rock flats dominate the surf zone.

227, 228   NAVY/SANDY POINT, SOUTH

Unpatrolled			
	Rating	No bar	Length
227 Navy/ Sandy Point	3	R + tidal flats	4 500 m
228 South	3	R + tidal flats	1 000 m
		Total length	5 500 m

Sandy Point is the northern equivalent of Phillip Island's Observation Point. It is a 4.5 km long sand spit that has been deposited by waves and tides across the entrance to the North Arm of Western Port. The point and beach are on Royal Australian Navy property, and it is therefore also called Navy Beach. There is public access to the roads and small car parks above South Beach, but only foot access along the beach to Navy Beach. **South Beach** is located on the eastern Somers foreshore, at the foot of 20 m high bluffs, containing a bushy foreshore reserve. The bluffs terminate at the end of the beach, where the long Sandy Point begins.

Both beaches face south and are exposed primarily to wind waves generated within Western Port, and occasional ocean swell. Both have relatively narrow, steep beaches with deep water off the beach at high tide, but tidal flats exposed at low tide. South Beach has intertidal rock flats off the beach, while Sandy Point is all sand.

Swimming: Best at mid to high tide, and relatively safe when the waves and wind are low. Strong westerlies will produce a drag along the beach.

Surfing: Only small wave chop.

Fishing: Better at high tide when there is deeper water off the beaches and the rocks at the western bluffs.

Summary: Two long sandy beaches, with only the western South Beach used primarily by the Somers locals.

229   SOMERS

Unpatrolled
No bar: R + tidal flats
Beach Hazard Rating: **3**
Length: 3 000 m

Somers is a holiday settlement on the bluffs overlooking the Western Passage to Western Port. **Somers Beach** is 3 km long, running between the eastern bluffs and rocks and the mouth of Merricks Creek. A foreshore reserve backs the entire beach and incorporates the 20 m high vegetated bluffs, with a low foredune between the bluffs and the beach. In places, the shoreline changes have cut into the dune and bluffs and a number of ineffective remedies including a seawall, groynes and car tyres have been used to counteract the changes.

The beach faces south into the Western Passage of Western Port and receives mainly low, choppy wind waves and only rare ocean swell. The beach is often narrow and steep with deep water at high tide, and sand and reef flats exposed at low tide. The Somers Yacht Club is located on the beach and small yachts and sail boats are commonly seen off the beach.

Swimming: Relatively safe under the usual calm conditions, with a drag down the beach in westerly winds when a chop is running.

Surfing: Only wind chop at best.

Fishing: Best at high tide off the beach or the eastern rocks.

Summary: A long, relatively calm beach, with most activity centred on the yacht club.

230, 231   BALNARRING, BALNARRING WEST

Unpatrolled			
	Rating	No bar	Length
230 Balnarring	**3**	R + tidal flats	2 500 m
231 Balnarring West	**3**	R + sand flats	1 300 m
		Total length	3 800 m

Balnarring is a small beach settlement located at the end of Balnarring Road. The road ends at the main **Balnarring Beach** (also called Tulum Beach), which has a foreshore reserve with picnic facilities. There is a small jetty crossing the beach, with boats moored offshore. Another reserve and picnic area are located behind the western beach.

The main beach is 2.5 km long and faces south, curving around at its western end to the south-east. It extends from the mouth of Merricks Creek, which runs along behind much of the beach, to sandy Balnarring Point, which is fronted by reef flats that separate the two beaches. The west beach is 1.3 km long and curves around to face the east by Palmer Bluff, its southern boundary. Both beaches receive low wind waves and are relatively narrow, with a moderately steep beach face and deep water off the beach at high tide. At low tide however, the sand and reef flats are exposed.

Swimming: Two relatively safe beaches protected from westerly winds and waves. Best at mid to high tide when the tidal flats are covered.

Surfing: Usually none, except in big outside swell, which can produce a small, well-shaped right hander around *Balnarring Point.*

Fishing: Best off the jetty or beach at high tide.

Summary: A relatively quiet holiday settlement with two attractive sandy beaches and foreshore facilities.

232, 233   MERRICKS, EAST CREEK

Unpatrolled			
	Rating	No bar	Length
232 Merricks	**3**	R + sand & reef flats	2 200 m
233 East Creek	**3**	R + sand & reef flats	1 500 m
		Total length	3 700 m

Merricks Beach is a small holiday settlement at the end of Merricks Beach Road. The road ends at a foreshore reserve where there is a yacht club and picnic facilities. The beach meanders for over 2 km south-west from Palmer Bluffs to a more prominent bluff. Beyond here, **East Creek Beach** continues to the bluff just past the mouth of East Creek, at Point Leo. Both beaches are backed by 10 to 20 m high, vegetated bluffs, with a low dune backing the northern section of Merricks Beach. Access is at Merricks in the north and on the south side of East Creek. A caravan park sits on the bluffs on the southern side of East Creek.

Both beaches receive low wind waves and very rare ocean swell. The beaches tend to be relatively narrow and steep, with deep water at high tide and exposed sand and reef flats at low tide.

Swimming: Most suitable at mid to high tide, as the tidal flats are exposed at low tide.

Surfing: Usually none, except when the swell is big. This produces a right hander along the southern point, known as *Merricks Point*.

Fishing: Better at high tide when you can fish the reef flats from the beach.

Summary: Merricks Beach is mainly used by locals and sailors, while East Creek Beach can only be reached on foot and therefore is little used north of the creek mouth.

234   POINT LEO

Unpatrolled
Single bar: R/LTT
Beach Hazard Rating: **3**
Length: 500 m

The small settlement of Point Leo is set back from the 20 m high point, which is covered by a wide foreshore reserve. Three beaches surround the point: Point Leo Beach, the Point Beach and Point Leo Surf Beach. **Point Leo Beach** is located on the east side of the point and is 500 m long. It faces east and is located between northern bluffs and Point Leo to the south, by which it is protected. A boat club and long boat ramp are located in the southern corner.

The beach receives low waves, and has a narrow high tide beach fronted by 200 to 300 m wide sand and rock flats, that are exposed at low tide.

Swimming: A relatively safe beach away from the boating activities. Better at mid to high tide.

Surfing: Usually none at the beach.

Fishing: Best off the point at high tide.

Summary: This is the quieter, safer and more sheltered of the Point Leo beaches, but is only suitable for swimming when the tide is in.

235   POINT LEO 'POINT'

Unpatrolled
Single bar: LTT + reefs
Beach Hazard Rating: **4**
Length: 200 m

Point Leo is a 20 m high, basalt bluff with a continuous sand beach at its base, and reef flats exposed at low tide. **The Point Beach** is 200 m long and faces south, but is partially protected by the extensive reef flats and

offshore reefs known to surfers as *The Peak*. The foreshore reserve covers the point and includes camping and picnic facilities.

Swimming: Not recommended, with the abundance of rock and reef on either side of the beach. Rips run out against the reefs when the surf is breaking. Swimming is safer at the adjoining bay or patrolled surf beaches.

Surfing: There are two breaks here: *The Point*, a right hander on the eastern side of the point; and *The Peak* offshore.

Fishing: A popular spot off the beach at high tide, or from the rocks and reef flats at low tide.

Summary: The focal point of Point Leo is the point itself. The beach is better suited for sunbathing, while the surf, when running, is for more experienced surfers.

236   POINT LEO SURF

Point Leo SLSC
Patrols: late November to Easter holidays
Surf Lifesaving Club: Saturday, Sunday and Christmas public holidays
Lifeguard: no lifeguard on duty, or weekday patrols
Single bar: R/LTT
Beach Hazard Rating: **4**
Length: 500 m
For map of beach see Figure 75

Point Leo is composed of low, basalt bluffs of the Older Volcanics. **Point Leo Surf Beach** lies at the base of the vegetated bluffs, with extensive intertidal rock platforms at each end and some rock reefs offshore. It is 500 m long and faces almost due south, however its location several kilometres inside the wide entrance to Western Port Bay affords considerable protection from high ocean waves. Waves average 0.5 m and combine with the sand to build a wide beach, fronted by a narrow, attached bar. At high tide, the waves usually surge up the beach without breaking, while a continuous bar with a shorebreak is present at low tide. Rips are rare, only occurring during and following high waves.

The beach is backed by Point Leo Foreshore Reserve, which contains a public park with most facilities required for a day at the beach, including a camping area and the surf lifesaving club, formed in 1955.

Swimming: A relatively safe patrolled beach. Water is deep close to the beach at high tide, with sand and rock flats exposed at low tide. The lifesavers rescue 13 people on average each year.

Surfing: Usually a low beach or shorebreak. However, during higher winter wave conditions, the reefs and point provide some excellent breaks. *The Point* (also known as *Suicide Point*) has a good right, while further out is a reef called *The Peak*. This can be surfed in moderate swell. Down the beach are two reef breaks known as *First* and *Second Reefs*, as well as the western *Honeysuckle Point*, which all provide right hand breaks.

Fishing: Best off the rocks at high tide, with access to both sand and rock reefs.

Summary: An attractive, well-maintained beach and reserve, offering safe family bathing in summer, with the chance of some good point and reef breaks during higher winter swell.

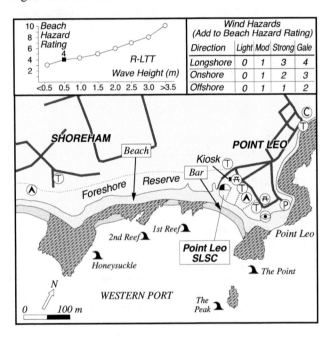

Figure 75. Point Leo Surf Beach is located on the southern side of the point and below the foreshore reserve. The beach usually receives low waves and has a shallow, attached bar. During bigger swell, waves break over the many reefs.

237, 238   SHOREHAM CAMPING, SHOREHAM

Unpatrolled			
	Rating	Single bar	Length
237 Shoreham Camping	3	R/LTT + reef flats	1 500 m
238 Shoreham	3	R/LTT + reef flats	700 m
		Total length	2 200 m

Shoreham is a small holiday settlement lying between Flinders Road and the coast. A foreshore reserve parallels both beaches with the Shoreham Foreshore Camping Reserve located behind the main beach (237). A separate Beach Road runs out to the second Shoreham Beach (238), where a long car park runs between the bluffs and the back of the low beach. There is also a picnic area located here. Stony Creek drains across the northern end of the beach.

Both beaches are very protected and usually have no waves. **Shoreham Camping Beach** is 1.5 km long and curves around to face south-east, while the 700 m long **Shoreham Beach** faces almost due east. Extensive sand and reef flats front both beaches and, as a result, the beaches have deep water only at mid to high tide. The tidal flats, and often seaweed, are exposed at low tide.

Swimming: Better at mid to high tide when the water is in.

Surfing: Usually no waves, however big swells can produce a right hand break over the reef that separates the two beaches. Known as the *Pines*, this break is very popular when working.

Fishing: Better at high tide, otherwise off the outer reef flats at low tide.

Summary: Two very accessible, attractive, low energy beaches.

239 - 242   PINES, BEACH 240, MUSK CREEK, MORNINGTON PARK

Unpatrolled			
	Rating	Single bar	Length
239 Pines	3	R/LTT + reef flats	1 200 m
240 Beach 240	4	R + reef flats	400 m
241 Musk Creek	4	R + reef flats	800 m
242 Mornington Park	4	R + reef flats	500 m
		Total length	2 900 m

Between Shoreham and Flinders, the coast trends south-south-west and consists of steep, 20 to 40 m high, grassy bluffs. The bluffs are fronted by 100 m wide intertidal reef flats, with a series of narrow, sandy, high tide beaches. The first four beaches are all backed by private farmland, and public access is by foot along the shoreline.

Pines Beach is 1.2 km long, backed by bluffs and fronted by the reef flat that narrows toward the centre. **Beach 240** is similar, but only 400 m long; while 800 m long **Musk** or **Manton Creek Beach** has lower bluffs, with the creek crossing the centre of the beach. **Mornington Park Beach** has high bluffs, with an 80 m wide, low dune between the bluffs and the beach, and 300 m wide reef flats.

Swimming: All these beaches have wide, exposed, continuous reef flats at low tide, so bathing is best at mid to high tide.

Surfing: None unless there is a very large swell; then a series of reef breaks begins working. These include one at *Pines Beach*, then *The Farm* at the mouth of Musk Creek, and finally *Lefts and Rights* off Mornington Park Beach.

Fishing: Better at high tide when the reef flats are covered.

Summary: Four relatively remote beaches backed by private property. They are only accessible from Shoreham Beach in the north, and by foot along the shore.

243, 244   DODDS CREEK, KENNON COVE

Unpatrolled			
	Rating	Single bar	Length
243 Dodds Creek	4	R + reef flats	400 m
244 Kennon Cove	3	R/LTT	1 400 m
		Total length	1 800 m

The town of Flinders lies behind the elongated Flinders Point. Beaches lie either side of the point, with two sheltered beaches on the east side called Dodds Creek and Kennon Cove. These sheltered beaches face east and are backed by 20 to 40 m high, vegetated bluffs. **Dodds Creek Beach** is 400 m long; has the creek draining across the southern end; houses perched on the bluffs; and 100 to 200 m wide reef flats in front.

Kennon Cove is 1.2 km long, with a 300 m long jetty in the centre and bluffs to either side. A car park, slipway and picnic area are located south of the jetty. Waves are usually very low and sand flats extend 200 m off the beach, beyond which there are boat moorings. Flinders Yacht Club and a Marine Rescue facility are located at the southern end of the beach.

Swimming: Two usually calm beaches, with bathing off the beaches only possible at mid to high tide.

Surfing: Usually none.

Fishing: The jetty is the most popular fishing spot in Flinders.

Summary: These are Flinders' two most sheltered beaches, with Kennon Cove providing the best access

and picnic facilities.

MORNINGTON PENINSULA NATIONAL PARK

Area:	2 686 ha
Coast length:	50 km
Beaches:	246 to 280 (34 beaches)

Mornington Peninsula National Park is a 50 km long coastal strip consisting of three parts: Flinders Point to Simmons Bay; Cape Schanck to Portsea; and the tip of Point Nepean. It is an exposed, high energy coast with both high, rugged cliffs and points, and more than 30 beaches. These range from little pockets of sand in amongst the rocks, to longer rip dominated sand beaches, usually containing a scattering of reefs.

Information:	Portsea:	(03) 5984 4276
	Rosebud:	(03) 5896 8987
	Dromana:	(03) 5987 3078

245, 246   PILOT STATION, OCEAN

Unpatrolled			
	Rating	Single bar	Length
245 Pilot Station	4	R+ rock flats	600 m
246 Ocean	5	LTT + rock & reef	800 m
		Total length	1 400 m

Point Flinders extends for 1.5 km south-east of the town of Flinders. The 30 m high point houses the golf course, a Navy Gunnery Range and, along the southern shore, the start of Mornington Peninsula National Park. The point is composed of ancient weathered basalt, which also lies along the shore and forms the offshore reef. West of Flinders, the rock platforms are composed of distinctive columnar basalt, resembling paving stones. The Golf Links Road runs around the point, but with no public access to the Navy land.

The two beaches lie on the south side of the point and are backed by high bluffs. **Pilot Station Beach** lies below the old Pilot Station. It is 600 m long; backed by 20 m high, grassy bluffs; composed of sand and basalt cobbles; and fronted by extensive reefs, that lower waves at the shore. **Ocean Beach** is 800 m long and backed by grassy bluffs, reaching 50 m in height. The beach is also narrow at high tide and includes sand and cobbles, with irregular reefs offshore.

Swimming: Best at high tide, otherwise in the inner tidal pools at low tide.

Surfing: None at the beach, however there are a few breaks on the outer reefs, that work during larger outside swells. These include *Meanos*, a heavy left and right reef break off the end of the point; *The Gunnery*, a right hand reef break out from the Pilot Station lookout; and *Cyrils*, another reef right hander off Ocean Beach.

Fishing: There are plenty of reefs and rocks to fish from the beach at high tide, however Mushroom Reef has been proposed as a Sanctuary under the Marine National Park legislation.

Summary: These are Flinders' ocean beaches; however extensive rocks and reefs dominate both, making them less suitable for safe bathing, but providing the surfers with some heavy reef breaks. A ramp is provided on the bluffs behind Ocean Beach for the use of hang gliders.

247, 248   TEA TREE CREEK, CAIRNS BAY

Unpatrolled			
	Rating	No bar	Length
247 Tea Tree Creek	8	R + rock & reef	200 m
248 Cairns Bay	8	R + rock & reef	100 m
		Total length	300 m

Tea Tree Creek and **Cairns Bay** are located 2 km west of the Blowhole lookout and 1 km south of the main Flinders Road. They are two cobble and boulder beaches, set in amongst the rugged, 60 to 80 m high, black basalt cliffs that run from Cape Schanck to Point Flinders. Access is via a 1 km walk from the road to the top of the cliffs, with a steep climb down to the beaches. Both are exposed to high waves, which break over the reefs and have built the narrow beaches in two small gaps in the cliffs, located either side of the end of the track.

Swimming: Neither beach is recommended, owing to the high waves, rocks, reefs and relatively isolated location.

Surfing: Not recommended.

Fishing: It is safer to fish from the beach, as the rocks are awash at high tide.

Summary: Worth the walk out to view the coast and the Blowhole, but unsuitable for bathing.

249   SIMMONS BAY

Unpatrolled
Single bar: LTT
Beach Hazard Rating: 5
Length: 100 m

Simmons Bay contains a 100 m long, south-east facing sand beach, located below 70 m high bluffs. The beach is surrounded by private land and is not in the National Park. Public access is only available by boat. It is sheltered by The Arch headland from high waves and usually has a narrow, attached bar, with rips against the rocks during higher waves.

Swimming: Moderately safe when waves are less than 1 m, but remote and little used.

Surfing: Usually a low shorebreak.

Fishing: The small bay can be fished from the beach or adjoining rock platforms.

Summary: An isolated but scenic pocket beach.

250   BUSHRANGER BAY

Unpatrolled
Single bar: LTT
Beach Hazard Rating: 5
Length: 300 m

Bushranger Bay Beach is a 300 m long, west facing beach lying inside the bay, and afforded some protection by headlands and reefs from the high westerly waves. A low foredune backed by sloping bluffs surrounds the back of the beach, and a creek drains across the western end. Waves average 1 m and have built a wide beach with a continuous bar, with rips against the rocks at either end.

The beach is privately owned but can be reached on foot along the 2 km long access track from Flinders Road.

Swimming: Moderately safe when waves are low, however stay on the bar and clear of the rocks, and rips against the rocks.

Surfing: During larger swell there are smaller beach breaks at the bay.

Fishing: You can fish the rip gutters from the beach or the adjacent rock platforms.

Summary: An attractive, isolated beach; well worth the walk for the views.

CAPE SCHANCK

Cape Schanck is a basalt headland blanketed in Pleistocene dune calcarenite, that is, successive layers of coastal sand dunes deposited during past high stands of sea level. The oldest dunes are at least several hundred thousand years old. The lighthouse, which stands 100 m above sea level, was built in 1859 to mark the eastern approach to Port Phillip.

There are excellent coastal views from the car park, lighthouse, and the coastal walk leading east to Bushranger Bay and west to Fingal Beach.

The Albert was wrecked on the cape in 1893, with the loss of 75 lives.

Surfing: During very big swell, the waves wrap around the cape and Pulpit Rock, to break over a reef located on the east side of the lighthouse. Stairs lead down to the rocks.

251, 252   FINGAL 1 & 2

Unpatrolled			
	Rating	No bar	Length
251 Fingal 1	7	R + rocks & reef	500 m
252 Fingal 2	7	R + rocks & reef	250 m
		Total length	750 m

The **Fingal Beaches** lie on the western side of Cape Schanck. They can be reached along the walking tracks from Cape Schanck and Fingal Picnic Area car parks. The two beaches are remnants of a once large beach, that supplied sand to the backing cliff-top dunes. Today they have been eroded down to the present strip of sand, fronted by rock platforms and reefs. They are backed by the 40 m high calcarenite cliffs with the active dunes on top.

Both beaches face south-west into the full force of the westerly winds and waves. The waves break on the offshore reefs and rocks; and the narrow beaches are only active at high tide.

Swimming: These are two dangerous beaches for bathing, with rocks and reefs dominating the high energy surf zone. Stay on the sand beach at high tide, and use the inner tidal pools at low tide.

Surfing: The only rideable break is a left hander over the reefs below the western side of the cape. A steep descent is required from Fingal Picnic Area car park to reach the rocks.

Fishing: There are plenty of opportunities for rock fishing, either from the beach at high tide, or off the rocks at low tide and in lower waves.

Summary: These are two narrow beaches dominated by cliffs, rocks and reefs. They are worth viewing from the coastal track, but are unsuitable for bathing.

253   GUNNAMATTA

Gunnamatta Beach SLSC

Patrols: late November to Easter holidays
Surf Lifesaving Club: Saturday, Sunday and Christmas public holidays
Lifeguard: 10 am to 6 pm weekdays, during Christmas holiday period

Single bar: TBR/RBB
Beach Safety Rating: **8**
Length: 3 000 m

For map of Gunnamatta Beach see Figure 76

Gunnamatta Beach is an exposed, high energy beach with a wide, rip dominated surf zone. It is located in the Mornington Peninsula National Park and is part of the 30 km long sandy and rocky coast that extends from Cape Schanck to Point Nepean. The Gunnamatta section is 3 km long, with extensive intertidal calcarenite reefs and rocks forming the boundaries, with some smaller reefs on the beach and in the surf. Truemans Road runs out through the dunes to the beach, where there is a large car park and the surf lifesaving club.

The beach faces south-west, exposing it to high westerly winds and waves. The waves average 1.9 m and combine with the medium sand to produce a 150 m wide single bar surf zone. The bar is cut by strong rips every 300 m, together with additional permanent rips next to major reefs and rocks. The rips intensify around low tide.

The Gunnamatta Surf Life Saving Club was founded in 1966. This is a very hazardous beach, with an average of 113 rescues a year, second only to its neighbouring Portsea Beach.

Swimming: This is a potentially hazardous beach for swimming, with usually high waves and strong rips close to shore. Definitely stay between the flags, on the bar and away from the rips, rocks and reefs.

Surfing: Gunnamatta offers the best beach breaks on the Mornington Peninsula, with consistency guaranteed by the high swell and reefs. Good breaks are found *Down the Beach* past the surf club, in front of the *First* and *Second Car Parks*, and up the beach at the *Pumping*

Station, which is, however, polluted by the sewerage outfall. Best conditions are with a low to moderate swell and north-easterly winds.

Fishing: Deep rip holes and gutters, together with rocks and reefs, are a permanent feature of this beach and make it a popular spot for beach and rock fishing.

Summary: A high energy, hazardous beach backed by extensive sand dunes. Best suited for experienced bathers and surfers.

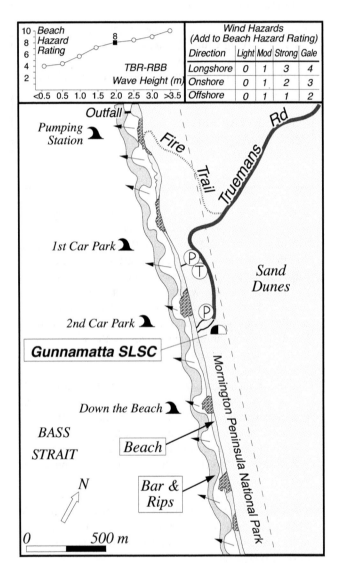

Figure 76. Gunnamatta Beach is an exposed, south-west facing beach nestled in the wide dunes of Mornington Peninsula National Park. The high surf and numerous reefs produce persistent and permanent rips, which result in good surf but hazardous bathing. Stay between the flags if swimming at this beach.

254   ST ANDREWS

Unpatrolled
Single bar: TBR/RBB + reefs
Beach Hazard Rating: **8**
Length: 2 000 m

St Andrews Beach is named after the backing holiday settlement and can be reached via a winding route through the houses and dunes. This terminates at a small car park in the dunes above the centre of the 2 km long beach. The Mornington Peninsula National Park runs in a strip between the houses and the beach.

Like all the beaches along this section, it is dominated by moderate to high waves, sand, reefs and rocks. St Andrews has rock along several hundred metres of the beach, with reefs also outcropping in the surf. Combined with the waves, which average 1.8 m, they produce several strong, permanent rips against the rocks and reefs.

Swimming: This is a hazardous, rip dominated beach. Stay close inshore, on the bar and well clear of the rocks, reefs and rips.

Surfing: There are always beach and reef breaks along St Andrews, with the best conditions occurring in a moderate swell with north-easterly winds.

Fishing: An excellent beach for beach fishing. Permanent rip holes exist against the reefs and gutters along the rocks.

Summary: An exposed, high energy beach and surf; most suitable for experienced surfers and beach fishers.

255   RYE OCEAN

Unpatrolled
Single bar: TBR/RBB + reefs
Beach Hazard Rating: **8**
Length: 4 000 m

Rye Ocean Beach lies 4 km due south of the more quiescent Rye Bay Beach. The beach is accessed via Sandy Beach Road, with a signposted road leading to a large car park. The beach is located inside the Mornington Peninsula National Park, and natural sand dunes back the length of this 4 km long beach. Foot tracks also lead to the beach from the St Andrews settlement at the eastern end, and at Orr Point at the western end.

The beach faces south-west and receives waves averaging 1.8 m. These produce large persistent rips every 300 m, with permanent rips against the rocks and reefs below the car park.

Swimming: A hazardous beach dominated by a wide surf zone with rips, rocks and reefs. Use extreme care if bathing here; stay close inshore on the bar and well clear of the above hazards.

Surfing: There are a number of reef breaks along the beach including at the rocks in front of the car park, which produces lefts and rights; at *Sunset Rocks* towards St Andrews; and *Snatches* toward Orr Point; together with beach breaks elsewhere.

Fishing: There are excellent permanent rip holes against the car park rocks, together with persistent rip holes and gutters along the length of the beach.

Summary: An exposed beach backed by sand dunes and fronted by a wide, energetic, rip and reef dominated surf. Only for experienced surfers.

256, 257   ORR POINT, NUMBER SIXTEEN

Unpatrolled			
	Rating	Single bar	Length
256 Orr Point	6	R	300 m
257 Number Sixteen	6	R	700 m
		Total length	1 000 m

Between Orr Point and Point Nepean, the calcarenite rocks and reefs increasingly dominate the beaches and surf. At Orr Point, intertidal rocks and reef flats front most of the beach, with only a small area of direct access to the surf. The beach is only active at high tide and is backed by both sand dunes and calcarenite bluffs. **Orr Point Beach** is 300 m long, while adjoining **Number Sixteen** is similar, but is 700 m long. It has a 200 m section that is free of reef, which provides access to the surf.

Waves break over the continuous outer reefs, causing the wave height at the shore to be lowered. This has resulted in two steep, high tide beaches, fronted by a surf entirely dominated by the reefs.

Access to the beaches is via a foot track over the dunes to Orr Point, and a car park behind Number Sixteen. The car park also marks the beginning of the 10 km long coastal walk through the National Park to Sorrento Ocean Beach and Sphinx Rock.

Swimming: Two beaches with lower waves at the shore, but with a surf zone dominated by rocks and reefs. If bathing, stay near the beach in the quieter inshore areas and do not venture out over the reefs, particularly at high tide.

Surfing: Only dangerous breaks over the near continuous rocks and reefs.

Fishing: There are plenty of rocks and reefs to fish here, however you need to find the deeper holes.

Summary: Two rock dominated beaches; best suited for viewing and fishing.

258 - 261   DIMMICKS 1 & 2, PEARSE'S 1 & 2

Unpatrolled			
	Rating	No bar	Length
258 Dimmicks 1	6	R + rock flats & reef	50 m
259 Dimmicks 2	6	R + rock flats & reef	100 m
260 Pearse's 1	6	R + rocks & reef	100 m
261 Pearse's 2	6	R + rocks & reef	100 m
		Total length	350 m

Dimmicks and **Pearse's Beaches** are four small pocket beaches, each consisting of a steep, sandy, high tide beach. They are backed by 10 to 20 m high, dune covered calcarenite bluffs, and are fronted by near continuous rock and reef flats. The beaches are awash at high tide, in lee of reef dominated surf, and fronted by exposed rock flats at low tide.

All four beaches can be reached along the coastal walk, and from a track across the dunes from the end of Pearse's Road.

Swimming: You can only bathe on these beaches at mid to high tide, however be very careful of the rocks and reef on and off the beach.

Surfing: There are no rideable waves at these beaches, however the offshore reefs can provide some waves at high tide. The breaks go by the names of the streets used to access the coast, with the break off Dimmicks called *Tibir Street*.

Fishing: This section is dominated by rock platforms and reefs, however you need to find the deeper gutters in the rocks.

Summary: These beaches are best suited for viewing from the coastal track, rather than bathing.

262, 263, 264   BRIDGEWATER BAY, FOWLERS, MONTFORTS

Unpatrolled			
	Rating	No bar	Length
262 Bridgewater Bay	6	R + rocks & reef	200 m
263 Fowlers	6	R + rocks & reef	250 m
264 Montforts	6	R + rocks & reef	500 m
		Total length	950 m

Bridgewater Bay, **Fowlers Beach** and **Montforts Beach** lie in the Mornington Peninsula National Park, either side of Koreen Point. St Johns Wood Road runs 2 km straight from Blairgowrie to a car park behind the point. All three beaches are backed by 10 to 20 m high calcarenite bluffs, capped by sand dunes, with some of the dunes reaching down to the beach at Fowlers and Montforts. The beaches are narrow and only active at high tide, being fronted by over 100 m of rocks and reef flats that dominate the surf zone. Waves average 1.5 m on the reefs, but are lower at the beaches.

Swimming: These are three dangerous beaches, with reef dominated surf at high tide and exposed rocks and reefs at low tide. Montforts has some sheltered tidal pools at low tide, which offer the most suitable spot for bathing.

Surfing: A very rocky surf zone, with the reef break at the end of *Kirkwood Street*, on Fowlers Beach, and *Central Avenue* on Montforts Beach being the best known spots.

Fishing: The best rock gutters are at the western end of Fowlers Beach, otherwise you can fish the reef flats at high tide from the beaches or rocks.

Summary: Three narrow, rock bound beaches; mainly used by rock fishers.

265   KOONYA

Unpatrolled
No bar: R + rocks & reefs
Beach Hazard Rating: **6**
Length: 750 m

Koonya Beach, while dominated by rocks and reefs, is one of the more open, sandy beaches in this section of Mornington Peninsula National Park. It is backed by a near continuous foredune with some calcarenite bluffs, but with extensive, generally deeper reefs off the beach. The 750 m long beach has good access via two car parks at the end of Hughes Road, with a loop walking track

from the road to the beach. The beach faces south-west and receives waves over the outer reefs averaging 1.5 m. The waves reform in the deeper channel and surge heavily up the beach face.

Swimming: This beach is moderately safe close inshore when waves are low. However, waves over 1 m will generate a heavy shorebreak and strong currents along the beaches. Also watch the rocks and reefs.

Surfing: Only a shorebreak at the beach, but there is a chance of waves over the outer reefs at high tide.

Fishing: There are wide rock platforms, particularly off Pelly Point, at the eastern end. However they are awash at high tide, so use caution if fishing from these. It is safer to fish straight off the beach into relatively deep water.

Summary: This beach has good parking and access and is popular for sun bathing and swimming when calm, however use care if there is any surf running and stay well clear of the reefs and rocks.

266, 267   DIAMOND BAY, ST PAULS

Unpatrolled			
	Rating	No bar	Length
266 Diamond Bay	5	LTT + reefs	80 m
267 St Pauls	6	R + rock flats & reefs	50 m
		Total length	130 m

Diamond Bay and **St Pauls** are two small pockets of sand lying at the base of 30 m high calcarenite bluffs. They are fronted by a reef and rock dominated surf zone. Diamond Bay Road and St Pauls Road both lead to car parks behind each beach. However access to the beaches requires a steep climb down steps.

Swimming: Diamond Bay Beach is set deep inside the bay with two points protecting it. It usually has low waves at the beach and is suitable for relatively safe bathing at high tide. St Pauls is not only difficult to reach, but has higher waves at high tide and more rocks and reefs along the beach.

Surfing: No rideable waves at either beach.

Fishing: Diamond Bay is a popular location as it offers a slightly more protected section of coast, with extensive rock platforms in the bay and on adjacent Jubilee Point.

Summary: Two small beaches, with only Diamond Bay offering some relatively safe bathing.

268   SORRENTO OCEAN (BACK)

Sorrento SLSC
Single bar: LTT
Beach Hazard Rating: **6**
Length: 400 m

Sorrento Ocean or **Back Beach** lies at the end of Ocean Beach Road, 1.5 km south of the bayside town of Sorrento. There is a large car park and a picnic area, as well as a new surf lifesaving club, on the bluffs behind the beach. The beach lies in a natural amphitheatre formed by the protruding western headland and the patchy offshore reefs. The beach faces south-west and waves average 1.6 m over the reefs, but are lower at the beach. Here they have built one of the wider beaches in this section, with a continuous attached bar.

Swimming: Moderately safe on the inner section of the bar at high tide and clear of the rocks; however, watch the rips and rocks at low tide. Stay between the flags.

Surfing: Usually a low shorebreak at the beach.

Fishing: At low tide, deep water can be fished from the rock platforms.

Summary: A popular beach for sun bathers and those wanting more action than at the Sorrento Bay Beach; and relatively safe under average waves.

269   PORTSEA (BACK)

Portsea SLSC
Patrols: late November to Easter holidays
Surf Lifesaving Club: Saturday, Sunday and Christmas public holidays
Lifeguard: 10 am to 6 pm weekdays, during Christmas holiday period
Inner bar: TBR + reefs   Outer bar: RBB
Beach Hazard Rating: **8**
Length: 2 500 m
For map of Portsea Beach see Figure 77 and Figure 10 for photograph

Portsea Beach is one of Victoria's most popular and infamous beaches, being the site of regular ironman contests, but also near the spot where Prime Minister Harold Holt disappeared in the surf in 1967. Portsea is a popular summer beach with extensive parking areas in the dunes, all of which provide a good view of the

beach, its wide surf zone and many rips and reefs. So take a good look before you go down and enter the surf. The beach is 2.5 km long, with extensive intertidal calcarenite rock platforms and reefs forming the boundaries, together with smaller reefs and rocks outcropping along the beach. It faces the south-west and receives the full force of the south-west waves and westerly winds. The waves average 1.8 m and combine with the finer sand and reefs to produce a 200 m wide surf zone, containing rip dominated inner and outer bars. Rips occur every 300 m along the inner bar, with strong permanent rips against major reefs. One permanent rip to the right of the surf club is known as *Huey's Reef.*

The Portsea Back Beach Road provides good access to the car parks and lookout above the beach, with a steep walk down to the beach and surf lifesaving club. The club was founded in 1949 and averages 143 rescues a year, the highest in Victoria.

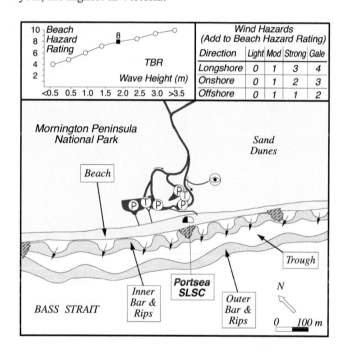

Figure 77.   Portsea Back Beach is a popular but hazardous beach. It faces south-west and receives high waves, which combine with the sand and reefs to produce a 200 m wide surf zone dominated by rips and currents, together with the reefs and rocks. If swimming, only bathe between the flags.

Swimming: A popular summer beach, however definitely stay on the bar and between the flags, and clear of the rips, rocks and reefs, particularly at low tide when the rips intensify.

Surfing: A popular surfing spot in summer, when Portsea is overrun with holiday makers. The beach breaks occur over the bars and reefs, with best conditions during low to moderate swell and northerly winds.

Fishing: A popular spot for beach and rock fishing, with deep rip holes a permanent feature, together with rocks and reef along the beach.

Summary: One of Victoria's better known and most popular summer beaches, however also one of its most potentially hazardous. Best suited for experienced bathers and surfers.

270, 271   **LONDON BRIDGE 1 & 2**

Unpatrolled			
	Rating	No bar	Length
270 London Bridge 1	6	R + rock flats & reefs	50 m
271 London Bridge 2	7	R + rock flats & reefs	100 m
		Total length	150 m

London Bridge is a well known, hollowed out sea stack that can be viewed from a lookout at the end of London Bridge Road. On either side of London Bridge are two high tide platform beaches, backed by 30 to 40 m high calcarenite bluffs and fronted by continuous intertidal rock flats, with reefs further offshore. Waves average 1.7 m on the outer reefs, with their height at the beach depending on the tide. The narrow sand beaches are awash at high tide, but fronted by exposed rock flats at low tide.

Swimming: Neither beach is suitable for safe bathing, owing to the high outer waves and dominance of rocks and reefs right up to the beach.

Surfing: No rideable waves along this section.

Fishing: The rock flats off the beach can be fished at high tide, or deeper water off the adjoining rock platforms at low tide. However, take care as waves wash over the rocks.

Summary: A popular tourist destination to see London Bridge, but only used by others for fishing.

POINT NEPEAN

Point Nepean became part of the larger Mornington Peninsula National Park in 1988. The entrance to the point is located 1 km west of Portsea at the Visitor Centre. It is open from 9 am to 5 pm daily. Private cars are not permitted past Gunners Cottage carpark, with the public being taken on a special Transporter to the point and other attractions, or you can walk or cycle the 6 km return distance. Access to the park by boat is prohibited.

Note: The beaches around Point Nepean are closed and no swimming is allowed.

Information: Portsea   (03) 5984 4276

HAROLD HOLT MARINE RESERVE

The Harold Holt Marine Reserve covers the shoreline and adjacent reefs from London Bridge to Point Nepean, and around into Nepean Bay as far as Observatory Point.

Fishing from the shoreline, collecting of marine species and spear fishing are all prohibited.

The reserve is named in honour of Prime Minister Harold Holt, who disappeared while swimming alone in heavy surf at Cheviot Beach, in December 1967.

Cheviot Beach was named after the wreck of the *Cheviot* in 1887, which resulted in the loss of 35 lives.

272 - 277   **BEACHES 272 to 275, RIFLE RANGE, CHEVIOT**

Unpatrolled			
	Rating	No bar	Length
272 Beach 272	6	R + rock flats & reefs	350 m
273 Beach 273	7	R + rock flats & reefs	100 m
274 Beach 274	7	R + rock flats & reefs	350 m
275 Beach 275	7	R + rock flats & reefs	350 m
276 Rifle Range	7	R + rock flats & reefs	250 m
277 Cheviot	7	R + rock flats & reefs	700 m
		Total length	2 100 m

Immediately west of London Bridge, the Mornington Peninsula National Park is separated from its western tip by 3.5 km of Commonwealth Land, which contains six beaches. All are platform beaches, consisting of narrow, high tide strips of sand. They are backed by 30 to 40 m high calcarenite bluffs and fronted by usually continuous intertidal rock flats, with patchy reefs offshore. They also face south-west and receive waves averaging 1.7 m on the outer reefs. The beaches are awash at high tide, but fronted by usually continuous rock flats at low tide, with reef dominated surf zones.

Access is restricted to all six beaches. The first four can only be reached on foot along the coast from London Bridge. A road runs to the top of the western end of **Rifle Range Beach**, and there is a viewing area on the 50 m high Cheviot Hill, behind **Cheviot Beach**. There is a steep descent to the beach from the road.

Swimming: BEACHES CLOSED - NO SWIMMING ALLOWED. Even if it were allowed, all six beaches are hazardous for bathing. At high tide the rocks and reefs lie immediately off the beaches and, as the tide drops, strong permanent rips intensify off the rocks and amongst the reefs.

Surfing: There are numerous breaks along here, but they are of poor quality, difficult to get to and rarely surfed.

Fishing: Line fishing is permitted along the numerous rocks and reefs here, but bait collecting is prohibited.

Summary: A cliff, rock and reef dominated section of coast, containing six strips of sand that are only awash at high tide.

SHIPWRECKS
at Point Nepean and Port Phillip Heads

Marmion	1853
Krisk	1853
Sea	1853
Nonpareil	1857
Amicus	1857
Sierra Nevada	1900

278, 279, 280   PEARCE HILL, NEPEAN HILL, POINT NEPEAN

Unpatrolled			
	Rating	No bar	Length
278 Pearce Hill	7	R + rock flats & reefs	500 m
279 Nepean Hill	7	R + rock flats & reefs	500 m
280 Point Nepean	6	LTT + rocks	100 m
		Total length	1 100 m

The western tip of Point Nepean, from Cheviot Hill out to the point and around to Observatory Point in the bay, is part of the Mornington Peninsula National Park. A road runs right out to the point with access to these three beaches from the road, via walking tracks down the backing bluffs. All three beaches are backed by 20 to 30 m high calcarenite bluffs and fronted by near continuous intertidal rock flats, with occasional deeper channels. **Pearce Hill** and **Nepean Hill Beaches** face south-west and have waves averaging 1.7 m breaking on the outer reefs. **Point Nepean Beach** wraps around the point and faces north-west, consequently receiving waves averaging only 0.5 m. However, the strong tidal currents of The Rip run just off the beach.

Swimming: BEACHES CLOSED - NO SWIMMING ALLOWED. All three beaches are potentially hazardous, owing to the usually high waves and predominance of rocks and reefs in the surf, together with strong tidal currents around the point.

Surfing: The best spot is off Point Nepean Beach, where the waves that wrap around just inside the point produce a left over the sand and reef flats, called *Quarantine*.

Fishing: Fishing is not permitted from the beaches or rock platforms.

Summary: These are the easternmost three beaches on Mornington Peninsula and are located in the Mornington Peninsula National Park. They provide a good view of this energetic and hazardous coast, as well as the strong tidal flows through The Rip.

6. PORT PHILLIP BAY

> **Port Phillip Bay**
> **(Point Nepean to Point Lonsdale)**
>
> Beaches: P1 to P132
> Coast length: 259 km
>
> *For maps of region see Figures 78 & 79*

Port Phillip Bay is 1 800 km^2 in size, with a maximum depth of 24 m. It has a maximum width of 67 km, with a total coast length of nearly 260 km. The 132 beaches occupy 177 km, or 68%, of the bay's coastline. Ocean tides penetrate into the bay, but are reduced slightly in size and take four hours to travel from the heads to Port Melbourne (Figure 21). The bay receives no ocean swell, apart from a few beaches near the entrance. As a result, many of the beaches may be considered low energy compared to the open ocean coast. However, the

strong westerly winds that periodically blow across the bay can generate waves up to 3 m high. Over time, these have built not only the beaches, but in some cases, wide multi-barred surf zones as well.

Unlike ocean beaches, the bars and surf zones of the bay are often inactive, particularly on calm, sunny days when most people frequent the bay beaches (Figure 80). There will only be waves on the beaches when the wind is blowing on- or along-shore. Consequently, while some of the eastern shore beaches have one, two or three bars, cut by rip channels and troughs, these systems are weak to inactive during low wave and calm periods.

These beaches do, however, pose a number of hazards and consequently are patrolled by 12 surf lifesaving and 15 Royal lifesaving clubs. During typical low wave conditions, the biggest problems are associated with the deep water close inshore, and variable water depth over the bars and troughs, particularly with the changing tide.

Victorian Beaches - Region 6 (Port Phillip)

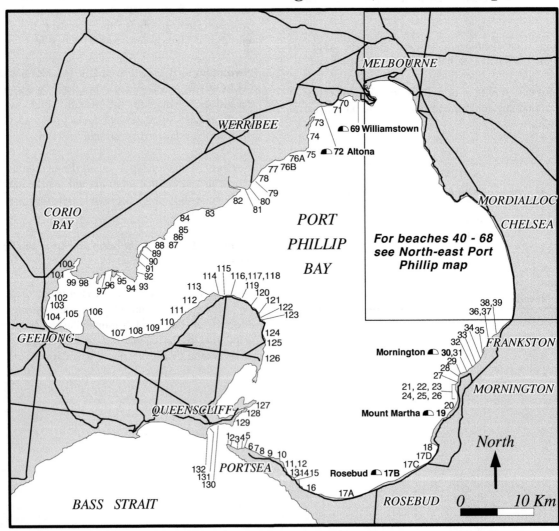

Figure 78. Region 6: Port Phillip Bay (Beaches P1 - P39, P69 - P132)

Victorian Beaches - Region 6 (North-east Port Phillip)

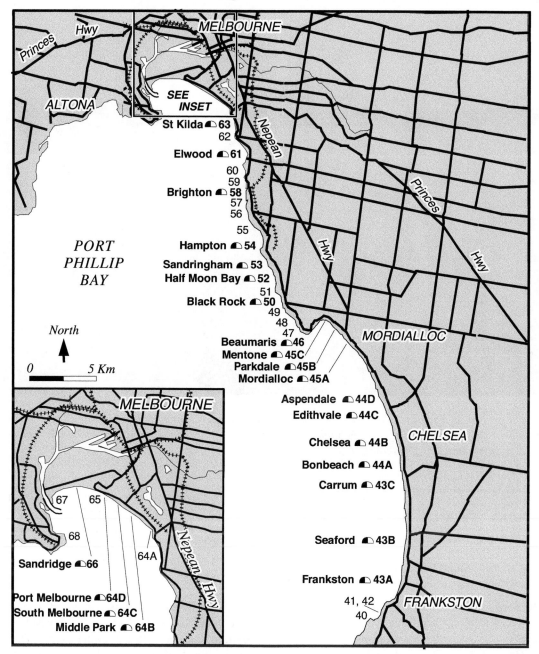

Figure 79.   Region 6 (north-eastern Port Phillip Bay): Frankston to Melbourne (Beaches P40 - P68)

Conditions can also change rapidly as soon as on- or along-shore winds start to blow, due to the dependence of waves on wind. This can cause waves, surf and strong currents in the rips and troughs.

The **beach hazard rating** given for the bay beaches refers to average wave conditions at the particular beach, not to the higher wave conditions that may have formed the bars and rips. When these conditions occur,

the rating rapidly increases, as indicated on the chart for all patrolled beaches.

Waves, and consequently surf, in the bay are entirely wind dependent. Strong winds produce short, 1 to 3 m waves that break along exposed beaches, producing usually sloppy beach breaks. These are all found between Mills Beach in the south and Port Melbourne in the north, with none on the Mornington, Western, Corio or Bellarine Bay shores.

Figure 80.   Port Phillip Bay has 132 beaches, that are for the most part dominated by low wind waves. The beaches are a tremendous recreational facility for Melbourne. This view from Rye Pier is typical of many bay beaches, with a popular, wide beach fronted by a shallow bar.

P1 - P4   POINT NEPEAN JETTY, NEPEAN BAY 1 to 3

Unpatrolled			
	Rating	No bar	Length
P1 Point Nepean Jetty	4	R + rocks, reefs & tidal currents	120 m
P2 Nepean Bay 1	3	R + tidal currents	120 m
P3 Nepean Bay 2	3	R + tidal currents	60 m
P4 Nepean Bay 3	3	R + tidal currents	400 m
		Total length	700 m

Point Nepean forms the eastern head of the entrance to Port Phillip. The point and the first 3 km of coast leading into Port Phillip Bay, as far as Observatory Point, are part of the Mornington Peninsula National Park. Unlike the beaches on the southern side of the point, the bay beaches receive only low to usually no ocean swell.

Away from the point, they are dominated by lower wind waves, generated by northerly winds within the bay.

The first four beaches form a narrow strip of sand below the steep, 20 to 30 m high bluffs that form the northern side of the point. The beaches average 20 m wide and have patchy reef flats offshore. The shoreline and adjacent waters are part of the Harold Holt Marine Reserve.

The **Jetty Beach** (P1) was the site of a jetty, that has since been removed. Nepean Point Road ends at the point, where there is a viewing area. The beach is 120 m long, faces north and has vegetated bluffs rising 27 m to the Nepean Trig behind it. The three **Nepean Bay Beaches** wind along below the crenulate bluffs, to where the bluffs terminate at the beginning of the broad Observatory Point beach ridge plain.

Swimming: BEACHES CLOSED - NO SWIMMING ALLOWED. All four beaches usually have calm to low wind waves, with very large swell only occasionally making it past The Rip to reach the beaches. However, very strong tidal currents run parallel to the beaches, making them potentially hazardous for swimming, even if it were allowed.

Surfing: None, once inside the bay.

Fishing: Line fishing, spear fishing and collecting of marine organisms are all prohibited in the reserve and park.

Summary: These beaches can only be reached through Commonwealth land. Public access is by a Transporter from the Orientation Centre, located at the Portsea entrance to the Commonwealth land.

P5, P6, P7   OBSERVATORY POINT, TICONDEROGA BAY, CADET

Unpatrolled			
	Rating	No bar	Length
P5 Observatory Point	2	R	1 200 m
P6 Ticonderoga Bay	3	R + tidal currents	1 600 m
P7 Cadet	3	R + reef	100 m
		Total length	2 900 m

Observatory Point is a 2 km long and up to 500 m wide accumulation of beaches deposited just inside Port Phillip over the past 6 000 years. They have been built by episodic swell, assisted by tidal currents penetrating the bay and moving sand along Nepean Bay, to accumulate on the point. Today they occupy nearly 100 hectares of low ridges. Much of the point is in the National Park; and the Marine Reserve extends to the point. The eastern section is in Commonwealth land. Access is either by boat or on foot from the Point Nepean Road.

Two long beaches lie on either side of the point, while the third, **Cadet Beach**, is a 100 m long, reef dominated extension of **Ticonderoga Bay**. The three beaches face north and usually receive only low wind waves, which build narrow, steep beaches fronted by subtidal flats that narrow toward the point, where tidal currents intensify.

Swimming: BEACHES CLOSED - NO SWIMMING ALLOWED. If swimming were allowed it would be relatively safe at the beaches, with deep water close inshore at high tide, and tidal flats off the Observatory Point Beach. However, there are deep tidal channels just off the Ticonderoga Bay beaches, with strong flows toward Point Nepean on a falling tide.

Surfing: None.

Fishing: Line fishing, spear fishing and collecting of marine organisms are all prohibited in the reserve and park.

Summary: Most of the point is in a natural state, with the two beaches backed by densely vegetated beach ridges. They make a fitting part of the National Park

and Marine Reserve, located next to the ocean entrance to Melbourne.

P8   WEROONA BAY (PORTSEA)

Unpatrolled
No bar: R
Beach Hazard Rating: **2**
Length: 1 200 m

Weroona Bay is the Portsea bay beach. It is bordered by Commonwealth land to the west and Point Franklin to the east, faces north and is 1 200 m long. The Nepean Highway and the town of Portsea parallel the back of the bay. The one hundred metre long Portsea Pier lies in front of the centre of the town, and numerous boats are moored in the bay.

Waters are usually calm off the beach, with strong northerly winds required to produce low wind waves. Strong tidal currents flow past Point Franklin, but reduce in strength toward the beach. The beach is relatively narrow and steep, with deep water close inshore at high tide.

Swimming: Relatively safe at the beach, with boat traffic and tidal currents increasing offshore.

Surfing: None.

Fishing: The pier is a very popular location, together with the rocks off each headland.

Summary: One of the more popular bayside beaches, with most people content to walk out on the pier, or in summer swim at the beach. The more adventurous can go out on the dive boats from the pier.

P9, P10   SHELLY, POINT MACARTHUR

Unpatrolled			
	Rating	Single bar	Length
P9 Shelly	2	R	1 200 m
P10 Point Macarthur	2	R	350 m
		Total length	1 550 m

Shelly Beach and **Point Macarthur Beach** lie between the 20 m high calcarenite bluffs of Point Franklin and Point King, with Point Macarthur forming the boundary. The Nepean Highway parallels both beaches and houses sit on top of the bluffs behind them both. Public access is restricted to side streets and through the reserve behind Shelly Beach. There are boat sheds below the bluffs at the back of the beaches, with several jetties extending across the beaches.

The beaches face north-north-east and usually receive low wind waves. They are both relatively narrow and steep, with deep water at high tide, but tidal flats at low tide. There are numerous boats moored off the beaches and the jetties.

Swimming: Relatively safe inshore and clear of the boats and jetties, however watch for strong tidal currents off the points.

Surfing: None.

Fishing: There are numerous private jetties along the beach, together with deep water off the headlands at each end.

Summary: Two protected beaches used by locals and boat owners, but with limited public access and facilities.

POINT KING

In March 1802, Lt John Murray landed at Point King and took possession of Port King, later renamed Port Phillip.

P11, P12   **POINT KING 1 & 2**

Unpatrolled			
	Rating	No bar	Length
P11 Point King 1	1	R + tidal flats	180 m
P12 Point King 2	1	R + tidal flats	200 m
		Total length	380 m

Point King is a 20 m high, calcarenite bluff, south of which lie two narrow strips of sand, fronted by 50 to 100 m wide sand and reef flats. The two beaches share the water with several boat sheds and private jetties. The beaches are very protected and are usually calm, with the tidal flats covered at high tide, but exposed at low tide.

Swimming: Best at high tide, however watch the boats, reef flats and jetties.

Surfing: None.

Fishing: There are plenty of rocks to fish from at high tide, and the jetties at low tide.

Summary: A bluff-top walking track runs south from Point King above these two beaches and the adjacent

bluffs. The beaches are primarily used by local residents and boat owners, with limited public access.

P13, P14   **SORRENTO PARK, SORRENTO FRONT**

Unpatrolled			
	Rating	No bar	Length
P13 Sorrento Park	1	R + tidal flats	900 m
P14 Sorrento Front	1	R + tidal flats	1 500 m
		Total length	2 400 m

Sorrento is the largest town on the Nepean Peninsula and the centre of numerous aquatic activities. The town sits on 20 m high, calcarenite bluffs, which outcrop at Policeman's Point. The point also divides the two Sorrento beaches. The **Park Beach** is backed by Sorrento Park, which has picnic facilities, a boat launching ramp and a number of private jetties and boat sheds. The 200 m long Sorrento Jetty runs out from Policeman's Point, with the **Front Beach** located on the eastern side. The beach is entirely backed by foreshore reserves with numerous facilities, including a second smaller jetty and a camping reserve toward the eastern end.

Both beaches face north-east and are sheltered from most waves, with the beach usually narrow and steep, and fronted by 100 to 150 m wide tidal flats. The water is deep inshore at high tide, while the flats are exposed at low tide.

Swimming: Two relatively safe beaches, with bathing best at mid to high tide. However, stay clear of the busy main jetty and boat ramp.

Surfing: None.

Fishing: The jetties are the most popular spots.

Summary: Two popular beaches with foreshore reserves containing most facilities for visitors.

P15   **SULLIVAN BAY**

Unpatrolled
No bar: R + tidal flats
Beach Hazard Rating: **1**
Length: 400 m

Sullivan Bay is the site of Victoria's first white settlement back in 1803. The settlement was located on the eastern point. The two calcarenite headlands that border the 400 m long bay are known as The Sisters. A foreshore reserve backs the western half of the bay and the historic section of the eastern Sister. Houses and some private jetties are located along the eastern half of the bay. There is a public parking area just off the Nepean Highway in the centre of the bay.

The beach faces north-east and receives usually low wind waves. It consists of a relatively narrow strip of sand, fronted by 100 to 200 m wide tidal sand flats.

Swimming: A relatively safe beach, best at mid to high tide.

Surfing: None.

Fishing: Best off The Sisters at high tide.

Summary: An attractive bay just off the highway, but bypassed by many.

P16   BLAIRGOWRIE

Unpatrolled
No bar: R + sand flats
Beach Hazard Rating: **1**
Length: 5 000 m

Blairgowrie is the first in a series of long, sandy beaches and wide sand flats that swing in a 24 km long, north facing arc from The Sisters to Martha Point. Most of the beaches are backed by a foreshore reserve, the Nepean Highway and Marine Drive, providing excellent public access and facilities.

Blairgowrie Beach is 5 km long and backed by a continuous foreshore reserve and the highway, all the way to the low White Cliffs that form the eastern boundary. The reserve contains all visitor facilities including a narrow, 200 m long jetty, boat ramps and a camping area. This section of the beach faces north-east and is usually calm, except during northerly winds when low wind waves are generated. The beach is usually narrow and fronted by 200 to 300 m wide, shallow sand flats. The flats have up to twelve very low amplitude, shore parallel ridges or 'bars', each ridge just a few centimetres high. Numerous wooden groynes cross the high tide beach, while scores of boats are moored in deeper water off the sand flats.

Swimming: A relatively safe beach, particularly on the sand flats. Best at mid to high tide when the flats are covered. Be careful off the flats as there is deep water and tidal currents.

Surfing: None.

Fishing: You can only fish off the beach at high tide, however the jetties provide the best access to deeper bay water.

Summary: A long, low beach, backed by a reserve and all facilities and fronted by shallow sand flats.

P17A   RYE

Unpatrolled
No bar: R + sand flats
Beach Hazard Rating: **1**
Length: 4 000 m

Between White Cliffs and Martha Point is a continuous 18.7 km long beach; the second longest in Port Phillip. The beach is backed by a foreshore reserve for most of its length, as well as by the Nepean Highway and Marine Drive. Several towns back the beach, which is divided into four sections based on these. The beaches are known as Rye, Rosebud, Dromana and Safety Beach; Beaches 17A, 17B, 17C and 17D, respectively.

Rye Beach begins at White Cliffs and runs past the 300 m long Rye Jetty, almost due east for 4 km. A foreshore reserve backs the entire beach which, beside the popular jetty, has two boat ramps, a yacht club and a camping reserve. The beach is backed by the low reserve and fronted by shallow sand flats that are between 300 and 400 m wide. The narrowest flats are found at the jetty. They are covered at high tide, but exposed at low. Waves are usually low and only exceed a few decimetres during strong northerly winds.

Swimming: Usually a safe beach close inshore. Depth varies over the flats with the tide, and deep water lies off the end of the flats.

Surfing: None, except for low wind waves during strong northerlies.

Fishing: Only off the jetty.

Summary: Rye Beach has a very popular and always full summer camping reserve, while the jetty area attracts day visitors. Do not jump or dive off the jetty as the water is very shallow.

P17B   ROSEBUD

Rosebud LSC
Single bar: R + sand flats
Beach Hazard Rating: **1**
Length: 9 000 m
For map of beach see Figure 81

Rosebud Life Saving Club is located 2 km east of the Rosebud Pier. It is readily accessible off Marine Parade and has a car park near the beach and lifesaving club. The club was founded in 1962. The entire 9 km long **Rosebud Beach** is backed by a 100 m wide foreshore reserve. The reserve contains numerous facilities including the 300 m long Rosebud Jetty, a boat ramp, swimming pool and camping area. The beach faces north-west, exposing it to westerly winds and waves, making it more suitable under easterly or calm conditions. The beach is 10 to 20 m wide, with a moderate slope to the shoreline. Irregular, shallow sand flats extend for up to 1 km off the beach.

Periodic wind waves have built a highly rhythmic beach, fronted by a 60 m wide inner bar. The bar attaches on the beach protrusions, with usually deeper rip holes and troughs in the embayments. Currents are only present when waves are breaking on the bar. Periods of higher waves will change the bar and rip configuration. Seaward of the shallow inner bar is a 1 km wide, deeper sand flat, that only drops off into deeper bay water at its outer edge. Depth on this bar depends on the tide height.

Movement of the shoreline protrusions, and subsequent beach erosion, has caused the council to respond by building a series of wooden groynes across the beach and onto the inner bar. There is also a low seawall along the car park and in front of the lifesaving club.

Swimming: A safe beach under low waves and calm conditions, apart from the change in water depth associated with the shoreline protrusions. Keep a watch on young children, particularly at high tide. Waves greater than 0.5 m will generate currents in the rip holes and gutters.

Fishing: The jetty is the most popular location, with little activity off the beach.

Surfing: None.

Summary: The only patrolled section of the long Rye-Rosebud-Dromana beach. It is usually not as crowded as the popular pier section.

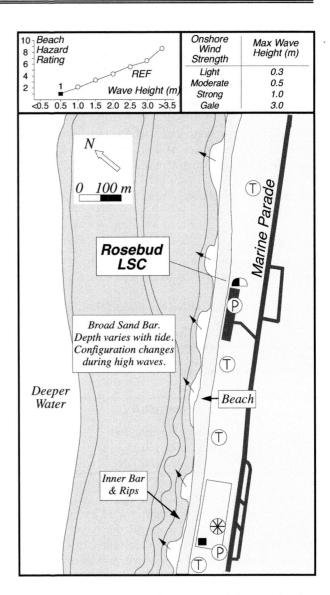

Beach Hazard Rating		Onshore Wind Strength	Max Wave Height (m)
		Light	0.3
	REF	Moderate	0.5
	Wave Height (m)	Strong	1.0
		Gale	3.0

Figure 81. Rosebud Life Saving Club patrols this section of the 18.7 km long beach. The approximate shape and location of the rhythmic inner bar and more variable outer bars and tidal shoals are shown here. The shoals extend up to several hundred metres off the beach, before dropping off into deeper bay water.

P17C & D   DROMANA, SAFETY

Unpatrolled			
	Rating	No bar:	Length
P17C Dromana	2	R + sand flats	3 000 m
		Double bar:	
P17D Safety	3	TBR/RBB	2 700 m
		Total length	5 700 m

Dromana and Safety Beaches occupy the eastern 6 km of the 18 km long Rye to Point Martha Beach. Both beaches are backed by a continuous foreshore reserve, the Nepean Highway and Marine Drive. The reserve contains numerous facilities including the Dromana

Camping Reserve, the Dromana Pier; and at Safety Beach, a launching ramp and sailing club.

Three kilometre long **Dromana Beach** faces north-west, while the 2.7 km long **Safety Beach** swings around to finally face west in lee of Martha Point. These beaches are more exposed to westerly winds and wind waves than the beaches to the west. Consequently, the high storm waves have steepened the nearshore. As a result, the sand flats that reach 1 km wide off Rosebud progressively narrow to 300 m at Dromana Pier, and 100 m at Safety Beach. Furthermore, the higher waves produce a double bar system at Safety Beach, characterised by alternating bars and rips every 200 m.

Swimming: Relatively safe during calms and low wind waves, however Safety Beach in particular will have higher waves and rips during strong winds.

Surfing: Wind waves can reach 1 m or more at Safety Beach during strong westerlies, otherwise it is usually flat.

Fishing: Best off the 200 m long Dromana Pier or the rocks at Martha Point.

Summary: Another very accessible beach, with a foreshore reserve with numerous facilities and a slightly more energetic surf. A Tourist Information Centre is located on Marine Drive just west of the pier.

P18   **MARTHA POINT**

+-------------------------------------+
| **Unpatrolled** |
| |
| No bar: LTT |
| Beach Hazard Rating: **3** |
| Length: 100 m |
+-------------------------------------+

On the southern side of **Martha Point** is a 100 m pocket of sand and cobbles, lying at the foot of steep, 40 m high, vegetated bluffs. It is fronted by a band of intertidal cobbles and rock, with a sandy bay floor lying beyond the rocks. Marine Drive passes around on the bluffs above the beach, while a walking track from Safety Beach to the point provides the best access.

The beach faces south-west and is well exposed to westerly winds. The beach is narrow and steep, with usually low waves at the base of the beach and some surf over the bay floor during strong south-westerlies.

Swimming: Moderately safe during low winds and waves, however watch the rocks at the base of the beach.

Surfing: Chance of a 1 m wind wave during strong westerlies.

Fishing: There are rip holes and gutters against the rocks, particularly following strong westerlies.

Summary: A small, pocket, rock bound and relatively isolated beach for those who want a change from the adjacent long beaches.

P19   **MOUNT MARTHA**

+---+
| **Mount Martha LSC** |
| |
| Inner bar: R Outer bar/s: RBB |
| Beach Hazard Rating: **3** |
| Length: 1 000 m |
| |
| *For map of beach see Figure 82* |
+---+

Mount Martha Beach is 2 km long, with Balcombe Creek mouth dividing it into two equal halves: Mount Martha (South) and Mount Martha North.

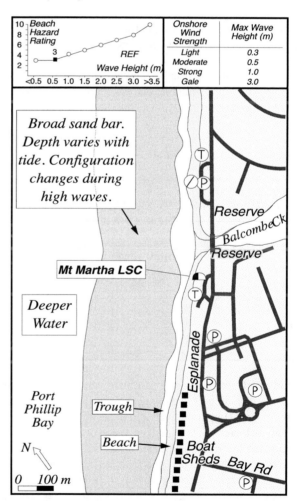

Figure 82.   Mount Martha Beach map showing the location of the Mount Martha Life Saving Club and the approximate shape and location of the broad bar that lies off the beach. A trough may lie between the beach and the bar.

The Mount Martha Life Saving Club and the yacht club are located at the northern end of the southern beach. South of the club a vegetated bluff backs the beach, with small boat sheds between the bluff and the 50 m wide beach. The beach slopes steeply to the shoreline, with usually deep water against the beach face. A trough often runs along the beach, with a broad, shallow, outer bar cut by deeper channels parallelling the beach. The beach faces north-west, exposing it to westerly winds which, when strong, produce the waves and currents to maintain and move the bars and troughs. Under normal calm to low wave conditions, and at high tide, waves will only break at the beach. However, during strong winds, higher waves break across the 100 to 200 m wide outer bar and shoals.

Swimming: The steep beach and drop-off close inshore are a problem for young children and poor swimmers. Use caution if bathing here, as waves will also generate rip currents along the beach and across the outer bar.

Surfing: During strong westerlies there are wind waves up to 1 m high breaking over the bars.

Fishing: A trough usually runs along the beach, while the rocks at the southern end also provide access to deeper water.

Summary: A popular beach for bathers, sailors and boaters, with facilities for all.

P20, P21, P22   MOUNT MARTHA NORTH, HAWKER, CRAIGIE

Unpatrolled			
	Rating	Single bar	Length
P20 Mount Martha North	3	RBB	900 m
P21 Hawker	3	RBB	600 m
P22 Craigie	3	RBB	120 m
		Total length	1 620 m

North of Balcombe Creek is a near continuous strip of beach, including 900 m long **Mount Martha North**, 600 m long **Hawker Beach** and the smaller, 120 m long **Craigie Beach**. They all face west-north-west and are separated by small sandstone bluffs, with bluffs behind the beaches that steepen and increase in height towards Hawker Beach. The Esplanade parallels the shore on top of the bluffs. There is a car park and easy access on the north side of the creek. However, to reach Hawker and Craigie Beaches requires a walk up the beach or a steep climb down the 20 m high bluffs. There are numerous boat sheds lining the rear of Mount Martha North and Hawker Beaches. At Hawker Beach, a seawall fronts the boat sheds to protect them from erosion that has narrowed the beach, as far as the seawall in some places.

The three beaches are exposed to westerly winds and waves, with storm waves building the rhythmic bar and beach. The rips are inactive under low waves, but will contain strong currents during strong winds and accompanying waves.

Swimming: Three moderately safe beaches when waves are low, however, still watch the variable water depth off the beach owing to the bars, channels and rip holes. Use care if there are any waves breaking, as the rips will be flowing.

Surfing: Only when there is a strong westerly, with a chance of wind waves over the bars.

Fishing: Usually a good, relatively deep trough along the beach with rip holes.

Summary: Three narrow beaches: the first two backed by boat sheds and bluffs, and the third, Craigie, offering a more isolated pocket beach

P23 - P26   BIRDROCK, DAVA, FOSSIL, FOSTERS

Unpatrolled			
	Rating	Single bar	Length
P23 Birdrock	3	LTT	550 m
P24 Dava	2	R + reefs	150 m
P25 Fossil	2	LTT + rocks	80 m
P26 Fosters	2	LTT + rocks	80 m
		Total length	860 m

Between Craigie Beach and Linley Point is a 3 km section of rocky coast dominated by 20 to 30 m high sandstone bluffs. In amongst the bluffs are four beaches: the longest being **Birdrock** at 550 m, while **Dava** is 150 m and **Fossil** and **Fosters** are both 80 m long. Although the Esplanade runs along the top of the bluffs behind the beaches, and there is a continuous foreshore reserve west of the road, access is restricted to a steep climb down the bluffs. There is a picnic area above Fossil Beach.

Birdrock Beach is embayed by headlands, with reefs off the centre; while the remainder are all bounded by steep bluffs and rocks, with a seawall also running off the northern end of Dava Beach. The four beaches face west and are exposed to westerly winds and waves which, during strong winds, produce waves high enough to form a bar and rips at Birdrock and Dava. Fossil and Fosters Beaches are more protected by the bluffs and tend to have narrower, steeper beaches.

Swimming: These are four relatively isolated beaches, so use care if bathing here. There is deep water close inshore at high tide; variable water depth owing to rip holes at Birdrock and Dava; and rips are present when waves are breaking.

Surfing: Usually nothing, though waves will break over the Birdrock and Dava sand and reefs during strong westerlies and accompanying waves.

Fishing: There are good rip holes and gutters at Birdrock and Dava, while deep water can be reached off the rocks along this entire section.

Summary: Four natural beaches lying below the busy Esplanade, but little used owing to the difficult access and limited parking.

P27   FISHERMANS

Unpatrolled
Single bar: RBB
Beach Hazard Rating: **3**
Length: 550 m

Fishermans Beach is located in lee of Linley Point. The Esplanade runs on 20 m high bluffs right behind the beach. Between the road and beach is a narrow foreshore reserve, with a seawall at the base of the bluffs and a few boat sheds on the beach. A car park has been built at the base of Linley Point, which also provides access to a double boat ramp.

The point only affords slight protection and the north-west facing beach receives sufficient wave energy during strong westerlies to build a one to two bar system. This is usually cut by rips at each end and one central rip. The rips are only active when waves are breaking.

Swimming: A moderately safe beach when waves are low. However, there is variable water depth owing to the presence of the bars and rip channels. Currents will intensify if waves are breaking. Best to bathe up the beach, away from the point where boats are launched.

Surfing: Only when strong westerlies are blowing. These can produce 1 to 2 m wind waves over the bar.

Fishing: Usually good rip holes along the beach, with gutters against the rocks at each end.

Summary: A very accessible and popular beach, which can provide a few waves to play in.

P28   ROYAL

Unpatrolled
No bar: R + rock flats
Beach Hazard Rating: **3**
Length: 150 m

Royal Beach is a small pocket beach, just 150 m long, lying between two 10 to 20 m high, red sandstone bluffs, and backed by grassy bluffs and the Esplanade. The narrow beach looks attractive at high tide, but at low tide exposes a continuous rock and reef flat, with deep water only out beyond the rocks.

Swimming: Only at high tide, but still watch the rocks.

Surfing: None.

Fishing: Best at high tide from the beach or off the rocks at each end.

Summary: A small, accessible beach dominated by the backing bluffs and fronting rocks and reefs.

P29   MOTHERS/SCOUT/SHIRE HALL

Unpatrolled
Single bar: TBR/RBB
Beach Hazard Rating: **3**
Length: 600 m

In lee of prominent Snapper Point and its Mornington Jetty is a continuous, 600 m long beach that begins as **Mothers Beach** in the western protected corner, then changes to **Scout Beach** in the centre and ends in the east as **Shire Hall Beach**. Beside the jetty at Snapper Point are a large car park, marina and boat ramp, with a 10 m high bluff backing Mothers Beach. At Scout Beach there is a scout hall and a few boat sheds below the bluffs, while at Shire Hall there are several boat sheds.

The beach faces north and is sheltered in the west by the point and jetty. A continuous bar fronts Mothers and Scout Beaches. However during strong north-westerlies, sufficient waves reach the beach to build a bar and two rips toward the eastern Shire Hall Beach.

Swimming: Safest at Mothers and Scout Beaches, so long as you are clear of the boat traffic. Water depth varies over the bars and troughs at Shire Hall, and care should be taken if children are using the beach.

Surfing: Strong west to north-west winds can produce waves over the bars at Shire Hall.

Fishing: Best off the jetty or the rocks at the end of Shire Hall.

Summary: An attractive, accessible beach with a foreshore reserve and picnic facilities. A choice of three beaches in one.

P30   MILLS

Mills Beach is a 700 m long, north to north-west facing beach, backed by boatsheds and a vegetated bluff for most of its length.

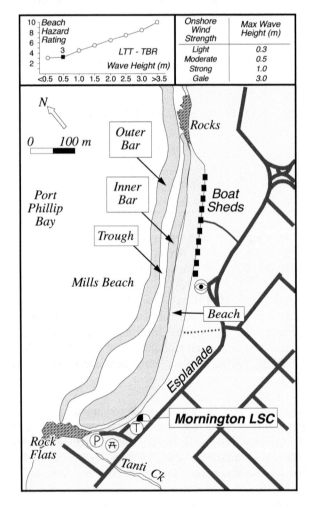

Figure 83.   Mills Beach, showing the location of the Mornington Life Saving Club and the approximate shape and location of the bars. Tanti Creek crosses the southern end, producing a broad, rocky delta. The remainder of the beach has two usually parallel bars, separated by a deeper trough.

The Mornington Life Saving Club and car park are located at the southern end. Tanti Creek also flows out at the southern end and has deposited a blanket of rocks under the sand. A low seawall curves around from the creek mouth to the club house. The lifesaving club's new premises provide a focal point for bathers.

The beach is exposed to wind and waves from the west and north-west. When these winds are blowing, they develop a moderately steep and narrow beach, fronted by a wide, shallow bar toward the south, and a deeper, shore parallel bar 100 m out. The waves also produce rip currents and channels that cut through both bars. The configuration is modified during strong winds and high waves, and so will change from year to year.

Swimming: Safest in front of the lifesaving club where the bar is usually shallowest. Deeper water may lie against the beach to the north. Be careful if walking out on the bars, as they are cut by deeper channels and troughs.

Fishing: Best off the southern rocks, or up the beach where the water is usually deeper.

Summary: A popular beach just off the Esplanade, with a grassy picnic area, lifesaving club and a kiosk.

P31 - P34   MILLS EAST, CARAAR, SUNNYSIDE, SUNNYSIDE NORTH

Unpatrolled			
	Rating	Single bar	Length
P31 Mills East	**4**	R + rocks	450 m
P32 Caraar	**3**	LTT + rocks	850 m
P33 Sunnyside	**3**	TBR	250 m
P34 Sunnyside North	**3**	TBR	500 m
		Total length	2 050 m

Between Mills Beach and Sunnyside North is a 2.5 km section of coast containing four beaches, that are either dominated or bounded by sandstone rocks, reefs and bluffs. Caraar Creek crosses the southern end of Caraar Beach. Access is limited to street parking on the bluffs behind **Mills East**, **Caraar** and **Sunnyside**, while **Sunnyside North** is only accessible on foot around the rocks.

The four beaches all face north-west and are exposed to westerly winds and accompanying waves. These produce a near continuous outer surf zone during storm conditions. During the more regular low waves, Mills Beach East and Caraar Beach have a steep, narrow beach and a surf zone dominated by rocks and reefs; and the two Sunnyside beaches have a bar cut by one to two rip channels, particularly against the rocks.

Swimming: Mills East and Caraar are only suitable for bathing at high tide in calm conditions, as there are many rocks off the beach. When waves are breaking, strong currents also occur amongst the rocks. The two Sunnyside beaches are relatively safe during low waves, however any surf will intensify the longshore and rip currents, particularly against the bluffs.

Surfing: Only at the Sunnyside beaches during westerly gales, when 1 to 2 m wind waves will break on the bars.

Fishing: Best off the rocks at Mills East and Caraar, and on the rocks either side of the Sunnyside beaches.

Summary: Four relatively natural beaches, with Sunnyside North also being an official Optional Dress (ie. nude) Beach.

P35   MOONDAH, EARIMIL

Unpatrolled				
	Rating	Inner bar	Outer bar	Length
P35 Moondah	3	TBR	RBB	1 500 m
P36 Earimil	4	R + rocks & reefs		250 m
			Total length	1 750 m

Moondah Beach is a 1.5 km long, west facing beach located at the end of Kuinyung St. Parking is limited to a small car park, and steps lead down the bluffs to the beach. Densely vegetated, 20 to 30 m high bluffs back most of the beach, with a few boat sheds near the steps. To the south, a small, red bluff separates it from Sunnyside North, while increasing rocks and reefs to the north finally form a boundary with the northern 250 m long **Earimil Beach**. Earimil Beach is dominated by the intertidal rock and reef flats.

The beaches are well exposed to westerly winds and waves which, during storm conditions, have built a 150 m wide double bar at Moondah Beach. The double bar is cut by rips every 200 m, with permanent rip holes existing against the rocks. Earimil Beach is dominated by the rocks and has no bars.

Swimming: Safest during low waves at Moondah, however there is usually a deep channel along the beach and rip holes, so use caution. Currents intensify if waves are breaking. Earimil is only suitable in calms at high tide.

Surfing: Strong westerlies will produce waves over the bars at Moondah.

Fishing: There are usually rip holes and gutters along Moondah, with Earimil best at high tide.

Summary: Two relatively natural beaches, off the main road and mainly used by locals and the boat shed owners.

P37   RANELAGH

Unpatrolled
No bar: R + rocks & reefs
Beach Hazard Rating: **2**
Length: 450 m

Ranelagh Beach is surrounded by bluffs and houses, with the best public access from southern Earimil Drive. The beach is 450 m long, with numerous boat sheds backing the crenulate beach. It faces west and, while exposed to westerly winds and waves, has a surf zone dominated by patchy reefs.

Swimming: Relatively safe during low waves and at high tide. However, watch the rocks and reefs at low tide and rips when waves are breaking.

Surfing: During strong westerlies there are some waves at high tide over the reefs.

Fishing: Good gutters run out along the reefs and rocks at each end.

Summary: A difficult to find beach, used mainly by locals and boat owners.

P38, P39   HALF MOON BAY, CANADIAN BAY

Unpatrolled			
	Rating	Single bar	Length
P38 Half Moon Bay	2	R	240 m
P39 Canadian Bay	3	TBR/RBB + rocks	850 m
		Total length	1 090 m

The Canadian Bay Road reaches the coast at a 10 m high point that separates Half Moon and Canadian Bays. There is a reserve on the point with picnic facilities and a car park. A seawall protects the southern side of the point while, on the northern side, is a large boat club, and a second car park behind Canadian Bay at the end of Williams Road. A walking track runs along the bluffs above Canadian Bay.

The two beaches face west and receive surf during strong westerlies. **Half Moon Bay** consists of over 200 m of high tide, sandy beach, which sweeps in three small arcs below the red bluffs and between more prominent reefs. At **Canadian Bay** there is a bar cut by rip channels along the southern half, with more rocks and reef along the northern half of the 850 m long, crenulate beach.

Swimming: Both are relatively safe when waves are low, with care required at low tide around the rocks and reefs. Strong westerlies and higher waves generate strong rips along both beaches, with permanent rips during westerly conditions against the rocks and reefs.

Surfing: Only during strong westerlies, when waves will break at high tide over the reefs at both beaches, and the bars at the southern end of Canadian Bay.

Fishing: Two good locations with permanent rip holes and gutters against the many rocks and reefs, and in the Canadian Bay rip channels.

Summary: Two crenulate, sandy beaches backed by red bluffs and dominated by reefs, with good access and parking.

P40, P41, P42   **PELICAN POINT, DAVEYS BAY, KACKERABOITE CREEK**

Unpatrolled			
	Rating	Single bar	Length
P40 Pelican Point	2	LTT	200 m
P41 Daveys Bay	3	TBR + rocks & reefs	250 m
P42 Kackeraboite Creek	3	TBR + rocks & reefs	650 m
		Total length	1 100 m

At Pelican Point, the coast turns east into Daveys Bay, then swings back to face west past Kackeraboite Creek. This 1.5 km long section is dominated by 20 to 30 m high bluffs, rocks and reefs, with the narrow beaches wedged in between them. Daveys Bay Road runs to the top of Pelican Point, with walking access down the bluffs to **Pelican Point Beach** and **Daveys Bay**, where a yacht club and jetty are located. Gulls Way provides access to the bluffs above **Kackeraboite Creek Beach**.

Pelican Point and Daveys Bay beaches face north and are partly protected from the westerlies, while Kackeraboite faces north-west and is more exposed. Strong westerly winds and waves produce a bar along the beaches. The bar is usually attached at Pelican Point, but in Daveys Bay and along Kackeraboite it is cut by rip channels, with rocks and reefs becoming more dominant to the north.

Swimming: Three relatively safe beaches during light winds and low waves. However, strong westerlies and accompanying waves will produce strong rip currents against either end of Pelican Point Beach, and along Daveys Bay and Kackeraboite Beaches.

Surfing: During strong westerlies there are waves over the bars and reefs of Daveys Bay and Kackeraboite Beaches.

Fishing: There are permanent rip channels and gutters against Pelican Point, and along Daveys Bay and Kackeraboite Beaches.

Summary: Three narrow beaches backed by high bluffs, mainly used by the yacht club and fishers.

**Port Phillip's Longest Beach
Frankston to Mentone
(Beaches P43 to P45)**

The longest beach in Port Phillip Bay runs for 20 km from Frankston in the south up to Mentone. The beach faces due west for most of its length, exposing it to the full force of the westerly winds and waves, which in turn have built a magnificent beach and backing sand dunes. These are fronted by two to three shore parallel bars, over which waves break during bigger seas. Three creeks cross the beach, two of which have been channelised and now break the run of the beach. The first is Kananook Creek, which flows across the beach at Frankston; the next is the channelised Patterson River mouth; and the third is Mordialloc Creek, where a breakwater has also been constructed. This long beach is backed by the Nepean Highway and the railway for much of its length, resulting in it being a readily accessible and popular beach. In response to this popularity, lifesaving clubs were established near all the railway stations: Frankston, Seaford, Carrum, Bonbeach, Chelsea, Edithvale, Aspendale, Mordialloc, Parkdale and Mentone.

In the following description, the beach is divided by the Patterson and Mordialloc Creeks into three beaches. These are from Frankston to Carrum (Beach P43), Bonbeach to Aspendale (Beach P44), and Mordialloc to Mentone (Beach P45). Each of these three is again subdivided into descriptions of the section adjacent to each of the ten lifesaving clubs.

P43A   **FRANKSTON**

Frankston LSC

Inner bar: LTT/TBR   Second bar: RBB   Outer bar: D
Beach Hazard Rating: **3**
Length: 2 500 m

For map of beach see Figure 84

Frankston Beach forms the southern end of the 9 km long beach that runs almost due north to the Patterson

River mouth at Carrum. The Frankston section is 2.5 km long and begins at the southern boat launching area. It includes the 500 m long Frankston Pier, Kananook Creek mouth and the beach to the north.

The beach averages 50 m in width, with a moderate to steep shoreline. Bayward of this are three bars. The inner bar is usually highly rhythmic, consisting of alternating shallow attached bars and shoals, separated by bays with deeper rip holes and channels, spaced about every 150 m. The second less rhythmic bar lies up to 100 m off the beach, while the third straight bar may be up to 200 m offshore. The three bars are clearly visible from the pier, though waves will only break on the outer two, when they are greater than 1 m.

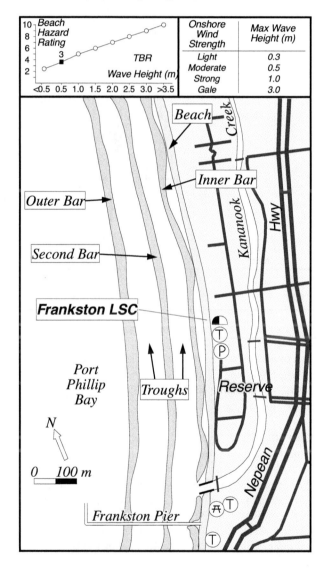

Figure 84. Frankston Beach map, showing Frankston Pier and Life Saving Club and the approximate shape and location of the three bars and troughs. During calms, waves and currents are usually absent, except near the creek mouth. However during strong winds, high waves will break over the bars and produce strong currents, as well as modifying the shape of the inner bar and rips.

Frankston Beach houses the lifesaving club, founded in 1924, and is backed by the town with all facilities, including the Frankston Pier. Parking is available at the boat ramp, pier and lifesaving club.

Swimming: Relatively safe on the inner bar and in the patrolled area. Watch children as deeper troughs and holes are common. Strong winds will rapidly increase the wave height and generate rip currents in the troughs. **Warning**: Do not jump off the pier, as the outer bars are shallow and this practice has been known to result in serious injuries.

Surfing: During strong westerly winds, wind waves up to 2 m high break over the outer bars, providing sloppy beach breaks.

Fishing: Best off the pier or the southern rocks.

Summary: A very popular beach, owing to the good road and rail access and full range of amenities.

P43B   **SEAFORD**

Seaford LSC
Inner bar: LTT/TBR   Second bar: RBB   Outer bar: D
Beach Hazard Rating: **3**
Length: 4 500 m
For map of beach see Figure 85

Seaford Beach, Life Saving Club and Pier are located just 250 m west of the railway station. This 4.5 km long section of the Frankston-Carrum Beach is not as popular as Frankston, but does attract a greater number of families, who prefer the more natural setting of a bush reserve that runs the length of the beach.

The 150 m long pier and the lifesaving club, founded in 1919, provide the focus for this beach. It essentially has identical beach and bar characteristics as its neighbour Frankston, with an attached inner bar, and deeper, more parallel second and outer bars. The outer bar usually lies just off the end of the pier.

Swimming: The beach is relatively safe under calm and low wave conditions, with only the deeper water in the rip holes and between the bars providing a problem for children and non-swimmers. The biggest danger is the pier, as people have been injured by jumping off it and landing on the shallow bar crests.

Surfing: During strong westerlies, wind waves up to 2 m high break over the outer bars, providing sloppy beach breaks.

Fishing: The pier is the most popular spot.

Summary: A long, sandy beach, backed by a naturally vegetated foredune and reserve, with the added attraction of the lifesaving club and pier. Very popular with families in summer.

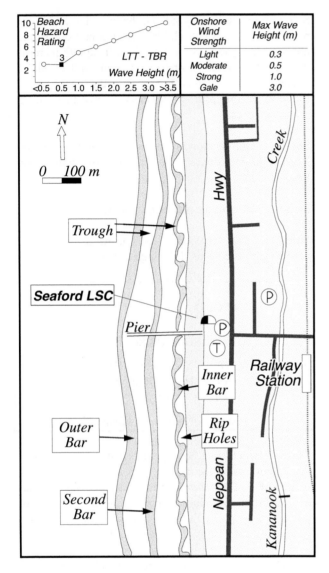

Figure 85.   Seaford Beach map, showing Seaford Pier, Life Saving Club and the railway station, and the approximate shape and location of the three bars. During calms, currents, waves and rips are usually absent. However during strong winds, high waves will break over the bars and produce strong currents, as well as modifying the shape of the inner bar and rips.

P43C   **CARRUM**

Carrum SLSC
Inner bar: LTT/TBR Second bar: RBB Outer bar: D
Beach Hazard Rating: **3**
Length: 2 000 m
For map of beach see Figure 86

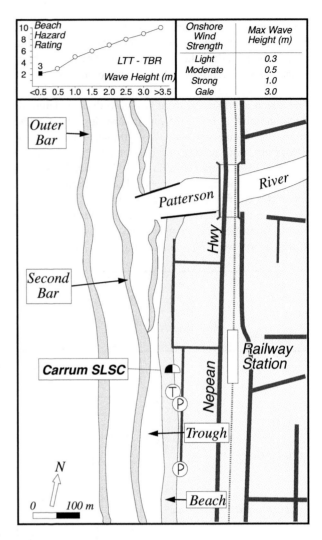

Figure 86.   Carrum Beach map, showing Carrum Surf Life Saving Club and railway station and the approximate shape and location of the three bars. During calms, waves and currents are usually absent, except near the river mouth. However during strong winds, high waves will break over the bars and produce strong currents, as well as modifying the shape of the inner bar and rips.

Carrum Beach forms the northern 2 km of the 9 km long Frankston-Carrum beach.   The beach, which is located 200 m west of Carrum railway station, houses the Carrum Surf Life Saving Club, Carrum Sailing Club and Patterson River Yacht Club. The latter is next to the Patterson River mouth and its short entrance walls. The Carrum Surf Life Saving Club patrols the northern section of the beach and was founded in 1937.

The beach faces west-south-west and is exposed to all westerly winds and waves. This produces a 200 m wide, three bar system, that is essentially identical to adjoining Seaford and Frankston Beaches.   The inner bar is usually highly rhythmic, alternating every 150 m between shallow shoals and rip holes. The second bar is less rhythmic and the outer bar is straighter still. The beach increases in width toward the river entrance walls, which act as a sand trap. Be careful bathing near the river mouth, as there is a deep tidal channel and strong tidal currents in the river and over the adjoining bars.

One kilometre south of the surf club is a short, concrete groyne, backed by a large park and car park. This is also a popular summer bathing area.

Swimming: Safest on the shallow parts of the inner bar. Watch children near the rip holes and troughs that parallel the inner bar. Stay clear of the river mouth, which has strong tidal currents and boat traffic. Best to bathe between the flags in front of the surf lifesaving club.

Surfing: During strong westerlies, wind waves up to 2 m high break over the outer bars, providing sloppy beach breaks.

Fishing: Best off the southern groyne and at the river mouth.

Summary: Readily accessible by train or car, with numerous aquatic facilities and the added safety of the surf lifesaving club.

P44A   BONBEACH

> **Bonbeach LSC**
>
> Inner bar: LTT/TBR   Second bar: RBB   Outer bar: D
> Beach Hazard Rating: **3**
> Length: 1 500 m
>
> *For map of beach see Figure 87*

Bonbeach Beach forms the southern end of the 7.7 km long beach that includes Bonbeach, Chelsea, Edithvale and Aspendale Beaches. Bonbeach is 1.5 km long, beginning at the northern Patterson River entrance wall and merging with adjoining Chelsea Beach to the north. All four beaches are located near their respective railway stations. Bonbeach has a lifesaving club, however access to the club is down a narrow street and parking is very limited. The club was formed in 1933 and is housed in an old, timber building.

The beach faces west-south-west and is exposed to the full force of westerly winds and waves, which interact with the fine sand to build a beach fronted by three bars extending 200 m bayward. The inner bar usually consists of alternating shallow bars and deeper rip holes, while the outer bars become increasingly straight. The outer two bars are usually inactive during low waves and calms. Strong winds and high waves are required to mobilise the bars and change their configuration. However near the river mouth, strong tidal currents can be encountered.

Swimming: Safest in front of the lifesaving club on the shallow sections of the inner bar. Stay well clear of the river mouth which has deep water, currents and power boats. Watch children, as deeper rip holes and troughs lie adjacent to the inner bar.

Surfing: During strong westerlies, wind waves up to 2 m high break over the outer bars, providing sloppy beach breaks.

Fishing: The river mouth is the most popular spot.

Summary: A beach for locals and train travellers, as parking is very limited.

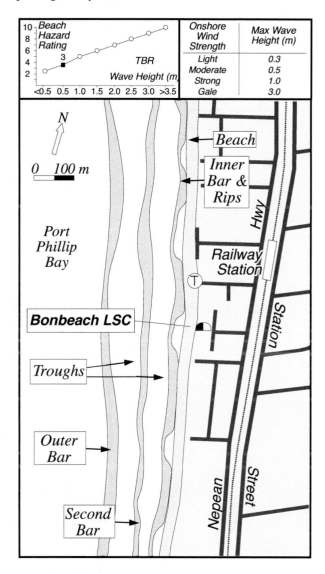

Figure 87. Bonbeach Beach map, showing Bonbeach Life Saving Club and railway station and the approximate shape and location of the three bars. During calms, waves and currents are usually absent, except near the river mouth. However during strong winds, high waves will break over the bars and produce strong currents, as well as modifying the shape of the inner bar and rips.

P44B   CHELSEA

Chelsea Longbeach LSC

Inner bar: LTT/TBR   Second bar: RBB   Outer bar: D

Beach Hazard Rating: **3**

Length: 1 500 m

For map of beach see Figure 88

Chelsea Beach is a 1.5 km long, straight stretch of sand between Bonbeach and Edithvale. The beach is located 250 m west of the Chelsea railway station and is home to the Chelsea Life Saving Club, founded in 1917.

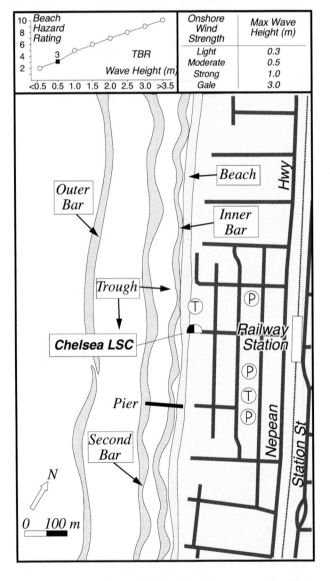

Figure 88.   Chelsea Beach map, showing Chelsea Life Saving Club, Pier and the railway station, and the approximate shape and location of the three bars. During calms, waves and currents are usually absent. However during strong winds, high waves will break over the bars and produce strong currents, as well as modifying the shape of the inner bar and rips.

Further up the beach, there is also the Chelsea Yacht Club. A short pier lies just south of the lifesaving club; and a foreshore park, picnic area and parking are all available immediately north of the club.

The beach has essentially identical beach, surf and bar characteristics as adjoining Bonbeach, ie. a usually attached and shallow inner bar, with two outer bars.

Swimming: Safest on the attached, shallow sections of the inner bar in the patrolled area. Avoid the deeper rip holes and troughs. Also, do not jump off the pier as it is surrounded by shallow water.

Surfing: During strong westerlies, wind waves up to 2 m high break over the outer bars, providing sloppy beach breaks.

Fishing: Best off the end of the pier.

Summary: A popular summer beach with good road and rail access, as well as parking, picnic and park facilities.

P44C   EDITHVALE

Edithvale LSC

Inner bar: LTT/TBR   Second bar: RBB   Outer bar: D

Beach Hazard Rating: **3**

Length: 1 500 m

For map of beach see Figure 89

Edithvale Beach is a 1.5 km long, straight section of sand located between Chelsea and Aspendale Beaches. The Nepean Highway and railway parallel the beach, with Edithvale railway station located 200 m east of the beach and Edithvale Life Saving Club. The club, which was founded in 1919, has a large club house on the beach; however parking is limited to the adjoining streets. On either side of the club are a few boat sheds, with beach-front houses backing most of the beach.

The beach faces west-south-west and, like adjoining Bonbeach and Chelsea, has a 50 m wide, sandy beach fronted by usually three bars. The inner bar is usually attached to the beach, while deeper troughs separate the second and outer bars. The bars and currents are only active during strong winds and high waves.

Swimming: Safest on the shallow sections of the inner bar in the patrolled area. Avoid rip holes and the outer troughs.

Surfing: During strong westerlies, wind waves up to 2 m high break over the outer bars, providing sloppy beach breaks.

Fishing: Best at high tide when you can cast into the deeper rip holes and troughs.

Summary: A long, sandy beach with good rail access but limited parking, more popular with locals for this reason.

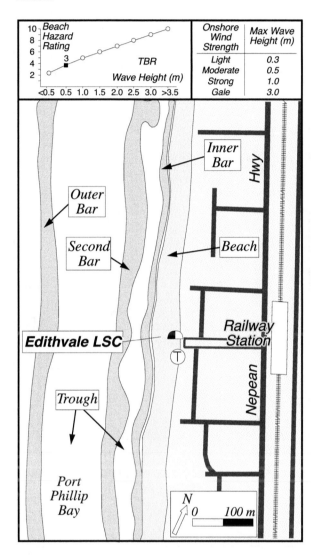

Figure 89. Edithvale Beach map, showing Edithvale Life Saving Club and the railway station, and the approximate shape and location of the three bars. During calms, currents, waves and rips are usually absent. However during strong winds, high waves will break over the bars and produce strong currents, as well as modifying the shape of the inner bar and rips.

P44D   **ASPENDALE**

Aspendale LSC		
Inner bar: LTT/TBR	Second bar: RBB	Outer bar: D
Beach Hazard Rating: **3**		
Length: 3 000 m		
For map of beach see Figure 90		

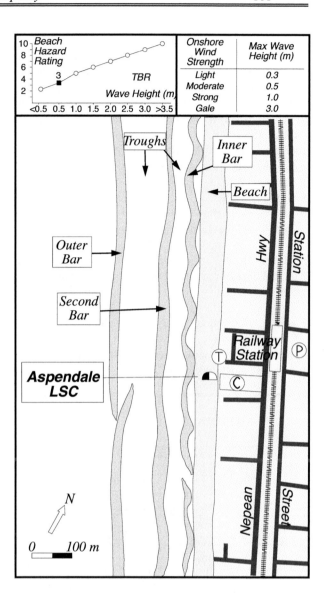

Figure 90. Aspendale Beach map, showing Aspendale Life Saving Club and the railway station, and the approximate shape and location of the three bars. During calms, currents, waves and rips are usually absent, except near Mordialloc Creek. However during strong winds, high waves will break over the bars and produce strong currents, as well as modifying the shape of the inner bar and rips.

Aspendale Beach forms the northern 3 km section of the 7.5 km long beach that stretches between the Patterson River mouth and the Mordialloc breakwater. The Nepean Highway and railway parallel the beach, with the Aspendale railway station being located 200 m east of the Aspendale Life Saving Club. The club was founded in 1927. There is limited parking at the club, as it is surrounded by beach-front houses and boat sheds, however a large car park is located 100 m to the south.

The beach faces the south-west and extends from adjoining Edithvale up to the Mordialloc drain, with the Mordialloc Sailing Club located at the northern end. The beach is exposed to west through south-westerly winds and waves, which have built a sandy beach fronted by three shore parallel bars. The inner bar alternates between shallow, attached sections and deeper rip holes approximately every 150 m, while the second and outer bars become increasingly straight, with deeper troughs in between. At the Mordialloc drain, tidal flows through the drain widen and deepen the channel. In addition, more than 50 wooden groynes have been built across the beach, supposedly to prevent beach erosion.

Swimming: Best in the patrolled area in front of the lifesaving club. Stay on the shallow section of the inner bar and avoid the deeper rip holes and troughs.

Surfing: During strong westerlies, wind waves up to 2 m high break over the outer bars, providing sloppy beach breaks.

Fishing: Best to look for the deeper rip holes.

Summary: A popular beach with locals, with rail access and some parking for visitors.

P45A   MORDIALLOC

Mordialloc LSC
Inner bar: LTT/TBR   Outer bar: RBB/LBT
Beach Hazard Rating: **3**
Length: 1 000 m
For map of beach see Figure 91

Mordialloc Beach is the southern section of a 4 km long beach running from Mordialloc Creek to the bluffs at Mentone. The beach has three sections, each with a lifesaving club, at Mordialloc, Parkdale and Mentone. The Mordialloc section is backed by an extensive foreshore reserve, that has a number of facilities including the pier, a large car park, park and picnic areas. It also has a lovely, curving promenade running behind the beach from the pier to the lifesaving club and beyond. The Mordialloc Life Saving Club was founded in 1919. Today it is fronted by fenced dunes and a wide beach. The good rail and road access, car parking and other facilities, make this a very popular beach.

The Mordialloc section is 1 km long, faces the south-west and extends to the north of the 300 m long Mordialloc pier and adjacent drain entrance wall. The beach has been nourished to help combat erosion toward the Parkdale and Mentone ends, and can be more than 50 m wide. It is fronted by two bars: the inner bar usually attached to the beach, with rips cutting it every 150 m; with a deeper trough separating it from the outer bar.

Swimming: Safest on the attached, shallow sections of the inner bar and clear of the rip holes and troughs, particularly in the patrolled area. **Warning**: Do not jump off the pier, as there is shallow water underneath and serious injuries can result.

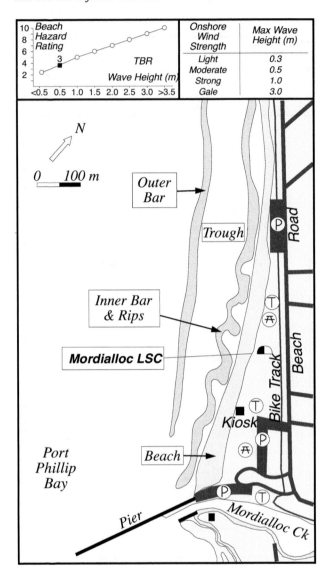

Onshore Wind Strength	Max Wave Height (m)
Light	0.3
Moderate	0.5
Strong	1.0
Gale	3.0

Figure 91. Mordialloc Beach map, showing Mordialloc Life Saving Club and Pier, and the approximate shape and location of the two bars. During calms, waves and currents are usually absent, except near Mordialloc Creek. However during strong winds, high waves will break over the bars and produce strong currents, as well as modifying the shape of the inner bar and rips.

Surfing: During strong westerlies, wind waves up to 2 m high break over the outer bars, providing sloppy beach breaks.

Fishing: The pier is the most popular spot, giving access to deeper water, while the adjoining breakwater is used to fish Mordialloc Creek.

Summary: A well developed and popular beach, with the added safety of the lifesaving club and the attraction of the pier.

P45B   PARKDALE

Parkdale LSC

Inner bar: LTT/TBR   Outer bar: RBB/LBT
Beach Hazard Rating: **3**
Length: 1 500 m

For map of beach see Figure 92

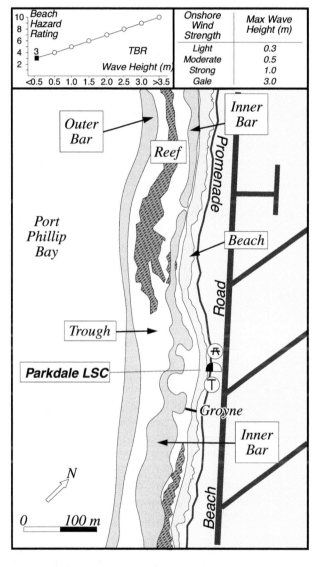

	Onshore Wind Strength	Max Wave Height (m)
	Light	0.3
	Moderate	0.5
	Strong	1.0
	Gale	3.0

Figure 92.   Parkdale Beach map, showing Parkdale Life Saving Club and the approximate shape and location of the two bars. The numerous reefs and rocks produce more rips, which result in a disjointed inner bar, while the outer bar is more continuous. During calms, waves and currents are usually absent. However during strong winds, high waves will break over the bars and produce strong currents, as well as modifying the shape of the inner bar and rips.

Parkdale Beach is located between Mordialloc and Mentone Beaches and 1 km south of the Parkdale railway station. Beach Road runs along the top of the bluff that backs the beach, the road providing most of the parking for the 1.5 km long beach. The beach has been severely eroded in the past, exposing the bluffs and rocks. As a result, the beach was renourished in the 1970s and 80s.

The Parkdale Life Saving Club was founded in 1922 and is located just off Beach Road. It stands on the edge of the 10 m high bluff, with a bike and walking path running along the base of the bluff. This path is part of the seawall, built in the 1960s to protect the eroding bluff. Also a small groyne is located just south of the club, while rocks are exposed to the north, providing a 'pocket' of sand on either side of the club house. The Parkdale Yacht Club is located 400 m north of the rocks.

The beach faces the south-west and is exposed to south-west to westerly winds and waves. These produce a beach of variable width, fronted by two bars. The shape of the beach and inner bar are also influenced by the presence of the bluff, and rocks and reef in the surf zone. The result of this is the formation of more rips and a more disjointed bar. The outer bar is more continuous.

Swimming: It is safest to stay in the pocket of sand below the lifesaving club, under the watchful eye of the lifesavers. Avoid the rocky areas, as deeper rip holes form next to the rocks and reefs; stay on the shallower, attached bars.

Surfing: During strong westerlies, wind waves up to 2 m high break over the outer bars, providing sloppy beach breaks.

Fishing: A popular section of beach owing to the easy parking and access.   There are good views and numerous holes next to the rocky sections.

Summary: The beach at the club house is 'small' and offers a change from the adjacent longer beaches, plus the facilities of the lifesaving club.

P45C   MENTONE

Mentone LSC

Inner bar: LTT/TBR   Outer bar: RBB/LBT
Beach Hazard Rating: **3**
Length: 1 600 m

For map of beach see Figure 93

Mentone Beach forms the northern boundary of the 20 km long stretch of beach that extends from Frankston to the bluffs at Mentone. The beach is backed by Beach Road and 20 m high bluffs. The once bare and eroding bluffs were the subject of some famous paintings by Charles Condor and Tom Roberts. Today the bluffs have been stabilised and vegetated, and a seawall runs along their base, topped by a promenade. In addition, the beach was nourished with sand in the 1970s and 80s, and several rock groynes were placed across the beach.

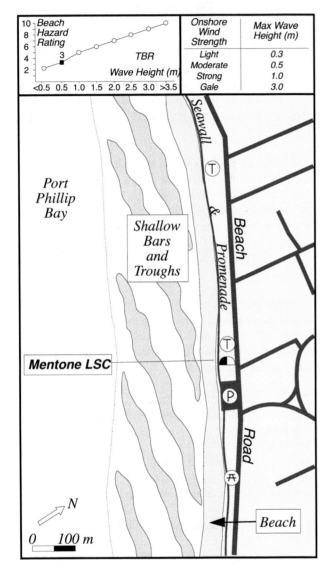

Figure 93.   *Mentone Beach map, showing Mentone Life Saving Club and the approximate shape of the shallow, transverse bars. During calms, currents, waves and rips are usually absent. However, strong winds and high waves will break over the bars and produce currents, as well as modifying the shape of the inner bar and rips.*

The beach in this section is 1.5 km long and curves around to face the south-west. The orientation, and Table Rock Point to the west, afford some protection from west and north-westerly wind and waves. As a result, the parallel, double bar system that has run all the way from Frankston transforms into a 200 m wide series of shallow, transverse bars. These cause large protrusions of the shoreline where they join the beach. At low tide, even low waves can be observed breaking on the many bars. In addition, the rock groynes cause changes in beach orientation, resulting in, at times, a very undulating shoreline. The Mentone Life Saving Club was formed in 1921. It is located in front of the bluffs, with a good view of the beach.

Swimming: A relatively safe beach owing to the generally shallow bars. However, numerous small rips and troughs also occur along the beach and on the bars, so it is best to bathe in the patrolled area in front of the lifesaving club.

Surfing: During strong westerlies, wind waves up to 2 m high break over the outer bars, providing sloppy beach breaks.

Fishing: The groynes are the most popular spot.

Summary: An easily accessible and popular beach offering usually low waves and a shallow surf zone.

P46   **BEAUMARIS**

Beaumaris LSC
Single bar: LTT/TBR + reefs
Beach Hazard Rating: **3**
Length: 650 m
For map of beach see Figure 94

Beaumaris Beach is a strip of sand located below the 20 m high bluffs of Table Rock Point, and surrounded by wide, intertidal rock reefs and flats. The beach has good vehicle access off Beach Road, with the road leading to the Beaumaris Yacht Club and a large picnic area. It then swings around to the lifesaving club and its large car park and picnic area.

The Beaumaris Life Saving Club was founded in 1956 and, apart from a few boat sheds, is the only structure behind the beach.

The beach is 650 m long and faces the south. The eastern end is backed by the bluff and fronted by reef flats. Only the section immediately east of the club house is free of reef, as extensive reefs surround the beach in front of the car park. Depending on the tide, these reefs will be covered or bare. Rip holes usually occur next to the reefs.

Swimming: Best at high tide on the bar in the patrolled area, just east of the club. Avoid the rip holes and reefs.

Fishing: A popular spot owing to the extensive reefs and rip holes.

Summary: An attractive, relatively natural beach with good access and parking, however there is as much reef flat as sand here, so watch where you bathe.

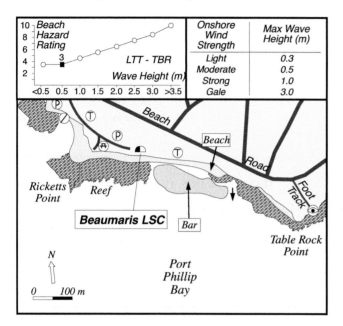

Figure 94. Beaumaris Beach is located below the bluffs of Table Rock Point, and is surrounded by extensive intertidal rock reefs and flats. It is easily accessible off Beach Road, with good parking and a picnic area. Bathe on the shallow bar and watch for the rip holes and reef flats. Strong onshore winds and waves produce rips and hazardous conditions at this beach.

P47, P48, P49   BEAUMARIS YACHT CLUB, BEAUMARIS NORTH, QUIET CORNER

Unpatrolled			
	Rating	Single bar	Length
P47 Beaumaris Yacht Club	3	LTT/TBR + rocks & reefs	700 m
P48 Beaumaris North	3	LTT + rocks & reefs	800 m
P49 Quiet Corner	3	R + rocks & reefs	100 m
		Total length	1 600 m

North of Ricketts Point, the coast trends north-west for 2 km to Quiet Corner. It is dominated by 10 to 20 m high, vegetated bluffs, fronted by extensive rocks and reefs. Between the bluffs and rocks are three narrow, crenulate beaches. **Beaumaris Yacht Club Beach** is 700 m long, with sand, then increasing reef, to the north.

Beaumaris North Beach is 800 m long and dominated by reefs throughout, while 100 m long **Quiet Corner** is hemmed in between two bluffs, with a seawall and walkway along the base of the bluffs. Beach Road runs along the top of the bluffs and parallels the beaches, however vehicle access is restricted to the yacht club, where there is a large car park and picnic area. Elsewhere, parking is on Beach Road, with a climb down the bluffs required to reach North Beach and Quiet Corner.

The three beaches face south-west, exposing them to periodic strong, south-west winds and waves. The waves break heavily over the reefs and only at Beaumaris Yacht Club Beach is there sufficient sand to build a generally attached bar, with rip channels running along the reefs.

Swimming: Best at high tide when the rocks and reefs are covered. Dangerous when there are wind waves, owing to the dominance of reefs and rip currents.

Surfing: Only during strong south-westerlies and at high tide over the reefs.

Fishing: There are numerous gutters in the reefs along this section, with best conditions at high tide.

Summary: Three relatively natural, narrow beaches, with a strip of sand and an increasingly rock and reef dominated surf.

P50, P51   BLACK ROCK, BLACK ROCK POINT

Black Rock Life Saving Club			
	Rating	Single bar	Length
P50 Black Rock	3	R + reefs	750 m
P51 Black Rock Point	3	R + reef	120 m
		Total length	870 m
For map of beach see Figure 95			

Black Rock Beach is a relatively narrow beach lying below 20 m high, vegetated bluffs. Beach Road runs along the top of the bluffs and the Black Rock Life Saving Club, founded in 1913, sits on top of the bluffs, with a good view of the beach. A car park and a picnic area are located on the bluffs just south of the club house.

The main beach is 750 m long and is composed of medium to coarse sand, which produces a steep beach face and usually no bar. As a result, deep water lies immediately off the beach. The beach narrows to the south, where the bluffs were stabilised in the 1930s with the construction of a seawall and walkway. The eroding bluffs left sandstone reefs off the beach, on which higher waves break, particularly at low tide. The beach faces south-west and is exposed to seasonal shifts in the wave climate. Summer southerlies tend to move the sand up the beach, while winter westerlies shift it back to the south. To the north, reefs increase toward Black Rock Point, with a bluff separating the main beach from the smaller, reef dominated **Point Beach**.

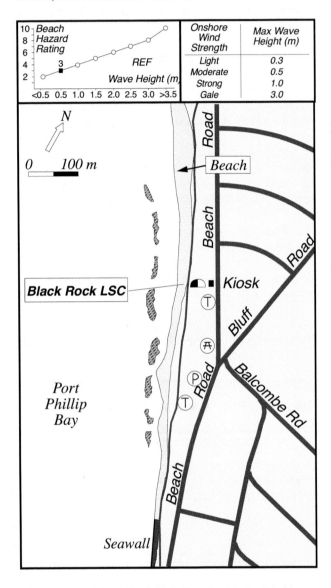

Figure 95. Black Rock Beach is located adjacent to Beach Road and the Black Rock shops, and below the 20 m high bluffs. It is a steep beach, with deep water close inshore. Relatively safe and popular in light winds, however waves will break over the patchy reefs that parallel the beach. High waves will also generate strong longshore currents against the beach.

Swimming: An exposed beach, with a steep beach face and deep water off the beach. This makes it good for swimming, but a hazard for young children and non-swimmers. Also watch the patchy reefs off the beach, and longshore currents if waves are breaking. The Point

Beach is only suitable at high tide, as the reefs are exposed at low tide.

Surfing: During strong westerly winds and accompanying waves, there is surf over the reefs off the beach.

Fishing: A popular spot for beach fishing with deep water and reefs off the beach, as well as rocks to either end.

Summary: A popular summer beach next to the main Beach Road, but shielded by the vegetated bluff, with the protection of the lifesaving club and shops across the road in Black Rock.

P52   HALF MOON BAY

Half Moon Bay SLSC
Single bar: TBR
Beach Hazard Rating: **3**
Length: 350 m
For map of beach see Figure 96

Half Moon Bay receives its name from its crescentic shape. The 350 m long bay faces north in the southern corner, then swings around to face west along the northern section. Due to the partially protected nature of the bay, it has long been a site for boating. It has a boat launching ramp, a 100 m long jetty, and off the jetty are the remains of HMVS Cerberus which was grounded in 1926, to provide additional shelter for the boats.

The beach is relatively protected in the southern corner and extensive sand shoals lie off the beach, necessitating the need for the long jetty. The bars narrow but continue up the beach, where they may be cut by deeper rip channels. High waves will produce strong currents in the rips and against the northern rocks at Red Bluff.

The Half Moon Bay Surf Life Saving Club was founded in 1910 and is the oldest in the state. It is located in the southern corner of the beach, next to a few boat sheds and the Black Rock Yacht Club. There is extensive parking for both clubs and the boat ramp on the point, with additional parking on the northern bluffs, as well as walkways down to the beach.

Swimming: A relatively safe beach in the southern corner, with deeper water and more rip channels toward the north. Best to stay close inshore in the patrolled area, just north of the surf club, as there is an average of 18 rescues a year at the beach.

Surfing: Waves break over the northern bars and reefs during strong westerly winds and accompanying waves.

Fishing: The pier is the most popular location, with the best beach fishing toward the northern end.

Summary: This beach has a very active and accessible southern corner, with a more natural and quieter northern section. Access and parking are good at both ends. Care should be taken if near the remains of the Cerberus, as it is deteriorating badly.

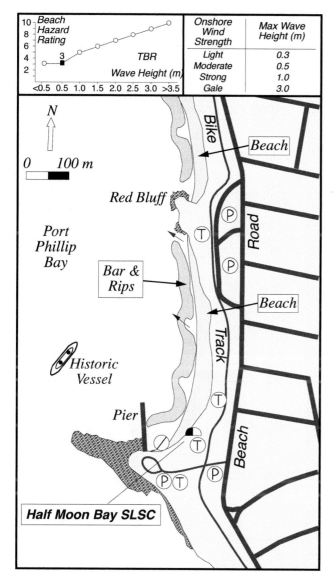

Figure 96.   Half Moon Bay is a partially protected, curving beach, with a full range of boating, yachting and bathing activities at the southern end, and a more natural beach toward the north. High waves will induce rip currents, especially at the northern end.

P53   SANDRINGHAM

Sandringham LSC
Single bar: TBR/RBB
Beach Hazard Rating: **4**
Length: 2 500 m
For map of beach see Figure 97

Sandringham Beach is a straight, 2.5 km long beach that faces the south-west. The southern half is backed by steep bluffs, which rise to 30 m at Red Bluff. This section has been eroding and is often narrow. Some seawalls and a groyne have been constructed to manage both the beach and bluff erosion. The eroding bluffs have also left shallow reefs in the surf zone.

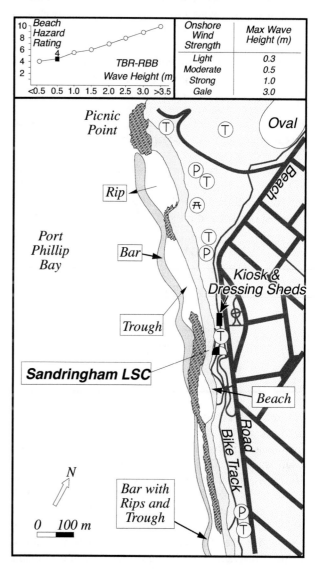

Figure 97.   Sandringham Beach is long and straight, however the rocks, reefs and waves combine to produce a highly crenulate shoreline, alternating between bars and rips. The northern patrolled section has the best access and amenities.

Beach Road parallels the beach. It is fronted by the bluffs in the south, with the best access between the Sandringham Life Saving Club and Picnic Point. As the name suggests, there is a foreshore reserve along this section with parking and picnic facilities, as well as the Brighton Amateur Swimming Club. The lifesaving club was formed in 1922.

The beach is exposed to westerly winds and waves and these, together with the numerous reefs and rocks, have produced a crenulate beach of varying width. A single bar attaches to the beach in places, and elsewhere is cut by deeper rip channels and longshore troughs. The deeper channels are more prevalent next to the rocks and reefs.

Swimming: Due to the high degree of variability in the nature of the beach, bar and surf, it is safer to bathe in front of the lifesaving club. Stay on the shallower bars and clear of the rocks and reefs.

Surfing: Usually calm, however strong south-westerly winds and waves will produce a break over the bar and reefs.

Fishing: The rocks, reefs and groyne all provide access to deeper water and rip holes.

Summary: A beach of two parts: a narrow, difficult to access southern half; and a very accessible and well-serviced northern half. Naturally the northern section is the most popular and is patrolled by the lifesaving club.

P54 HAMPTON

Hampton LSC
Single bar: LTT
Beach Hazard Rating: **3**
Length: 900 m
For map of beach see Figure 98

Hampton Beach was once part of a continuous strip of sand running from Picnic Point to Green Point, including Brighton Beach. The Hampton Life Saving Club was formed in 1913, to patrol the southern end. However, construction of the boat harbour in the early 1950s trapped sand from the beach and erosion resulted. Beach nourishment was attempted in the 1970s, with little success; and the New Street groyne was built in 1986. This has trapped sand on the northern side and helped Brighton Beach, however Hampton remains virtually a shadow of its former self.

Today the 900 m long beach is broad, but stagnant in the boat harbour. It consists of a few patches of sand at the

Hampton Life Saving Club, and only a seawall, rocks and groyne up to the New Street groyne.

Swimming: The safest place is on the patch of sand in front of the lifesaving club, but even here, keep away from the rocks and groynes.

Fishing: The groynes and seawall may not have saved the beach, but they have provided a good fishing spot, and access to relatively deep water.

Summary: The southern boat harbour is perfect for boating, however the beach has been heavily degraded by the lack of sand and numerous rock structures.

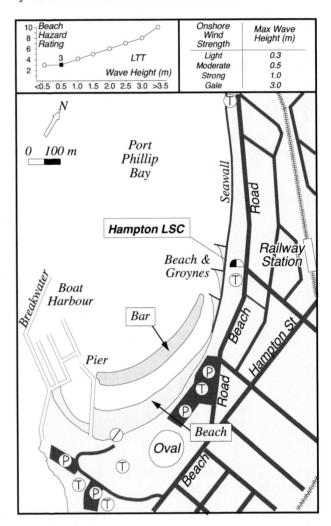

Figure 98. Hampton Boat Harbour dominates Hampton Beach and has resulted in the erosion and loss of much of the beach. The boat harbour breakwater, jetties and pier dominate the southern corner, while seawalls, groynes and rocks occupy the rest.

P55   BRIGHTON

Unpatrolled
Single bar: TBR
Beach Hazard Rating: **3**
Length: 1 000 m

Brighton Beach is a 1 km long, south-west facing beach, lying between the New Street groyne and Green Point. It is paralleled by a seawall and walkway, foreshore reserve and Beach Road. It is backed by Brighton shops and the railway station. The reserve has a large parking area and picnic facilities.

The beach is exposed to strong south-west winds and waves and, when they occur, they maintain a 60 to 100 m wide bar. This is either separated from the beach by a trough and/or cut by rip channels. In addition, rocks outcrop on parts of the beach, with more extensive reefs toward Green Point.

Swimming: Relatively safe when waves are low. However, water may be deep off the beach and there are rocks and reefs in places. During periods of strong winds and waves, longshore and rip currents are generated.

Surfing: Strong south-west winds will produce surf over the bars and reefs at high tide.

Fishing: There are usually good rip holes against the rocks and often a trough along the beach, plus the southern groyne and rocks at Green Point to fish from.

Summary: A very accessible and popular beach with foreshore facilities. However, be aware of the often deep water and reefs.

P56, P57   HOLLOWAY, DENDY STREET

Unpatrolled			
		Rating	Single bar	Length
P56	Holloway	2	R	200 m
P57	Dendy Street	3	LTT	500 m
			Total length	700 m
For map of beaches see Figure 99				

At Green Point, the low rocks and reefs have been encased in a 300 m long seawall, which is backed by Brighton Beach Gardens. On the north side of the wall are Holloway and Dendy Street Beaches, both tied to cuspate forelands in lee of more prominent rock reefs. **Holloway Beach** curves around for just 200 m and is backed by a reserve and scout hall. Pulses of sand move along Green Point and onto Holloway, often forming a small lagoon between the sand pulse and the beach.

Dendy Street Beach is 500 m long and tied to two reefs, with the famous series of colourful boat sheds lining most of the beach. The Esplanade parallels both beaches and there are large car parks in the Brighton Gardens, and at the northern end of Dendy Street.

The two beaches face due west and are exposed to strong westerly winds and waves. However, extensive reefs lie off both beaches, resulting in deeper rip channels against the reefs, and only a bar along the northern section of Dendy Street Beach.

Swimming: Both beaches are relatively safe under the usual low waves, with the inner section of Holloway and northern Dendy offering a sandy bar, while rocks and reefs occur off the remainder of both beaches.

Surfing: Dendy Street has waves over the bar during very strong westerly winds.

Fishing: The southern seawall and the reef points are the best spots to fish the adjacent deeper rip channels.

Summary: Dendy Street, with its boat sheds, is one of Melbourne's most photographed beaches. It is also very popular, with good parking and access.

P58   MIDDLE BRIGHTON

Brighton LSC
Single bar: LTT/TBR
Beach Hazard Rating: **3**
Length: 500 m
For map of beach see Figure 99

Middle Brighton Beach is the home of the Brighton Life Saving Club, which was formed in 1917. Unlike its boat shed lined neighbour (Dendy Street), Middle Brighton has a seawall and promenade running the length of the 500 m long beach. The beach is backed by The Esplanade, with a large car park next to the lifesaving club at the southern end of the beach.

The beach is bordered by rock flats and reefs at each end. In addition, a groyne has been built across the northern end, while the seawall and low grassy bluffs back the beach. The beach tends to widen to the north, while the surf zone alternates between the reefs and an irregular, shallow, usually attached bar.

Swimming: It's best to go a little north of the club house, away from the reefs. However, stay in the patrolled area.

Surfing: During strong south-west winds and waves, there are waves over the scattered reefs and bar.

Fishing: The northern groyne offers the best access to deeper water and the adjoining reefs.

Summary: A very accessible patrolled beach right next to the main road, with a promenade running the length of the beach. Best bathing is toward the centre, away from the rocks and reefs at each end.

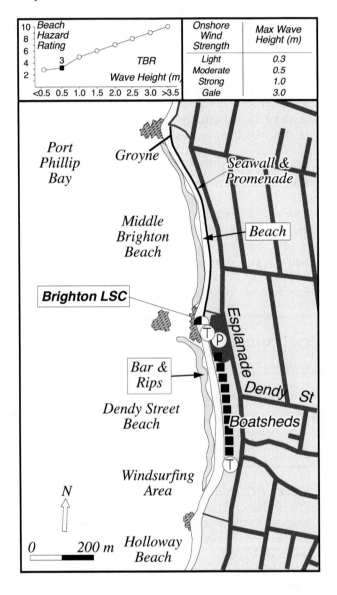

Figure 99. Middle Brighton and Dendy Street Beaches are very similar in size and shape. However, whereas Dendy Street has the photogenic row of colourful boat sheds, Middle Brighton has a low seawall and promenade. The Brighton Life Saving Club is located between the two beaches.

P59, P60 MIDDLE BRIGHTON BATHS, SANDON STREET

Unpatrolled				
		Rating	Single bar	Length
P59	Middle Brighton Baths	1	R/LTT	200 m
P60	Sandon Street	1	R + sand flats	500 m
			Total length	700 m

Past Middle Brighton Beach, the coast becomes increasingly dominated by artificial structures. A groyne forms the northern boundary of Middle Brighton, beyond which the beach has receded and a seawall has been built. This continues on to the large Middle Brighton pier and breakwater, in lee of which is Sandon Street Beach. Two rock groynes also cross the northern end of Sandon Street Beach.

The **Middle Brighton Baths** and adjacent yacht club both have steep, narrow strips of sand; remnants of once wider beaches. **Sandon Street Beach**, on the other hand, is protected by the breakwater, and the shoreline has built out up to 100 m in its lee, while erosion to either end has exposed the seawall.

Both beaches face west and are exposed to westerly winds and waves, however waves are reduced by the structures. The beaches are usually calm, or have a low shorebreak when waves reach the beach.

Swimming: The Baths are probably the safest place to bathe in Port Phillip, while Sandon Street is usually calm in lee of the breakwater. The best bathing is at high tide, as sand flats are exposed at low tide.

Surfing: Waves are surfed next to the Baths in big south-westerlies.

Fishing: The jetty and breakwater are the best spots, with the exposed section of seawall also giving access to deeper water.

Summary: Two heavily modified beaches: one now used for pool bathing and sailing; the other enclosed by the pier and breakwater, with boats moored offshore.

P61 ELWOOD

Elwood LSC
Inner bar: TBR   Outer bar: RBB
Beach Hazard Rating: **3**
Length: 1 300 m
For map of beach see Figure 100

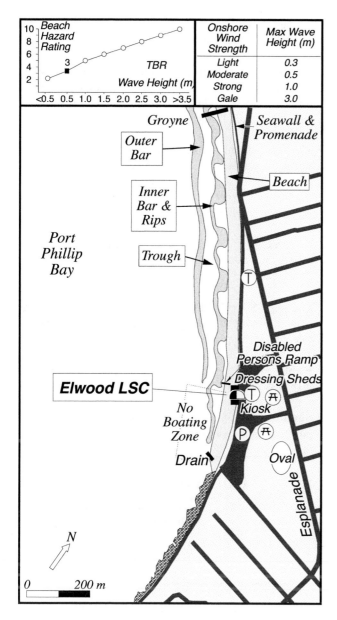

	Beach Hazard Rating					Onshore Wind Strength	Max Wave Height (m)
						Light	0.3
			3	TBR		Moderate	0.5
				Wave Height (m)		Strong	1.0
	<0.5 0.5 1.0 1.5 2.0 2.5 3.0 >3.5					Gale	3.0

Figure 100. Elwood Beach is backed by numerous amenities and a large car park. A large amenities block houses the lifesaving club, kiosk and dressing sheds. The beach usually has an inner bar with alternating, shallow bars and deeper rip holes. The rips are only active during breaking waves.

Elwood Beach (and the backing foreshore reserve) is a popular area used for a range of activities. Besides the Elwood Life Saving Club, founded in 1911, the beach is the site of the Elwood Bowling Club and Sea Scouts and Sailing Club. Extensive car parking and a park, picnic area and oval back the beach. The lifesaving club is located in a modern multi-purpose building, and a seawall and promenade run the length of the low beach. A launching ramp for the sailing club and a disabled access ramp cross the beach just north of the club house.

The beach itself is 1300 m long, extending from the Head Street diversion drain up to a groyne on Point

Ormond. It faces the south-west and receives sufficient waves to produce a double bar system. The inner bar alternates between a shallow, attached section and deeper rip channels, with a trough separating it from the rhythmic outer bar. Low waves break on the inner bar, particularly at low tide, while strong winds and higher waves are required to activate the outer bar, during which time the rip currents intensify.

Swimming: The safest location is on the shallow sections of the inner bar, away from the rip channels and in front of the lifesaving club, where boats are prohibited.

Surfing: Strong southerly winds will produce waves high enough to break over the bars.

Fishing: The southern drain and northern groyne provide access to deeper water, while the beach usually has several deeper rip holes.

Summary: A very accessible beach that has been highly, but attractively, developed for a range of recreational activities, including water access for the disabled. It also has good parking and facilities. Wind surfing is very popular off Point Ormond.

P62   ST KILDA MARINA

Unpatrolled
Single bar: LTT
Beach Hazard Rating: **2**
Length: 150 m

Out towards the end of the large St Kilda Marina is an artificial beach, that has formed since the construction of the marina. It is just 150m long and located between two rock groynes, and is backed by a seawall. It faces south-west and is well exposed to southerly winds and waves. These have built a continuous, narrow bar.

Swimming: A relatively safe, if unattractive, beach, surrounded by seawalls and groynes.

Surfing: Only during very strong southerlies, and then only a short beach break.

Fishing: There are excellent fishing spots off the groynes.

Summary: A small, very accessible, but unattractive artificial beach.

P63   ST KILDA

St Kilda LSC

Single bar: LTT/TBR
Beach Hazard Rating: **3**
Length: 650 m

For map of beach see Figure 101

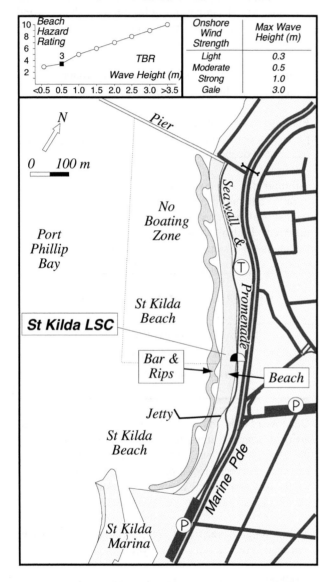

Figure 101. St Kilda Beach lies between the large marina and the pier, with a jetty crossing the southern end. The patrolled beach usually has a single bar, cut by deeper rip holes. The rips are only active when waves are breaking over the bar.

St Kilda Beach has been modified by foreshore development. Today the main beach is 650 m long. The beach abuts the side of St Kilda Marina in the south, with Brooks Jetty also crossing the southern end, while the St Kilda Pier and breakwater form the northern boundary. As a result of these structures, the beach has widened at each end.

The beach is backed by the busy Esplanade, however parking is limited. The St Kilda Life Saving Club was formed in 1914. The two-storey club house is located next to the main beach access point.

The beach faces the south-west and receives sufficient storm waves to build a single bar, consisting of predominantly shallow, attached sections, cut by deeper rip channels every 100 m. The rips are inactive unless waves are breaking over the bar.

Swimming: The safest section is from the lifesaving club to the northern pier, which is also a No Boating Zone. Stay clear of the southern pier and breakwater.

Surfing: Strong south-west winds will produce large enough waves to break over the bars, producing a rideable, if sloppy, surf.

Fishing: The two piers are very popular locations and give easy access to deeper water.

Summary: A popular beach with locals and visitors, however parking and amenities are limited.

P64A   ST KILDA PIER

Unpatrolled

No bar: R + sand flats
Beach Hazard Rating: **1**
Length: 1 000 m

In lee of 1 km long **St Kilda Pier** and breakwater is a protected beach. The beach has built out over 100 m in lee of the breakwater, at the expense of the southern section, which has eroded back to the seawall. The beach is backed by a continuous seawall and walkway, with Catani Gardens behind the southern half. This is a protected and usually calm beach. As a result, sand flats extend out from the shore.

Surf Life Saving Victoria's State Centre is located here on Beaconsfield Rd.

Swimming: Better at high tide when the sand flats are covered.

Surfing: None.

Fishing: Most fishing is done from the pier, breakwater and along the exposed parts of the seawall.

Summary: A wide, sheltered beach.

P64B   MIDDLE PARK

> ### Middle Park LSC
>
> Single bar: LTT/TBR
> Beach Hazard Rating: **3**
> Length: 1 000 m
>
> *For map of beach see Figure 102*

Middle Park is part of a 4 km long beach that runs from the Cowdroy Street drain to the large Station Street Pier. There are three lifesaving clubs on the beach: at Middle Park, South Melbourne and Port Melbourne; together with the Kerferd Road Pier and a small boat harbour.

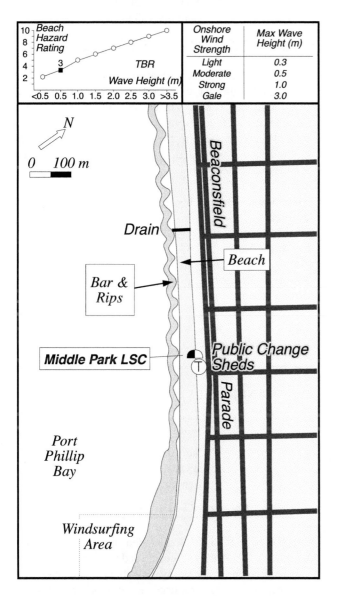

Figure 102. Middle Park Beach and Life Saving Club are located off the busy Beaconsfield Parade. Parking is available on the parade. The beach usually has an attached bar, often cut by small rip holes.

The Middle Park section is a straight, 1 km long, south-west facing beach, backed by a low seawall, promenade and busy Beaconsfield Parade. Access is available along the length of the beach, with parking limited to the road. The Middle Park Life Saving Club was founded in 1913. It also houses a kiosk and dressing rooms.

The beach is exposed to any south through westerly winds and waves, and these usually build a 50 m wide beach, with a moderate slope and a single bar about 50 m wide. The bar is predominantly shallow and attached to the beach, with rip channels sometimes cutting across the bar every 100 m. The rips are only active when waves are breaking across the bar.

Swimming: The lifesaving club patrols the beach in front of the club house and this is the best place to bathe. Avoid the rip channels and stay on the shallow bars.

Surfing: During very strong south-westerlies there is surf along this beach, with the Kerferd Street Pier being the most popular location.

Fishing: Beach fishing is best when there are rip holes, otherwise the drain north of the club house and Kerferd Road Pier are used to access deeper water.

Summary: A readily accessible beach, with amenities at the lifesaving club. This is a popular windsurfing area.

P64C, D   SOUTH MELBOURNE, PORT MELBOURNE

> ### South Melbourne SLSC
> ### Port Melbourne LSC
>
> | | | Rating | Single bar | Length |
> |---|---|---|---|---|
> | P64C | South Melbourne | 3 | TBR | 800 m |
> | P64D | Port Melbourne | 2 | TBR/LTT | 300 m |
> | | | Total length | | 1 100 m |
>
> *For map of beaches see Figure 103*

South Melbourne and **Port Melbourne** Life Saving Clubs patrol a 1.5 km section of beach between the Kerferd Road and Lagoon piers. The reason for having two clubs is that the boundary of South and Port Melbourne Shires crosses the beach midway between each club. The low beach is backed by Beaconsfield Parade, as well as a low seawall and promenade. While the South Melbourne Surf Club was established in 1927, the Port Melbourne Life Saving Club was formed in 1913, but it now occupies a new building. Both clubs incorporate dressing rooms and kiosks.

The beach faces the south-south-west, and can receive moderate waves during strong southerly winds. When these occur, they maintain a 100 m wide beach fronted by a 50 m wide bar, that is usually attached to the beach, with occasional rip channels. The rips are only active when waves are breaking over the bar.

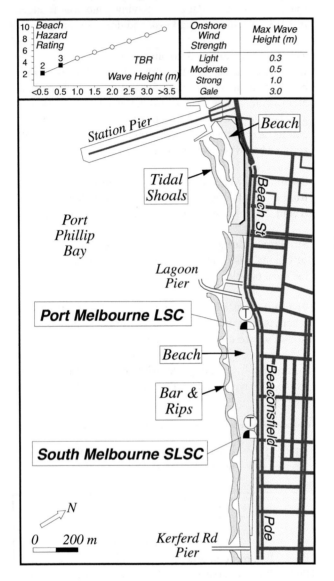

Figure 103. *South Melbourne and Port Melbourne Beaches are part of one continuous beach, bordered by the Kerferd Road and Lagoon Piers. The beach is patrolled by the two clubs. It usually consists of a continuous, attached bar, cut by small rip channels. The rips are only active when waves are breaking over the bars. Beyond Lagoon Pier, waves are lower and sand flats front Station Pier Beach.*

Swimming: A relatively safe beach when waves are low, still it is best to stay on the bar and in the patrolled area in front of each club, as there is an average of 5 rescues each year.

Surfing: There is some surf during very strong southerly winds along these two beaches, with waves breaking over the bars at high tide.

Fishing: The piers at each end provide the best location for bay fishing, with easy access to deeper water.

Summary: These are the two closest patrolled beaches to Melbourne city. However, the most accessible parking is restricted to Beaconsfield Parade, and hence these beaches are more popular with locals than visitors.

P65   STATION PIER

Unpatrolled
No bar: R + sand flats
Beach Hazard Rating: **1**
Length: 700 m
For map of beach see Figure 103

Wedged in between the Port Melbourne Yacht Club and the large Station Pier, and backed by busy Beach Street, is a 700 m long beach. The eastern end of **Station Pier Beach** narrows to an exposed seawall, while it broadens toward the pier. Due to the relatively protected nature of the beach it is fronted by irregular sand flats, which are exposed at low tide.

Swimming: Best at high tide when the sand flats are covered.

Surfing: None.

Fishing: Better off the adjacent piers, as it is shallow off the beach.

Summary: A heavily modified beach, surrounded by a number of structures.

P66   SANDRIDGE

Sandridge LSC
No bar: R + sand flats
Beach Hazard Rating: **2**
Length: 500 m
For map of beach see Figure 104

Sandridge Beach is surrounded by piers and port facilities. However, recent redevelopment of the surrounding area, including new beach amenities and landscaping, have produced a reasonably attractive beach. The beach is very accessible, with good parking. The Sandridge Life Saving Club, founded in 1927, now has a relatively new club house, that houses dressing rooms and a kiosk. A seawall backs the beach with a promenade running the length of the beach.

Construction of Webb Dock to the west has resulted in realignment of the beach. It has been built out at the western end, but eroded toward the east. Today the beach is 500 m long, and faces the south-south-east. It is a relatively protected location, however strong southerly winds generate enough wave activity to build a shallow, 50 m wide sand flat, which is cut by some channels. Surf currents are usually inactive, unless a strong southerly wind and resulting waves are present.

Swimming: A relatively shallow and safe beach. However, seaweed accumulation on the beach and shipping pollution can be a problem when present.

Surfing: None.

Fishing: It is usually shallow off the beach, with the adjoining seawall and breakwaters offering the best access to deeper water.

Summary: A heavily modified beach, that has been recently rejuvenated and is now a reasonably attractive place for a picnic and a swim in harbour waters.

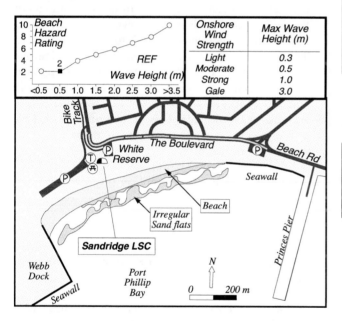

Figure 104. Sandridge Beach is located in the heart of Port Melbourne and is bordered by Princes Pier and Webb Dock. The beach reserve has, however, been redeveloped and provides good access, parking and beach amenities.

P67   WEBB DOCK

Unpatrolled
No bar: R
Beach Hazard Rating: **2**
Length: 250 m

Webb Dock is part of the Port of Melbourne. Deep inside the dock, a 250 m long sand beach has formed. The beach is narrow and relatively steep, with deep dredged water off the beach. As the beach is located in the middle of a major port, public access to the area is restricted.

West Port Phillip Bay Beaches
The 65 beaches on the western side of Port Phillip Bay are, for the most part, very low wave beaches. Most of them face east and the prevailing westerlies, that are required to build waves in the bay, blow offshore. As a result, many of these beaches consist of narrow, high tide beaches backed by low or non-existent foredunes. They are fronted by shallow, intertidal sand or rock flats of varying width.

P68   POINT GELLIBRAND

Unpatrolled
No bar: R + rocks
Beach Hazard Rating: **2**
Length: 500 m

Point Gellibrand is a low, basaltic point in lee of which is a narrow, high tide, sand beach fronted by basalt boulders and sand flats. A road parallels the back of the 500 m long, crenulate beach, and historic Fort Gellibrand occupies the backing point.

The beach faces south-east and consequently (coupled with the point and rocks) receives very low waves. As a result, there are sand flats amongst the rocks.

Swimming: Best at high tide when the rocks and sand flats are covered.

Surfing: None.

Fishing: There is shallow water right along the beach, even off the rocks.

Summary: A narrow, winding beach dominated by basalt rock flats and boulders.

P69 WILLIAMSTOWN

> **Williamstown Swimming and Surf Life Saving Club**
>
> Single bar: R + sand flats
> Beach Hazard Rating: **2**
> Length: 550 m
>
> *For map of beach see Figure 105*

Williamstown Beach is located just south of the Williamstown Railway Station. It is also backed by the Esplanade, making it a very accessible beach. A foreshore reserve is located at the eastern end and contains the Anglers Club, the pier, a car park, a park and a picnic area. The Williamstown Swimming and Surf Life Saving Club, formed in 1922, is located at the western end next to the Baths.

The beach is 550 m long and faces almost due south. It receives waves during strong southerlies. These waves have built a wide, sandy beach fronted by an attached bar that widens to the west. At low tide, it can be a 100 m wade before you can swim off the bar.

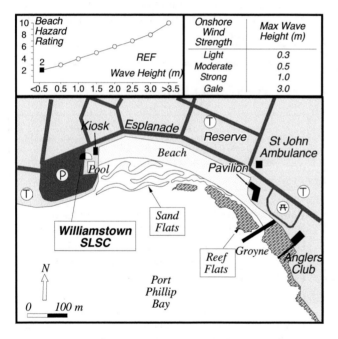

Figure 105. Williamstown has a nice, wide, sandy beach, fronted by a shallow bar and sand flat, as well as a number of facilities at either end of the beach. Water depth over the bar varies with the tide.

Swimming: Safest at the western end, where the surf club patrols the beach and the bar is relatively shallow. There are many rocks at the Pavilion end of the beach, and the holes off the end of the rock groyne are a danger to non-swimmers.

Surfing: Only during south-westerly winds and waves.

Fishing: The eastern pier and rocks to either end are the best locations, providing access to deeper water.

Summary: An attractive, well laid-out beach and reserve, with all the usual public amenities, except for a lack of public toilet facilities.

P70, P71 ALTONA (BURNS RESERVE), ALTONA (CRESSER)

Unpatrolled			
	Rating	No bar	Length
P70 Altona (Burns Reserve)	1	R + sand flats	700 m
P71 Altona (Cresser)	1	R + sand flats	400 m
		Total length	1 100 m

The Altona Coastal Park lies at the mouth of Kororoit Creek. Most of the beach is fronted by 500 m wide sand flats, while the backing shoreline is composed of marsh and patches of sand. Only toward the southern end of the park, in the P.A. **Burns Reserve**, does a continuous beach form, still fronted by 300 to 400 m wide flats. The reserve terminates at an open, 30 m wide drainage easement, on the other side of which is **Cresser Park**. The park contains a large car park, boat ramp, yacht club and a small jetty. The narrow beach alternates with seawalls, while the tidal flats deepen off Cresser Park and are covered by seagrass, with numerous boats moored off the beach.

Swimming: Only possible at high tide, and even then better suited for wading than swimming.

Surfing: None.

Fishing: Better at Cresser Park, off the jetty or seawall at high tide.

Summary: Two very low energy beaches, both backed by large foreshore reserves.

P72 ALTONA (MAIN)

> **Altona LSC**
>
> Single bar: R + sand flats
> Beach Hazard Rating: **1**
> Length: 3 100 m
>
> *For map of beach see Figure 106*

Altona Beach forms the southern boundary of the town of Altona, with the Esplanade paralleling the beach.

The Altona Railway Station is located 500 m north of the beach, and the beach is crossed by the 500 m long Altona Pier. The Altona Life Saving Club, formed in 1927, is located just east of the pier and patrols a bathing and No Boating Zone, extending 200 m east of the pier.

The beach runs roughly east-west for 3 km from the point at Frazer Reserve, past the town to the western drainage channel, called Laverton Creek. Most facilities are located near the pier and lifesaving club. There are also picnic areas in foreshore reserves at either end of the beach.

The beach is widest near the pier, but narrows west of the groyne, which is located 100 m past the pier. Intertidal and subtidal sand flats front the beach, requiring a long wade to reach deep water.

Swimming: It is best to bathe in the patrolled area east of the pier. The depth over the sand flats varies with the tide, and is best at high tide for a swim, or low tide for a wade.

Summary: A long, accessible beach, with good facilities at the pier and reserves at each end.

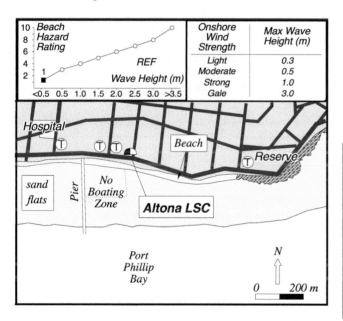

Figure 106. Altona Beach and Pier lie adjacent to the centre of town, with the lifesaving club located just east of the pier. The beach is fronted by wide, shallow sand flats.

P73, P74   SKELETON CREEK, POINT COOK NORTH

Unpatrolled			
	Rating	No bar	Length
P73 Skeleton Creek	1	R + sand flats	1 500 m
P74 Point Cook North	1	R + sand flats	3 000 m
		Total length	4 500 m

Past Laverton Creek drainage channel, the coast continues to the south. It consists of a low, narrow beach (**Skeleton Creek**), backed by low, swampy ground. The northern end is backed by the Altona Explosives Reserve, while the southern half has swampy lakes that drain into Skeleton Creek, whose mouth forms the boundary between the two beaches. Both beaches are fronted by 200 m wide, low gradient, intertidal sand flats. There is presently no public access to these beaches, although they may become part of the adjacent Point Cook Coastal Park.

Swimming: Only possible at high tide when the tidal flats are covered.

Surfing: None.

Fishing: Only over the tidal flats at high tide.

Summary: Two very low energy beaches, presently inaccessible to the public.

POINT COOK COASTAL PARK AND MARINE RESERVE
Coastal Park: Area: approximately 400 ha
Coastal Park and Marine Reserve: Coast length: 3.5 km Beaches: P75 & P76 (2 beaches)
Information: (03) 9395 1198

P75   POINT COOK HOMESTEAD

Unpatrolled
No bar: R + sand & rock flats Beach Hazard Rating: **2** Length: 700 m

Point Cook lends its name to the original homestead built in 1849. It is adjacent to the RAAF base, and to the new Coastal Park. The point is named after John

Cook, who was the mate on the HMS Rattlesnake, which chartered Port Phillip in 1836. Today the original homestead and its buildings are part of the Coastal Park and are open to the public.

Point Cook Homestead Beach lies between the actual Point Cook, a low basalt promontory, and a protrusion 700 m to the south-west. It is a narrow, 10 m wide beach fronted by sand and basalt rock flats up to 200 m wide. The old Homestead Jetty crosses the beach, but is now in ruins.

The beach and adjacent tidal flats are part of the Point Cook Marine Reserve. Line fishing is permitted, but collecting of marine organisms is prohibited.

Swimming: Only possible at high tide on the shallow sand flats. Watch out for the numerous rocks.

Surfing: None.

Fishing: Better at high tide when the sand flats are covered.

Summary: The park and homestead are well worth a visit, and have extensive shady park and picnic facilities.

P76A, B   POINT COOK SOUTH 1 & 2

Unpatrolled			
	Rating	No bar	Length
P76A Point Cook South 1	1	R + sand flats	1 800 m
P76B Point Cook South 2	1	R + sand flats	2 800 m
		Total length	4 600 m

South of the Point Cook Homestead, the coast swings to the west, then south-west. The first beach (**Point Cook South 1**) is part of the Coastal Park and has extensive parking and picnic facilities. The second beach (**Point Cook South 2**) forms the boundary of the Point Cook RAAF base and is off limits to the public.

Both the beaches receive low waves, which build a low, narrow beach fronted by shallow, intertidal sand and rock flats up to 200 m wide. There is a 400 m long jetty extending out from the RAAF base.

Swimming: Only at high tide when there is water over the flats, and even then watch out for the rocks.

Surfing: None.

Fishing: Better at high tide when the flats are covered by water.

Summary: The park provides excellent parking and picnic facilities and good access to the beach and bay.

P77, P78   CAMPBELLS COVE NORTH & SOUTH

Unpatrolled			
	Rating	No bar	Length
P77 Campbells Cove North	1	R + sand flats	1 200 m
P78 Campbells Cove South	1	R + sand flats	700 m
		Total length	1 900 m

Campbells Cove is a crenulate, south-east facing section of low coast, located at the end of Cunninghams Road and backed by the rough Campbells Cove Road, which runs along the 5 m high basalt bluffs. Numerous fishing shacks back and, in some cases overhang, the beaches, many fronted by small jetties and slipways. The north beach begins at the fenced southern boundary of Point Cook RAAF base and extends south for 1.2 km to Cunningham Road and the basalt outcrops. Beyond this the south beach, with more shacks, extends for another 700 m to a seawall reinforced point.

The northern end of the north beach is an official Optional Dress (ie. nude) Beach. The beaches are fronted by shallow, 100 to 200 m wide sand and rock flats. In places where the beaches have eroded, basalt boulders and seawalls replace the beach.

Swimming: Only at high tide when the sand flats are covered.

Surfing: None.

Fishing: There are many fishing shacks here, however most fishing is done from boats in the bay.

Summary: A rough and unattractive beach, that is popular with the shack owners and sunbathers (at the northern end), but with little to attract others.

P79, P80   DUNCANS ROAD NORTH & SOUTH

Unpatrolled			
	Rating	No bar	Length
P79 Duncans Road North	1	R + sand flats	2 500 m
P80 Duncans Road South	1	R + sand flats	150 m
		Total length	2 650 m

Past Campbells Cove, the coast continues due south. It consists of low, basalt bluffs fronted by a narrow, high tide beach and 100 to 200 m wide intertidal sand flats. The only access is along **Duncans Road**, that reaches the coast at the southern end of the 2.5 km long north beach. The road then parallels the coast as far as Werribee South. The small, but wider, south beach is located between low basalt bluffs and seawalls.

Swimming: The south beach is the more accessible and attractive of the two, and is best at high tide when the tidal flats are under water.

Surfing: None.

Fishing: The southern seawall and bluffs can be fished at high tide only.

Summary: The north beach is narrow and is largely inaccessible by car, while the small south beach is the more popular with picnickers and bathers.

P81   WERRIBEE SOUTH

Unpatrolled
No bar: R + tidal flats & channel
Beach Hazard Rating: **2**
Length: 1 200 m

The small settlement of **Werribee South** on the northern banks of the Werribee River marks the last public access to the coast, until Beacon Point boat ramp 11 km to the south-west. Beach Road runs on the low bluffs behind the beach, and the beach curves around into the river mouth, where there is a park, picnic area, boat ramp and jetties.

The 1.2 km long beach is narrow and fronted by 200 to 300 m wide sand flats on the open, east facing coast, with a deep tidal channel along the south facing river mouth section.

Swimming: It is safest on the open beach at mid to high tide. Be careful in the river mouth because of the deep water, strong tidal currents and boat traffic.

Surfing: None.

Fishing: This is a very popular spot to go out into the bay, as well as fishing from the bluffs, river banks and jetties.

Summary: An accessible and popular beach with good park facilities, a large boat ramp and relatively safe away from the river channel.

P82, P83   WERRIBEE SEWER FARM NORTH & SOUTH

Unpatrolled			
	Rating	Single bar	Length
P82 Werribee Sewer Farm North	1	R + sand flats	4 500 m
P83 Werribee Sewer Farm South	1	R + sand flats	5 100 m
		Total length	9 600 m

The **Werribee Sewer Farm** occupies a large area south and west of the Werribee River. The low coastal plain is cris-crossed by the farm paddocks, and is fronted by a low beach and wide, tidal sand flats. The whole area has restricted access and the only way to reach the beaches is by boat.

The northern section extends south-west for 4.5 km from Werribee River mouth to a low point, from where the south section runs west for 5.1 km to the mouth of Little River. The entire section has a low, narrow beach and 300 to 500 m wide sand flats. The Werribee River has a deep channel, while Little River is just a small creek.

Swimming: Not recommended, owing to the proximity of the sewer farm. The sand flats are covered at high tide and exposed at low tide.

Surfing: None.

Fishing: Only at the river mouth, or along the beach at high tide.

Summary: Two very low energy, crenulate beaches, fronted by wide tidal flats and backed by even more extensive sewer paddocks.

P84 - P87   LAKE BOWRIE, BEACON POINT, BEACON POINT BOAT RAMP, KIRK POINT

Unpatrolled			
	Rating	No bar	Length
P84 Lake Bowrie	1	R + sand flats	2 300 m
P85 Beacon Point	1	R + sand flats	1 000 m
P86 Beacon Point Boat Ramp	1	R + sand flats	500 m
P87 Kirk Point	1	R + sand flats	1 700 m
		Total length	5 500 m

Between the Little River mouth and **Kirk Point** is a 5.5 km section of south-east facing coast containing four low, crenulate, sandy beaches. The first two beaches between the river and **Beacon Point** are backed by sewer farms. Low basalt and shelly points and reefs form the boundaries between the beaches, with the most prominent being Beacon and Kirk Points. The beaches are low and narrow, and are fronted by intertidal sand flats. These are 500 m wide at the river mouth, but decrease in width to 200 m off Kirk Point.

The only public access to the area is a boat ramp located 1 km north of Beacon Point. The remaining area is restricted due to the sewer farm.

Swimming: All of these beaches are fronted by shallow flats, the water depth of which depends on the tide. They are covered by high tides and exposed at low tide.

Surfing: None.

Fishing: Most people fish from boats along here, with high water required to fish from the beaches or points.

Summary: A largely off limits section of the bay, used only by fishers launching their boats from the Beacon Point ramp.

P88 - P91   KIRK POINT WEST, SAND HUMMOCKS NORTH, MID & SOUTH

Unpatrolled			
	Rating	No bar	Length
P88 Kirk Point West	1	R + sand flats	1 700 m
P89 Sand Hummocks North	1	R + sand flats	2 500 m
P90 Sand Hummocks Mid	1	R + sand flats	1 500 m
P91 Sand Hummocks South	1	R + sand flats	300 m
		Total length	6 000 m

At the low, basaltic **Kirk Point**, the coast swings to the west and breaks into a series of low, shelly barrier islands, called the **Sand Hummocks**. The first section west of Kirk Point is backed by sewer farms, with a drain forming its western boundary. From there the two major islands (north and mid) partly enclose a shallow lagoon, with the coast swinging to the south at the lagoon inlet. The islands are low and irregular, with major storms often overwashing the beaches and cutting new channels, as well as reshaping the beaches and islands. The south beach is backed by slightly higher land. The entire area is part of the sewer farm in the north and the Murtcaim Wildlife Sanctuary in the Sand Hummocks area, and is off limits to the public.

The entire 6 km section consists of low, narrow beaches fronted by sand flats ranging from 50 to 100 m wide, with some deeper tidal channels along the lagoon section of the Sand Hummocks.

Swimming: The low beaches are fronted by sand flats, with water depth depending on the tide.

Surfing: None.

Fishing: Mainly done from boats off the Sand Hummocks and Kirk Point.

Summary: This is a relatively natural section of low coast and lagoons. It is now preserved as a wildlife sanctuary, and is off limits to the public.

P92, P93   POINT WILSON NORTH & EAST

Unpatrolled			
	Rating	No bar	Length
P92 Point Wilson North	1	R + sand flats	2 000 m
P93 Point Wilson East	1	R + sand flats	500 m
		Total length	2 500 m

Point Wilson is a low basalt point that protrudes south into Port Phillip Bay and forms the northern entrance to Corio Bay. The point is backed by the Commonwealth Explosives Department and is off limits to the public.

Between the Sand Hummocks and the point is a 2.5 km section of east facing coast, composed of low, sandy, beach ridges and narrow beach. The beaches are fronted by seagrass meadows. The 2.5 km long Point Wilson jetty separates the two, with the 500 m long eastern beach terminating at the point.

Swimming: These are two narrow beaches, with narrow sand flats and relatively deep, seagrass covered sand off the beaches. However, the public is prohibited from landing in this explosives area.

Surfing: None.

Fishing: The seagrass beds are popular areas for catching whiting.

Summary: Two low energy beaches fronted by seagrass flats, but backed by Australia's largest explosive depot.

P94 - P97   POINT WILSON WEST, SNAKE ISLAND, SNAKE ISLAND WEST 1 & 2

Unpatrolled			
	Rating	No bar	Length
P94 Point Wilson West	1	R	500 m
P95 Snake Island	1	R + sand flats	800 m
P96 Snake Island West 1	1	R + tidal flats	500 m
P97 Snake Island West 2	1	R + tidal flats	400 m
		Total length	2 200 m

At Point Wilson, the coast turns west into Corio Bay and then continues in an irregular manner for 9 km to Point Lillas. Between the two points are four very low energy, narrow beaches and tidal flat areas. The first is immediately west of **Point Wilson** and is on Commonwealth land. The second is on the east side of land-locked **Snake Island**; and the remaining two are inside an open bay to the west of Snake Island.

All four beaches are usually calm, with the beach fronted by shallow, irregular rocks; sand tidal flats; and seagrass beds. There is no public access to Point Wilson and restricted access to the Snake Island beaches, which are backed by salt evaporators.

Swimming: Only possible at high tide on these essentially tidal flat beaches.

Surfing: None.

Fishing: Most fishing is done from boats over the sea grass beds.

Summary: Four very low energy beaches and tidal flats, all with restricted public access.

P98   AVALON

Unpatrolled
No bar: R + sand flats
Beach Hazard Rating: **1**
Length: 2 000 m

Avalon Beach is the first public beach since Werribee South. It lies on the north side of Corio Bay and faces due south. The Avalon Beach Road runs due south from the Princes Highway for 6 km to reach the beach. There is a foreshore reserve and a number of houses, fishing shacks and numerous jetties and rock groynes across the low beach. Tidal sand flats extend out into the bay and are exposed at low tide.

Swimming: Only at mid to high tide, when the sand flats are covered.

Surfing: None.

Fishing: Most fishers go out onto the bay in boats, with fishing off the jetties and groynes only possible at high tide.

Summary: This is a popular fishing settlement, with the houses, jetties and moored boats dominating the very low energy beach.

P99   POINT ABEONA

Unpatrolled
No bar: R
Beach Hazard Rating: **2**
Length: 500 m

Point Abeona is a 500 m long, sandy and shelly spit, that partly blocks the entrance to Limeburners Bay. The spit has been built by low southern waves running along and wrapping around the point. The low spit is backed by the bay and fronted by shallow, seagrass covered tidal flats.

Swimming: You can swim on both sides of this beach, with deeper water on the back (bay) side, or off the tip of the spit. Shallow tidal flats front the beach.

Surfing: None.

Fishing: The tip of the point is the most popular spot, with deep water and tidal flows.

Summary: A long, narrow spit surrounded by usually calm water, with tidal currents off the tip.

P100   GEELONG GRAMMAR

Unpatrolled
No bar: R + sand flats
Beach Hazard Rating: **1**
Length: 1 500 m

Geelong Grammar School is located in the north-west corner of Corio Bay. Running parallel to Foreshore Road, which fronts the school, is a 1.5 km long south-east facing beach. The beach is located between the western entrance to Limeburners Bay and the large Shell Corio oil refinery. The beach has good access, with a car park, two boat ramps and a small jetty at the bay entrance. Waves are usually absent and the narrow beach slopes gently into the bay.

Swimming: A relatively safe, usually calm beach, with deeper water and tidal currents off the northern end.

Surfing: None.

Fishing: More popular off the jetty and toward Limeburners Bay.

Summary: A quiet, low energy beach, mainly used for launching boats and fishing.

P101 MOORPANYAL PARK

Unpatrolled
No bar: R + sand flats
Beach Hazard Rating: **1**
Length: 200 m

Moorpanyal Park Beach is surrounded by a bluff-top park of the same name. It lies at the junction of The Esplanade and Seabeach Parade. The bluff-top park backs the entire 200 m long beach, with ample parking and other facilities in the park, and shops across the road.

The beach faces south-east and is backed by the 10 m high bluffs and a seawall along the southern end. Two drains cross the beach and there is a gravel road that leads down to the beach for boat launching. The high tide beach is less than 10 m wide, with a 50 m wide, intertidal sand flat running parallel. Rocks lie on the flats below the bluff headlands at either end.

Swimming: Better at high tide when the sand flats are covered.

Surfing: None.

Fishing: Most fishers use the beach for boat launching, with the beach itself best off the rocks at high tide.

Summary: Behind this quiet, little beach is a nice park for a picnic.

P102, P103 ST HELENS, RIPPLESIDE

Unpatrolled			
	Rating	No bar	Length
P102 St Helens	**1**	R + sand flats	150 m
P103 Rippleside	**1**	R + sand flats	300 m
		Total length	450 m

St Helens and **Rippleside** are two small but popular beaches located in North Geelong. Both are backed by formal parks of the same names. These contain picnic facilities and parking areas. Jetties extend out from the end of both beaches, with a larger working jetty next to Rippleside.

Swimming: Usually calm conditions, with water depth dependent on the tide.

Surfing: None.

Fishing: Better off the jetties.

Summary: Two small but popular beaches located in the heart of North Geelong.

P104 WESTERN

Unpatrolled
No bar: R + sand flats
Beach Hazard Rating: **1**
Length: 500 m

Western Beach is a narrow beach located below The Esplanade and its foreshore reserve, between Griffiths Gully Jetty and the Western Beach Motor Boat Club. A walkway and bike track run behind the low, narrow beach.

Swimming: The beach slopes gently into the calm waters of the bay, with depth dependent on the tide.

Surfing: None.

Fishing: Better off the jetties than the shallow beach.

Summary: A quiet beach backed by Geelong's foreshore reserve, and bordered by jetties and boat activity.

P105 EASTERN

Unpatrolled
No bar: R + sand flats
Beach Hazard Rating: **1**
Length: 700 m

Eastern Beach is the main swimming beach for Geelong. It is backed by the grassy slopes of Eastern Beach Reserve and has a 200 m long, circular swimming enclosure toward its eastern end. There are numerous facilities in the reserve and at the tidal pool, including a children's pool.

Swimming: This is the safest place to swim in Geelong, with a choice of two pools and a range of other aquatic activities.

Surfing: None.

Fishing: Only off the outside of the pool enclosure.

Summary: Geelong's premier beach area, with a full range of facilities for picnickers and bathers.

P106   **POINT HENRY**

Unpatrolled
No bar: R + sand flats
Beach Hazard Rating: **1**
Length: 1 000 m

Point Henry forms the southern entrance to Corio Bay, as it protrudes 3 km northward into the bay. Most of the point is occupied by salt works and the Alcoa processing plant. A sealed road runs out past the plant to the point. There is a reserve with parking and limited facilities on the point, including beach boat launching.

The 1 km beach lies on the north-eastern tip of the point, between the point and the 1 km long Alcoa jetty. The northern half is a low, narrow spit extending northward into the bay. The beach is low and shelly and is fronted by 100 to 200 m wide sand flats.

Swimming: Better at high tide when the sand flats are covered.

Surfing: None.

Fishing: The water is shallow off the beach, with the deepest water off the back of the spit.

Summary: A low energy beach and spit at the entrance to Corio Bay.

P107, P108   **SEABRAE-SANDS, GRAND SCENIC**

		Rating	Single bar	Length
107	Seabrae-Sands	1	R + sand flats	2 000 m
108	Grand Scenic	1	R + sand flats	1 800 m
			Total length	3 800 m

The northern side of the Bellarine Peninsula between Point Henry and Point Richards contains 22 km of north facing, low energy shores. For the most part, these are backed by sandstone and clay bluffs between 5 and 40 m high. At the base of the bluffs are 15 km of narrow, crenulate beaches, usually fronted by seagrass covered sand flats that are between 200 and 400 m wide.

The beaches begin at the Seabrae Caravan Park located at the end of Clifton Avenue. The first very low energy beach (**Seabrae-Sands Beach**) runs from here to the Sands Caravan Park 2 km to the east. **Grand Scenic Beach** runs from some rocks just past the Sands, to a low point 1.8 km to the east. Both high tide beaches are shell rich and are only 5 m wide, with grassy, intertidal sand flats extending 200 m into the bay.

Swimming: Only at high tide when the sand flats are covered.

Surfing: None.

Fishing: Most fishers launch at high tide to fish in the bay.

Summary: A low energy, low gradient beach, with good access at the caravan parks, and via Grand Scenic Drive.

P109   **HERMSLEY**

Unpatrolled
No bar: R + sand flats
Beach Hazard Rating: **1**
Length: 5 000 m

Hermsley Beach is the name given to a series of near continuous ribbons of sand, below 5 km of crenulate, 20 to 40 m high bluffs. Much of the beach is backed by private land, with the best access via Hermsley Road. At this point the beach consists of a steep, narrow, shelly beach, backed by grassy slopes and fronted by 200 m wide sand flats. There are no facilities at the beach.

Swimming: A usually calm and relatively safe beach, which is better at mid to high tide when the sand flats are covered.

Surfing: None.

Fishing: Best at high tide.

Summary: A relatively natural, but isolated and little used beach.

P110, 111, 112 CLIFTON SPRINGS WEST, MAIN & EAST

Unpatrolled			
	Rating	Single bar	Length
110 Clifton Springs West	1	R + sand flats	800 m
111 Clifton Springs Main	1	R + sand flats	500 m
112 Clifton Springs East	1	R + sand flats	4 500 m
		Total length	5 800 m

The **Clifton Springs** section of coast is characterised by near continuous bluffs averaging 20 m high, and is fronted by a narrow, crenulate beach and wide sand flats. The first section lies to the immediate west of the boat ramp, which provides the best access, and has 400 m wide sand flats.

The central section is the site of the main Clifton Springs recreational beach. It is the most popular of the northern Bellarine Peninsula beaches. It is located below 20 m high bluffs and is backed by an extensive foreshore reserve, with ample parking and picnic facilities. Most facilities and parking are on the bluffs, with more limited facilities at the beach. The narrow, 500 m long beach faces north-west and is fronted by tidal flats that extend several hundred metres into the bay. The width of the flats can be gauged by the length of the ruins of the 400 m long jetty.

The eastern beach is a 5 km long, narrow, crenulate beach lying below the bluffs, which slowly decrease in height to the east. It fronts the Clifton Springs golf course and terminates 1 km west of Point Richards. While all of this beach is backed by a foreshore reserve, access is limited to the golf course or Beacon Point Road and bluff-top Water Drive. It has no facilities.

Swimming: Three relatively safe beaches, with best bathing on the main beach at mid to high tide.

Surfing: None.

Fishing: Only at high tide from the beach.

Summary: A usually quiet section of coast with the main beach being the most accessible and popular.

P113, P114, P115 POINT RICHARDS WEST, PORTARLINGTON, PORTARLINGTON JETTY

Unpatrolled			
	Rating	No bar	Length
P113 Point Richards	1	R + sand flats	300 m
P114 Portarlington	1	R + sand flats	2 000 m
P115 Portarlington Jetty	1	R + sand flats	200 m
		Total length	2 500 m

Point Richards is a large accumulation of sand that forms the northern tip of the Bellarine Peninsula. The point is still growing slowly to the west, while in the east it is attached to the bedrock at **Portarlington**. There are three beaches along this 2.5 km section of coast. The first is a 300 m long, low energy section west of the point, which grades into tidal flats. Between the point and the **Portarlington Jetty** is a 2 km long, north facing beach, backed by a large reserve and a caravan park. The third beach runs for 200 m east of the jetty to the bluffs.

All three beaches have good access, with a large car park on the point servicing a boat ramp, and a second boat ramp on Portarlington Beach.

Swimming: These are three relatively safe beaches, with usually calm to low wave conditions. Due to the extensive sand flats, bathing is best at mid to high tide. However, watch the boat traffic near the boat ramps and jetty.

Surfing: None.

Fishing: The jetty is the most popular location.

Summary: A very popular section of coast, particularly during the summer holidays when the backing caravan park is full.

P116, P117, P118 STEELES WEST, STEELES, STEELES EAST

Unpatrolled			
	Rating	No bar	Length
P116 Steeles West	1	R + sand flats	200 m
P117 Steeles	1	R + sand flats	150 m
P118 Steeles East	1	R + sand flats	100 m
		Total length	450 m

East of Portarlington, The Esplanade follows the coast for 10 km to The Bluff, with a continuous foreshore reserve between the road and the shore. There are ten low energy beaches in this section, all with excellent access and are backed by a grassy reserve.

The first three are small pockets of sand below 5 m high bluffs, with low, rocky reefs bordering each beach. They are located 1 km east of Portarlington and lie either side of Steele Trig station, and all face north. The first beach (**Steeles West**) is 200 m long and is backed by sloping grassy bluffs. **Steeles Beach** is 150 m long, with a boat ramp and an informal parking area on flat land below the bluff; while **Steeles East** is narrow and backed by eroding bluffs. All three are usually calm or have low waves, and are fronted by scattered rock reefs and 100 to 200 m wide rock and sand flats.

Swimming: Three relatively safe beaches, with the best bathing at mid to high tide. Watch the occasional rocks and reefs and the boats using the middle beach.

Surfing: None.

Fishing: The rocks are the best location at high tide.

Summary: Three small beaches not visible from the road, with good access but no facilities.

P119, P120, P121   GRASSY POINT WEST, GRASSY/POINT GEORGE, POINT GEORGE SOUTH

Unpatrolled			
	Rating	No bar	Length
P119 Grassy Point West	1	R + sand flats	2 000 m
P120 Grassy/ Point George	1	R + sand flats	2 800 m
P121 Point George South	1	R + sand flats	900 m
		Total length	5 700 m

East of Steele Trig, the coast trends east for 2 km to **Grassy Point**, then turns to the south-east for 2.8 km to **Point George**. Here it turns again, to the south-south-east, for 900 m down to Half Moon Bay. There are three beaches located between these points. The Esplanade and foreshore reserve back the beaches, and there is a camping area on the eroding beach front south of Point George.

The three beaches are all narrow and crenulate, with low (1 to 2 m high) backing bluffs, and 100 to 200 m wide sand flats. Seagrass grows on the deeper parts of the sand flats and washes up onto the beaches.

Swimming: Three relatively safe beaches, with usually calm to low wind waves. Water depth over the shallow

sand flats depends on the tide.

Surfing: None.

Fishing: You can only fish from the beach at high tide.

Summary: Three very accessible beaches, with Point George housing a popular, beach front camping reserve.

P122, P123   HALF MOON BAY, HOOD BIGHT

Unpatrolled			
	Rating	No bar	Length
P122 Half Moon Bay	1	R + sand flats	400 m
P123 Hood Bight	1	R + sand flats	1 000 m
		Total length	1 400 m

Half Moon Bay and **Hood Bight** are two east facing beaches, bordered by low rocky points. The Esplanade runs behind both beaches, with a narrow foreshore reserve between the road and the beaches. The reserve increases in width on the points. There is a camping reserve in Batman Park at Half Moon Bay, and a yacht club and picnic area in Hood Bight. Several boat sheds back both beaches and there is a large car park, boat ramp, jetty and picnic area on Indented Head, at the southern end of Hood Bight. Both beaches are narrow and steep and fronted by 100 m wide sand flats.

Swimming: Two relatively safe beaches, with usually calm to low wave conditions. Depth over the shallow sand flats depends on the tide.

Surfing: None.

Fishing: Generally shallow water off the beaches, with the points being the best location at high tide.

Summary: Two curving, embayed beaches, with good camping and picnic facilities.

P124, P125   ST LEONARDS, ST LEONARDS PIER

Unpatrolled			
	Rating	No bar	Length
P124 St Leonards	2	R + sand flats	2 700 m
P125 St Leonards Pier	2	R + sand flats	300 m
		Total length	3 000 m

At Indented Head, the coast turns and runs due south for 3 km down to the low bluffs at St Leonards. The Esplanade runs right behind the beach and low foredune.

There are two picnic areas behind the main beach, with a camping reserve toward St Leonards, and a foreshore reserve with numerous facilities backing the bluffs and **St Leonards Pier Beach**. Toward the southern end of **St Leonards Beach**, there are several wooden groynes, as well as the breakwater and pier, that form the boundary with the 300 m long St Leonards Pier Beach. This beach terminates at a low, rocky point and reef flats.

Both beaches are low and narrow and fronted by shallow, 100 to 200 m wide sand flats, containing low amplitude bars and runnels. The flats are exposed at low tide.

Swimming: Two relatively safe beaches fronted by shallow sand flats and low bars. Bathing is better at mid to high tide when the flats are covered.

Surfing: None.

Fishing: The rocks on the south side of Indented Head and the St Leonards Pier are the two best places to reach deeper water.

Summary: A long, relatively natural beach, with good access and numerous facilities, plus the small town of St Leonards at the southern end.

P126   RED BLUFF

Unpatrolled
No bar: R + sand flats
Beach Hazard Rating: **1**
Length: 5 200 km

On the south side of the low St Leonards bluffs, the coast trends almost due south for over 5 km to a series of recurved spits at Edward Point. The Esplanade runs behind the first 2 km, with a camping area on the beach just south of St Leonards. A picnic area, boat ramp and a second camping area are situated just before the 10 m high bluff section. On the south side of **Red Bluff** are another boat ramp and the yacht club, with several wooden groynes crossing the crenulate beach. The southern 3 km are part of Point Edward Wildlife Reserve and are backed by Swan Bay.

Swimming: The low, narrow beach is fronted by 100 to 200 m wide sand flats, with low amplitude bars. Bathing is better at mid to high tide, with the inner portions of the flats exposed at low tide.

Surfing: None.

Fishing: Better at high tide when the flats are covered.

Summary: A long beach with numerous facilities between Red Bluff and St Leonards, and a rich wildlife reserve in the south.

P127   SWAN ISLAND

Unpatrolled
No bar: R + tidal flats and shoals
Beach Hazard Rating: **2**
Length: 2 800 m

Swan Island is a dynamic collection of beaches and recurved spits that have been deposited by waves and tidal currents over the past 6 000 years. The island now covers several hundred hectares and is reached via a causeway from Queenscliff. However, the western side of the island belongs to the Queenscliff Golf Club, while the eastern half is Commonwealth land, resulting in most of the island and the beaches being off limits to the public.

The beach begins in lee of Sand Island (another collection of recurved spits) and runs north-east for 2 km to Swan Point, where it curves around to the west for a few hundred metres. The beach is low and narrow and is fronted by highly variable tidal flats, shoals and channels associated with tidal flow through The Rip. Unfortunately a seawall has been built along much of the beach which, together with a scattering of groynes, destroys the natural integrity of this dynamic island.

Swimming: The beach is relatively safe, however care should be taken if swimming off the beach and flats owing to the strong tidal currents and deep water.

Surfing: None.

Fishing: There are highly variable tidal channels along the island and at Swan Point, which offer the best fishing locations.

Summary: A restricted island maintained in a relatively natural state by the Commonwealth.

P128   SAND ISLAND

Unpatrolled
No bar: R + tidal shoals
Beach Hazard Rating: **3**
Length: 1 000 m

Sand Island has come into being since dredging of the Queenscliff harbour began in the 1940s. Most of the island has formed in the last 30 years. The dredged sand has been washed northward to form a 1 km series of recurved spits. The dredged Queenscliff harbour channel and entrance wall form the southern boundary. There is a deep channel, with ebb tide shoals extending a few hundred metres into Port Phillip Bay.

The beach runs north-east for several hundred metres before curving around to the west. It is steep and narrow, with considerable variation in the extensive offshore tidal shoals.

Swimming: Relatively safe at the beach, but beware of tidal currents through the harbour channel and further out associated with the larger flows through The Rip.

Surfing: None.

Fishing: The breakwater along The Cut at the southern end of the island provides good access to the harbour channel, while the northern spits are prograding into deeper water.

Summary: A dynamic, new island that can only be reached by boat.

P129   QUEENSCLIFF

Unpatrolled
Single bar: LTT
Beach Hazard Rating: **3**
Length: 800 m

Queenscliff Beach fronts the town of the same name. It is 800 m long, faces south-east, and is backed by a large foreshore reserve with numerous facilities. The Queenscliff harbour channel and breakwater form the northern boundary, with the vegetated slopes of 20 m high Shortland Bluff forming the southern boundary. Two long jetties cross the beach, one servicing the passenger ferry to Portsea; and the other is the old Pilot Jetty. There are several boat sheds below the bluffs and the Queenscliff Lighthouse on top of Shortland Bluff.

The beach is low and flat, with a continuous, wide, shallow bar and no rips. Shallow reef flats extend east of the bluff.

Swimming: Relatively safe, with a wide, shallow bar.

Surfing: None.

Fishing: The harbour channel, the two jetties and the seawall around the base of the bluff all provide excellent fishing locations.

Summary: A very accessible beach, with numerous facilities in the foreshore reserve and the attractive town of Queenscliff behind.

P130, P131, P132   LONSDALE BAY, FRONT, POINT LONSDALE EAST

Unpatrolled			
	Rating	Single bar	Length
P130 Lonsdale Bay	**3**	LTT	2 500 m
P131 Front	**3**	LTT	1 000 m
P132 Point Lonsdale East	**3**	LTT	600 m
		Total length	4 100 m

Between Shortland Bluff and Point Lonsdale is a curving, south-east facing, 3 km long series of three beaches. The entire section is backed by foreshore reserves and the Queenscliff and Point Lonsdale Roads. There are camping and picnic areas in the reserve, just west of Shortland Bluff and further west at Golightly and Royal Parks. The central section of Lonsdale Bay Beach is backed by natural, vegetated dunes, with a walking track linking the camping areas. On the bluffs above Point Lonsdale are car parks, picnic areas and lookouts to view The Rip and passing ships. The lighthouse was built in 1863 and stands 37 m above sea level. Steps lead down the bluffs to the beaches below.

Lonsdale Bay Beach is 2.5 km long and faces south. It consists of a low, wide beach and shallow, attached bar, usually with waves of less than 0.5 m. The southern end of the beach is backed by a seawall and promenade, with several wooden groynes crossing the beach.

Front Beach is 1 km long, faces due east and has more than 15 groynes crossing the low gradient beach. A shallow bar and patchy rock reef form the surf zone.

Point Lonsdale East Beach is 600 m long, lies at the base of the 20 m high bluffs, and is fronted by a mixture of patchy rock flats and a shallow sand bar. The 200 m long Point Lonsdale Jetty is located at the southern end.

Swimming: During normal low waves, these beaches are relatively calm with wide, shallow surf zones. Watch the groynes and patchy rocks and reefs to the south, and strong tidal currents off the point.

Surfing: Usually low beach breaks. However during large winter swells, there is a left and right that breaks over the reefs off the *Lighthouse*, and a right hander that runs over the reef at Front Beach.

Fishing: The seawall around the bluff, the southern rocks and the jetty are the best locations.

Summary: An open bay that forms the western entrance to Port Phillip Bay and receives increasing ocean wave energy toward the point. The beach has good access and numerous facilities.

7. POINT LONSDALE TO CAPE OTWAY

Point Lonsdale to Cape Otway

Beaches: 281 - 410
Coast length: 139 km

For maps of region see Figures 107 & 116

The coast between Port Phillip Heads and Cape Otway is the prime surfing coast of Victoria. Located within one to two hours of Melbourne, this coast has long attracted people wanting to view the scenery and/or recreate along some of Australia's most attractive coastline. There are 130 beaches along this 139 km of coast. Twelve of these have surf lifesaving clubs and an additional two have summer lifeguards. Consequently, this area has the highest concentration of patrolled beaches in Victoria. Furthermore, the coast boasts some of Victoria's top surfing breaks, including Bells Beach: Australia's first surfing reserve and site of the world's longest running surfing contest. It is also the home of malibu surfing at Torquay, which houses Australia's first surfing museum.

In addition to the surf, the coast (particularly south of Anglesea) is backed by some of the most spectacular coastal scenery in Australia. The Great Ocean Road certainly lives up to its name, as it winds along between the Otway Ranges and the Southern Ocean. The numerous headlands and valleys are matched by the beaches, rock platforms, reefs and surf (Figure 108)

Note: The Surf Coast Shire (Breamlea to Lorne) publishes free Visitor and Fishing Guides that indicate all surfing and fishing spots.

---

281   **LONSDALE**

Unpatrolled

Single bar: R/LTT + reef flats
Beach Hazard Rating: **6**
Length: 500 m

Point Lonsdale and its lighthouse form the boundary between Port Phillip Bay and the open Bass Strait coast. As soon as the point is rounded, waves pick up and energetic surf dominates the beaches.

Victorian Beaches - Region 7: East (Point Lonsdale to Fairhaven)

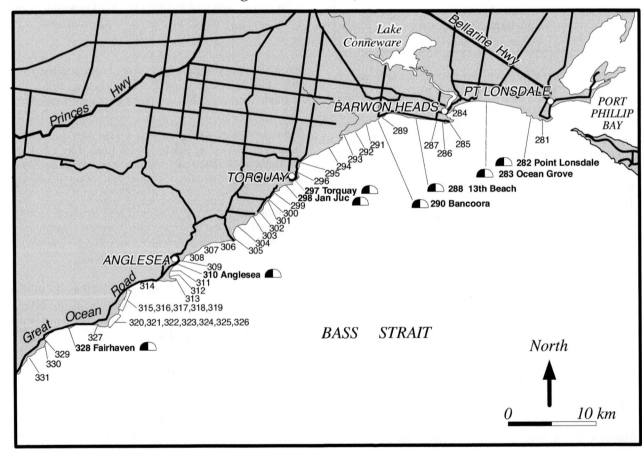

Figure 107.   Region 7: (eastern section) Point Lonsdale to Fairhaven (Beaches 281 to 331)

Figure 108. Swell Point, just south of Apollo Bay, consists of a series of small sand beaches, separated by sandstone shore platforms, and backed by vegetated bedrock slopes. The sequence of rock and sand is typical of the Point Lonsdale to Cape Otway coastal region. Along the 139 km of coast, there are 130 beaches, each averaging only 590 m in length.

Below the lighthouse and extending either side are extensive intertidal, calcarenite rock platforms. To the east they form reefs that shield the jetty, while to the west they are backed by the 20 m high, dune covered bluffs and a crenulate, 500 m long, south facing beach. There is a car park beside the lighthouse, with steps down to **Lonsdale Beach** and a camping area in the reserve behind the bluffs.

At high tide, waves break over the reefs and reach the beach as lower waves, while at low tide the reefs are exposed, with surf and deep holes off the beach.

Swimming: Be careful here as rocks, reefs and deep holes dominate the surf. Best close inshore at mid to high tide, when waves are lowered by the reef and some protected tidal pools lie off the beach.

Surfing: Best at high tide over the reefs on a low to moderate swell.

Fishing: Excellent deep rip and reef holes right off the beach at high tide, or from the rock flats at low tide.

Summary: A beach viewed by all who go to the lighthouse and mainly used by fishers and sunbathers. Use care if bathing, as reefs dominate the surf.

282　POINT LONSDALE SURF (BACK)

Point Lonsdale SLSC

Patrols: late November to Easter holidays
Surf Lifesaving Club: Saturday, Sunday and Christmas public holidays
Lifeguard: 10 am to 6 pm weekdays, during Christmas holiday period

Single bar: TBR + reefs
Beach Hazard Rating: **8**
Length: 900 m

For map of beach see Figure 109

Point Lonsdale forms the western side of Port Phillip Heads, with The Rip separating it from Point Nepean. The town of Point Lonsdale has a protected bay beach and more exposed ocean beaches. The main ocean beach is known as the **Surf** or **Back Beach** and is the site of Point Lonsdale Surf Life Saving Club, founded in 1947. A walking track leads from the surf club over the dunes to the beach.

Surf Beach extends for 900 m from a wide, intertidal rock platform, located just east of the surf lifesaving club, to where more rocks and reefs outcrop in the surf.

In fact, low tide rock flats dominate this beach and are clearly visible at low tide.

The beach faces south-west and receives waves averaging 1.4 m, which produce a single attached bar, cut by strong rips every 250 m. In addition, strong permanent rips run out against some of the reefs, the worst being *The Escalator* to the left of the club house. These rips have been responsible for many rescues, with an average of 30 each year. There have also been drownings at the beach, so be very wary and stay between the flags.

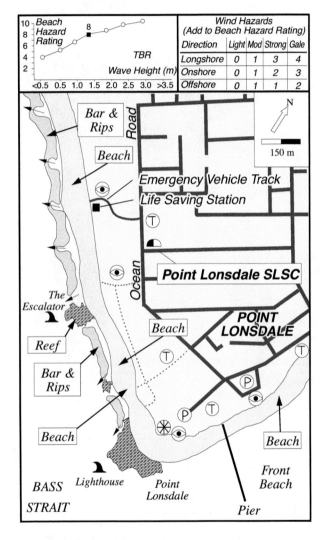

Figure 109. Point Lonsdale Beach and Surf Life Saving Club are located 1 km west of Point Lonsdale, which forms the western entrance to Port Phillip Bay. The beach receives moderate waves and is dominated by reefs, which produce some strong permanent rips, including The Escalator. It is best to bathe at mid to high tide and between the flags on this potentially hazardous beach.

Swimming: A reasonably hazardous beach, owing to the mixture of moderate waves and strong permanent and shifting rips, together with rocks and reefs.

Definitely stay on the bars, clear of the rips and rocks and between the flags.

Surfing: Beach breaks are common over the numerous reefs, with the best known being *Glaneuses*, located at the end of Glaneuse Road, and next to The Escalator rip, which offers a good left over the reef. Surfing is best with northerly winds and a low to moderate swell at mid to high tide, as the reefs are exposed at low tide.

Fishing: A popular spot offering permanent rips and gutters, particularly adjacent to the reefs and rocks.

Summary: This is the surf beach for the popular holiday town of Point Lonsdale. It is very popular with bathers in summer and surfers year round. However, it is a hazardous beach with strong permanent rips, so use extreme care.

283A   COLLENDINA

Unpatrolled
Single bar: TBR
Beach Hazard Rating: **7**
Length: 6 000 m

Collendina Beach occupies most of the open bay between Point Lonsdale and Barwon Heads. It is 6 km long, extending from the reefs west of Point Lonsdale Beach to 1 km west of the Collendina Beach car park. The only public access is at the car park, together with tracks over the foredune from the caravan park.

The beach faces south-south-east and for the most part is backed by 10 to 20 m high, vegetated dunes, with a few blowouts. It receives waves averaging between 1 and 1.5 m, which break over a wide, low gradient surf zone and occasional reefs and rocks. Persistent rips occur every 250 m, with some permanent rips against the more prominent reefs. During bigger seas, waves break on outer, deeper reefs.

Swimming: Be careful on this beach as there are usually deep rip holes and strong currents along the beach. Stay inshore on the attached section of the bars and well clear of the rips and reefs.

Surfing: There are many beach breaks and a few reef breaks along the beach, with best conditions occurring in a low to moderate swell and northerly winds.

Fishing: There are excellent persistent rip holes and occasional gutters along the beach, together with scattered reefs.

Summary: A long, relatively natural beach offering plenty of sand, a low gradient inner surf for bathing, rip holes for fishing and numerous beach breaks for surfing.

283B   OCEAN GROVE

Ocean Grove SLSC

Patrols: late November to Easter holidays
Surf Lifesaving Club: Saturday, Sunday and Christmas public holidays
Lifeguard: 10 am to 6 pm weekdays, during Christmas holiday period

Single bar: TBR
Beach Hazard Rating: **6**
Length: 2 000 m

For map of beach see Figure 110

Ocean Grove Beach is located in the centre of the 9.5 km long beach that curves in a broad, south facing arc from Point Lonsdale to the Barwon River mouth. The Ocean Grove section is 2 km long and faces southeast. Some protection is offered by Barwon Heads and the beach receives waves averaging 1.4 m. These waves interact with the fine beach sand to produce a wide, low gradient beach face, fronted by a 300 m wide surf zone that contains strong rips every 250 m. During moderate waves, the rips increase in size and intensity toward Collendina, while decreasing toward Barwon Heads. At low tide, the beach and exposed bar can be over 100 m wide, with the deeper rip channels clearly visible.

The town of Ocean Grove backs the beach, with a wide, well-arranged foreshore reserve between the town and the beach. It provides extensive parking, together with most beach amenities. The good parking and easy access, together with the surf club patrols and slightly lower waves make this a popular summer beach. The Ocean Grove Surf Life Saving Club was formed in 1948 and performs an average of 8 rescues each year.

Swimming: A moderately safe beach, particularly during average summer conditions, when extensive bars dominate. Best at high tide, however watch the rips, particularly at low tide. Best to stay between the flags.

Surfing: Usually has wide, moderate to low beach breaks; more popular with summer surfers.

Fishing: Best to go up the beach away from the summer crowds, and where rip holes are more common.

Summary: A popular summer beach, which can hold a large crowd. It has a wide, shallow surf zone with rips increasing up the beach, so it is best to stay near the surf club and bathe in the patrolled area.

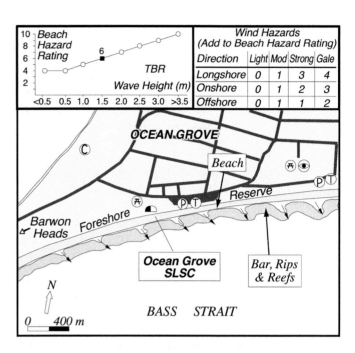

10	Beach					Wind Hazards (Add to Beach Hazard Rating)			
8	Hazard					Direction	Light	Mod	Strong	Gale
6	Rating									
			TBR			Longshore	0	1	3	4
		Wave Height (m)				Onshore	0	1	2	3
	<0.5 0.5 1.0 1.5 2.0 2.5 3.0 >3.5					Offshore	0	1	1	2

Figure 110.   Ocean Grove Beach is a popular, southeast facing beach located in front of the Ocean Grove business centre. It has good access, parking and facilities and is a popular summer beach, with a wide beach and surf zone. During moderate to high waves, rips dominate the outer and low tide surf, so stay close inshore in the patrolled area.

The annual **Surf Coast Fun Run** is held every January, over an 11.5 km course between Point Lonsdale and Barwon River. It attracts over 1 000 runners.

Ocean Grove Beach, west of the surf lifesaving club, becomes a horse racing track each February. The **Collendina Cup** is held on the beach over distances from 600 m to 1 000 m. The Cup, which commenced in 1985, is now an annual event.

283C   RAFFS

Unpatrolled

Single bar: LTT
Beach Hazard Rating: **5**
Length: 1 500 m

Raffs Beach occupies the western 1.5 km of the 11 km long stretch of sand and reef that runs from Point Lonsdale to the Barwon River mouth. The beach extends from the western side of Ocean Grove out to the river mouth. It is backed by a 200 m wide, vegetated sand spit with the Barwon River on the other side. The Barwon Heads Road parallels the beach. There is a foreshore reserve between the road and the beach, that has a camping area on both sides of the road. It also has a car park and numerous beach access tracks over the low dunes.

The beach swings around in lee of Barwon Heads to face south, then south-east. Waves are lowered by refracting around the head, and average 0.6 m at the beach. This produces a wide, low gradient, continuous bar, which is usually free of rips.

Swimming: A moderately safe beach that is better at high tide, with a greater chance of rips at low tide. Be careful near the river mouth as it has deep tidal channels and strong currents.

Surfing: There are beach breaks, which decrease in size, all the way to the river mouth. Best conditions are in a moderate swell and north-west to westerly winds. This is a good place to check out when westerlies are blowing.

Fishing: The river mouth is the best location, as the beach tends to be wide and shallow.

Summary: A popular summer camping area, with people using both the beach and backing river for various recreational pursuits.

284   BARWON RIVER

Unpatrolled
Single bar: R/LTT + river
Beach Hazard Rating: **4**
Length: 700 m

The **Barwon River** enters the sea in lee of Barwon Heads. The river flows along the eastern side of the heads and the reefs, that run for another 1 km to the east. The tip of the head is called Point Flinders, which is a 20 m high sand dune surrounded by rock platforms and reefs. In behind the point, a narrow sand beach runs up along the river to the bridge. The beach is initially backed by a seawall and foreshore reserve, and then the Barwon Heads caravan and camping area. Closer to the bridge there are two car parks and a jetty.

The beach is narrow, steep and usually calm. However, shifting tidal shoals, deeper channels and strong currents parallel the beach.

Swimming: This beach is a popular spot for people from the backing caravan park and those looking for calmer waters. Take care as water depth varies with the tide and the shifting banks, and strong tidal currents flow past the beach.

Surfing: None.

Fishing: A very popular spot, with fishing from the jetty or, at high tide, straight off the beach or rocks into the river.

Summary: A very accessible and quiet beach. Very popular in summer, however care must be taken with the deep channels and tidal currents.

285, 286, 287   POINT FLINDERS, BARWON HEADS 1 & 2

Unpatrolled			
	Rating	Single bar	Length
285 Point Flinders	4	R + rocks	50 m
286 Barwon Heads 1	7	TBR + reefs	800 m
287 Barwon Heads 2	6	TBR + reefs	1 000 m
		Total length	1 850 m

At Barwon Heads, the coast trends due west for 7 km to Black Rocks. The first 2 km are dominated by calcarenite rocks and reefs, which outcrop on the beach and in the surf. These divide the coast into three beaches. The first (285) is below **Point Finders** and is a 50 m pocket of sand facing south-east and bordered by rock platforms and reefs. The two **Barwon Heads** beaches (286, 287) face south and are more exposed, with higher waves and patchy reefs. These conditions result in a wide, low gradient beach, rock flat and surf, with persistent and some permanent rips against the reefs.

All three beaches are easily accessible. There is a car park and a lookout on Point Flinders, and car parks on the Torquay Road, which parallels the two Barwon Heads beaches.

Swimming: Point Flinders is relatively safe close inshore, however there are rocks and reefs off the beach. The Barwon Heads beaches are both potentially hazardous, owing to the higher waves, reefs and strong permanent rips.

Surfing: There are several breaks along this section, mostly reef breaks that work best at higher tide, with a low to moderate swell and north winds. Those immediately west of Point Flinders are called *The Hole*.

Fishing: There are excellent rip holes and gutters next to the reefs, together with rocks and reefs to fish from at low tide.

Summary: A reef dominated section of coast, most suitable for beach fishing and experienced surfers.

288   THIRTEENTH

Thirteenth Beach SLSC

Patrols: late November to Easter holidays
Surf Lifesaving Club: Saturday, Sunday and
Christmas public holidays
Lifeguard: no lifeguard on duty or weekday patrols

Single bar: TBR
Beach Hazard Rating: **6**
Length: 4 500 m

For map of beach see Figure 111

Thirteenth Beach is part of the 7 km long section of
coast between Barwon Heads and Black Rock.   It
occupies the western 4.5 km and faces essentially due
south.   The beach receives waves averaging 1.5 m, is
moderately sloping and is fronted by a single bar,
dominated by rips every 250 m.   The beach is backed by
a vegetated foredune for most of its length, and the
Barwon Heads to Torquay Road.   The best access is
provided at the surf lifesaving club, with additional car
parks and access tracks located along the road.   The surf
club, founded in 1961, is the only development on what
is a relatively natural beach.   Its members rescue 5
people on average each year.

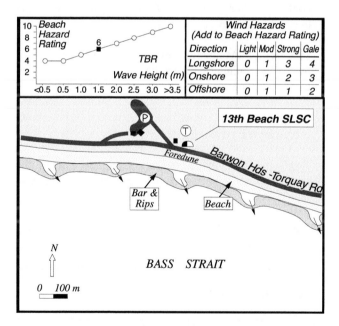

*Figure 111.   Thirteenth Beach is part of the longer
beach that runs 7 km from Barwon Heads to Black
Rock.   The patrolled section has waves averaging 1 to
1.5 m.   Bars and rips are usually spaced every 250 m
along the beach.*

Swimming: Rips are a common feature of the beach,
with safest bathing on the bars in the patrolled area.
Strong permanent rips lie east of the surf club.   The
western end is adjacent to the Black Rock sewer outlet
and should be avoided.

Surfing: A popular surfing beach with low to moderate
swell providing numerous beach breaks, all readily
accessible from the main road.   One of the more popular
areas is in front of the shipping beacon, known as *The
Beacon*.   Best with northerly winds.

Fishing: A good, natural spot for beach fishing, with
good road access to the numerous rip holes that persist
along the beach.

Summary: A relatively undeveloped beach, more
popular with surfers and bathers who want a patrolled
beach away from the crowds.

289   BLACK ROCK

Unpatrolled

No bar: R + sewer outlet
Beach Hazard Rating: **10**
Length: 150 m

Black Rock is a low basalt point, one of four such
points between here and Noble Rocks 3 km to the south-
west.   The Torquay Road takes a right hand turn at the
end of Thirteenth Beach, to bypass the Geelong Sewer
Works, the effluent from which is discharged at Black
Rock.

The beach itself is a narrow strip of sand, bordered and
fronted by basalt rocks and boulders.   Waves break over
the boulders and only reach the beach at high tide.

Swimming: A polluted and hazardous beach with a
rock dominated surf, that should be avoided.

Surfing: There is none on the beach, however surfers
who do not mind the smell, surf the appropriately named
Turd or *Black Rock*.   This right hand break works best in
a large swell.

Fishing: There are plenty of nutrients in the water to
attract fish, and rocks to fish from.

Summary: The beach is off limits owing to the sewer
outlet.

290   BANCOORA

Bancoora SLSC

Patrols: late November to Easter holidays
Surf Lifesaving Club: Saturday, Sunday and
Christmas public holidays
Lifeguard: no lifeguard on duty or weekday patrols

Single bar: TBR
Beach Hazard Rating: **5**
Length: 1 000 m

For map of beach see Figure 112

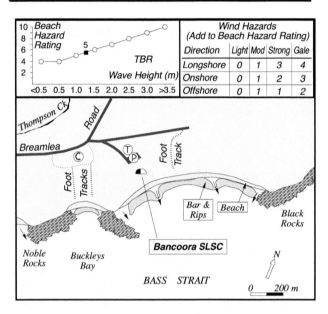

Figure 112. Bancoora Beach and Buckleys Bay are both nestled between low basalt rocks and boulders (Black Rock and Noble Rocks) and are backed by natural foredunes. On Bancoora in particular, watch the rips against the rocks and at times along the beach.

Bancoora Beach is a 1 km long, south-east facing beach located between low, basaltic, rocky points and reefs, and backed by a natural, vegetated foredune. The Bancoora Surf Life Saving Club and car park are located behind the foredune, leaving the beach in an attractive natural state.

The beach receives waves averaging 1.3 m, which usually cut three rips across the 80 m wide single bar and surf zone. Higher waves intensify the rips, with strong permanent rips running out against the rocks at each end. On average, 10 people are rescued here each year.

Swimming: An attractive, moderately safe, patrolled beach, particularly during lower summer swell. Stay on the bars in the patrolled area, and avoid the strong rips near the rocks.

Surfing: Usually a low to moderate beach break, with a right hand point break out on the southern point during higher swell.

Fishing: Popular in summer with campers. It offers both beach fishing and some rip holes and rock fishing off the points.

Summary: An out of the way, relatively natural beach, more popular in summer when the nearby caravan park is full and the beach is patrolled. It is only used by surfers in winter.

291   BUCKLEYS BAY

Unpatrolled

Single bar: LTT
Beach Hazard Rating: **5**
Length: 150 m

For map of beach see Figure 112

Buckleys Bay is a small, attractive, 150 m long beach, bordered by a low basalt point and boulders and backed by 10 to 20 m high dunes. There is a caravan park behind the dunes with foot access to the beach. The beach faces south-east and receives waves averaging 1 m. These waves usually maintain a shallow, continuous bar, with rips forming against the rocks during higher waves.

Swimming: Moderately safe when waves are 1 m or less. When higher, stay to the centre of the beach and clear of the rocks and rips.

Surfing: Usually a low beach break, which closes out during higher waves.

Fishing: There is deep water along the side of the rocks, however take care as they are washed by waves at high tide.

Summary: A small beach used by holiday makers from the backing caravan park.

292   BREAMLEA

Unpatrolled

Single bar: TBR
Beach Hazard Rating: **6**
Length: 2 000 m

Breamlea is a small holiday settlement lying between the banks of Thompson Creek and **Breamlea Beach**. The beach faces south-south-east and runs for 2 km from the low basalt rocks at Noble Rocks to the mouth of the creek at Point Impossible. There is road access to the back of the foredunes, with foot tracks crossing the 20 m high foredune to reach the beach.

The beach receives waves averaging just over 1 m, which usually produce an attached bar cut by rips every 250 m. At the creek mouth, both a tidal channel and shoals are present.

Swimming: A moderately hazardous beach, owing to the persistent rips and creek mouth. Stay on the attached section of the bars and clear of the rips, rocks and creek.

Surfing: Usually low to moderate beach breaks along the length of the beach.

Fishing: This beach has rocks at one end, the creek at the other and usually rip holes and gutters along the beach.

Summary: A natural beach, mainly used by the Breamlea locals for bathing, surfing and fishing.

293, 294   POINT IMPOSSIBLE, WHITES

Unpatrolled			
	Rating	Single bar	Length
293 Point Impossible	**5**	LTT + creek	100 m
294 Whites	**5**	LTT/TBR	4 500 m
		Total length	4 600 m

Point Impossible is a low, calcarenite point, capped by 10 m high foredunes and bordered by the mouth of Thompson Creek. The gravel road from Torquay runs out to the point, where there is a large car park. A small beach (**Point Impossible Beach**) lies in front of the car park and forms the western boundary of Thompson Creek. A foreshore reserve and the road back the 4.5 km long **Whites Beach**, with car parks and access tracks across the dune. The eastern section of the beach, just back from the point, is an official Optional Dress (nude) Beach.

The beach faces south-east and is protected to the south by Point Danger, and along the central-eastern section by extensive rock reefs. As a result, waves average 1 m at the beach and usually produce a continuous, shallow bar only cut by rips during and following high seas.

Swimming: The small Point Impossible Beach varies with wave and tide conditions. Take care if swimming here and watch the deeper tidal channel and currents. Whites Beach is a moderately safe beach close inshore, in lee of the reefs. Watch for rips during higher waves, particularly near the reefs and rocks.

Surfing: Usually low shorebreaks along the beach. However during big winter swell, many surfers head for Point Impossible, where there are two breaks. These are *Insides* against the car park and creek, when waves are up to 1.5 m; and *Outsides* on the outer reef, when waves are higher.

Fishing: The point and creek mouth are the most popular spots, with the beach tending to be shallow.

Summary: A natural beach next to the popular town of Torquay, used by those who want to get away from the more crowded (and clothed) town beaches.

295   FISHERMANS

Unpatrolled
Single bar: LTT
Beach Hazard Rating: **4**
Length: 1 000 m

Fishermans Beach, as the name suggests, is a low energy beach traditionally used to launch fishing boats. This is still true today with a boat ramp on the beach, as well as sailing, yacht, and motor boat clubs all located behind the western end of the beach. The beach lies in Zeally Bay and runs south-west for 1 km from the mouth of the small Deep Creek, then south to the 10 m high limestone bluffs at Yellow Bluff. The entire beach is backed by a foreshore reserve and The Esplanade. It has parking areas and other facilities.

Swimming: A relatively safe beach with a wide, shallow bar and usually no rips. Stay clear of the boating activity near the ramp and boat clubs.

Surfing: Usually a low shorebreak. Big winter swells do however break over the shallow reefs and bars to produce reasonable waves, when everything else is closed out.

Fishing: Best off Yellow Bluff at high tide where you can reach the reef. At low tide, shallow water and exposed reefs dominate.

Summary: This is Torquay's most protected beach and is very popular in summer with those who are looking for quieter surf conditions.

Cliff and rock hazards
Steep and unstable cliffs and in some cases slippery rocks dominate much of the rocky coast between Torquay and Fairhaven. Beware of all cliffs and rocky sections in this region. Only access the coast via designated access points and use extreme caution if having to access the coast via rocks.

296   FRONT

Lifeguard

Patrols: late November to Easter holidays
Lifeguard: 10 am to 6 pm, weekdays and weekends

Single bar: LTT
Beach Hazard Rating: **4**
Length: 1 000 m

For map of beach see Figure 113

Torquay's **Front Beach** fronts the town centre. It is a well-appointed beach with a well-maintained foreshore reserve between The Esplanade and the beach. There are numerous facilities in the reserve, including a tourist information centre. A seawall and a row of tall Norfolk Island pines back the beach, and several wooden groynes cross the beach.

The beach faces due east and runs for 1 km from Yellow Bluff to Point Danger. The point and its reefs protect the beach, which receives waves averaging less than 1 m. These maintain a shallow, continuous, attached bar.

Swimming: This is Torquay's most popular family beach, with usually low waves, a shallow bar and no rips, plus the added safety of a summer lifeguard patrol.

Surfing: Usually a low beach break used by learners. During big swell, waves can make it around *Point Danger* to break as right handers off Front Beach.

Fishing: The best location is on Point Danger. However, watch the waves and tides, as it is awash at high tide.

Summary: Torquay's showpiece beach with good access, facilities, a lifeguard and usually low, safe surf.

297   TORQUAY (SURF)

Torquay SLSC

Patrols: late November to Easter holidays
Surf Lifesaving Club: Saturday, Sunday and Christmas public holidays
Lifeguard: 10 am to 6 pm weekdays during Christmas holiday period

Single bar: TBR
Beach Hazard Rating: **6**
Length: 800 m

For map of beach see Figure 113

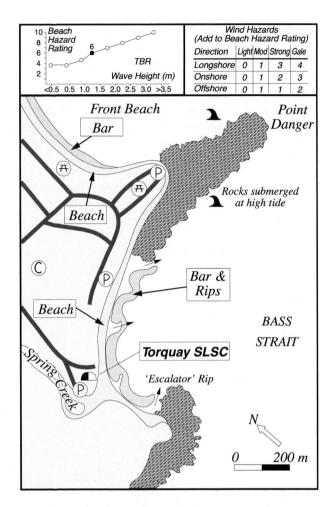

Figure 113.   Torquay is one of the state's most popular holiday destinations and offers a range of beaches, including the more protected Front Beach, and Torquay (Surf) Beach where the surf club is located. The Surf Beach usually has a wide surf zone with rips, and permanent rips against the rocks, so be careful and bathe in the patrolled area.

Torquay is promoted as the 'Surfing Capital of Australia'. It is definitely the commercial surfing capital. **Torquay Beach** was the site of the first malibu board demonstration in Australia, back in 1956. Today Torquay is more important for being the closest town to the famous Bells Beach, and the stepping-off point for a number of surfing locations along the Great Ocean Road. The Torquay Surf Life Saving Club, founded in 1945, has also hosted state, national and international surf lifesaving carnivals. More recently, a number of major surfing companies and an excellent Surf World exhibition have been located at Torquay's Surf Coast Plaza.

Torquay Beach is 800 m long and faces south-east, with some protection provided toward the southern end by Rocky Point. Extensive intertidal rock reefs lie off Point Danger at the northern end, and Spring Creek drains across the beach just west of the surf club. Waves average 1.2 m and usually cut three rips across the single bar, with additional permanent rips against the rocks at each end. The southern rip, known as the 'Escalator' is particularly strong during easterly conditions. The beach itself is moderately steep and is backed by extensive parking areas, particularly along the eastern half.

Swimming: A very popular summer beach bolstered by its name, good accessibility and surf lifesaving club. The beach is moderately safe on the bars in the patrolled areas, however avoid the rocks and strong rips, particularly toward Point Danger as, on average, 27 people are rescued here each year.

Surfing: The site of the first short board riding in Australia and still a very popular, if crowded, location year round. The beach offers a wide beach break, which is moderately protected during westerlies, though best in a north-westerly, with a left hander off Point Danger.

Fishing: Both beach and rock fishing are available, with the best rip holes toward the northern end. Take care on the rocks, as they are awash at high tide.

Summary: One of Victoria's best known and most popular summer surfing beaches. The adjacent town offers all facilities, while the patrolled beach is popular with bathers and surfers.

298　**JAN JUC**

Jan Juc SLSC

Patrols: late November to Easter holidays
Surf Lifesaving Club: Saturday, Sunday and Christmas public holidays
Lifeguard: 10 am to 6 pm weekdays during Christmas holiday period

Single bar: TBR/RBB
Beach Hazard Rating: **7**
Length: 1 200 m

For map of beach see Figure 114

Jan Juc Beach is located immediately south of Torquay and is a little more exposed, receiving waves averaging 1.4 m. It extends for 1.2 km between Rocky Point and Bird Rock and faces almost due south, resulting in larger waves. The waves combine with the fine to medium sand to produce a single bar cut by three to four rips, with permanent rips against the rocks at each end.

The northern half of the beach is backed by low bluffs, partly covered by dunes. The surf lifesaving club, parking and access, together with Jan Juc Creek, are in the centre, while the narrow, southern half of the beach is backed by 20 m high cliffs. The Jan Juc Surf Life

Saving Club was founded in 1963 and annually rescues an average of 30 people.

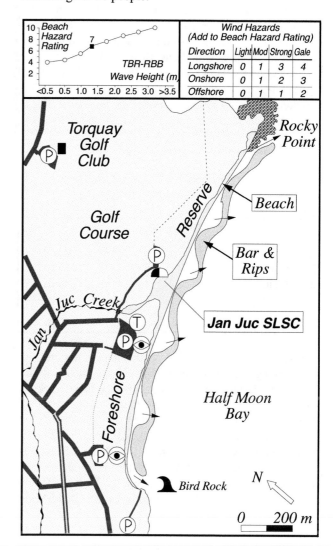

Figure 114. Jan Juc Beach has good access in the centre, with bluffs behind the northern half and higher cliffs to the south. The beach usually has several rips, particularly against Rocky Point and Bird Rock.

Swimming: A potentially hazardous beach, owing to the high waves and persistent rips. More suitable for experienced bathers and surfers. Stay between the flags and away from the rips and rocks.

Surfing: Usually variable beach breaks, however Bird Rock can provide excellent rights with a moderate swell and high tide.

Fishing: Best toward the northern end, where rip holes are more persistent.

Summary: Jan Juc is Torquay's second and more exposed surfing beach. Still popular in summer for those escaping the Torquay crowds, however the variable beach and surf conditions warrant extra care.

SURFING'S GOLDEN MILE

The 3 km section of coast from Bird Rock to Bells Beach is rated by many surfers as Victoria's 'Golden Mile of Surf'. The coast is composed of limestone bluffs for the most part, with Bells Beach being the longest beach at 300 m. However, all the surfing is done on the limestone reefs that lie along the coast. The predominantly right hand breaks include *Bird Rock, Sparrows, Steps, Evo's, Boobs, Winkipop* and *Bells*.

299, 300, 301   PRINCES ST, BEACH 300, DEAD MAN

Unpatrolled			
	Rating	Single bar	Length
299 Princes St	**5**	LTT	250 m
300 Beach 300	**7**	TBR	200 m
301 Dead Man	**6**	R + reefs	80 m
		Total length	530 m

South of Bird Rock, the coast is dominated by steep, 30 to 40 m high, limestone bluffs, which wind their way for 2.5 km to Bells Beach. The original vehicle track to Bells ran along the top of these bluffs. At the base of the bluffs are three small beaches, each surrounded by rocks and reefs. The three beaches are all relatively short and narrow, with the one below **Princes St** awash at high tide. Access is difficult, requiring a walk around the rocks from Princes St, or down the bluff to **Beach 300**, and down the gully of the same name to **Dead Man Beach**.

The three beaches receive waves averaging 1 to 1.5 m, with the reefs lowering waves at Dead Man and producing a steep, barless beach. At both Dead Man and Beach 300, permanent rips persist against the reefs.

Swimming: None of these beaches are recommended for bathing. They are difficult to access, are dominated by rocks and reefs and are backed by steep slopes.

Surfing: There are four recognised surfing breaks along this section, and all are right handers breaking over reefs. *Sparrows* is off the southern end of Princes St, with an inner and outer reef break; *Steps* is out from Beach 300; while *Boobs* on the southern side of Dead Man has a left and right; and finally, the more famous *Winkipop* is on the northern side of Bells Beach.

Fishing: Access is difficult, however there are plenty of deep holes to fish from the rocks along this section.

Summary: Three difficult to reach beaches, used mainly by the more experienced surfers and fishers.

---

302   BELLS

Unpatrolled
No bar: R
Beach Hazard Rating: **6**
Length: 300 m

Bells Beach is Victoria's most famous surfing beach and one of the world's great surfing breaks. The excellent break is due to a combination of clean waves, that have refracted around the Otways, and particularly a gently sloping limestone reef off the southern point that produces one of the world's best right handers. It can handle anything from 1.5 up to 7 m.

The world's longest running surfing contest began at Bells in 1961 and continues every Easter. In recognition of its surfing status, the Victorian government proclaimed it the state's and Australia's first Surfing Reserve in 1971.

Bells is now serviced by a good road, a large car park, viewing areas and facilities. The walk to the beach is still, however, down the gully at the southern end of the beach. The beach is just 300 m long and faces south-east, with prominent, 40 m high, limestone headlands at each end. Waves average 1 to 1.5 m and the beach is composed of coarse sand which, even under high waves, stays steep and barless. A normally low shorebreak becomes very heavy in high waves.

Swimming: During low waves, the beach is relatively safe close inshore, however any surf will produce a rip running along the beach and out toward the northern point. During big seas, this rip becomes an express ride for the surfers.

Surfing: *Bells* is a world class right when above 1.5 m. When smaller, the waves break close in to the headland and produce a right called *Rincon*. Further around the head are two more reef breaks which work below 2 m, called *Centre Side* (a right) and *Southside* (a left).

Fishing: The water is deep right off the beach, while at low tide you can fish from the reefs at each end.

Summary: One of the meccas of surfing and well worth a visit, if only to view the beach and surf from the bluffs.

---

303 SOUTHSIDE

Unpatrolled
Single bar: TBR/RBB
Beach Hazard Rating: **7**
Length: 1 000 m

Southside Beach is located on the southern side of Bells Headland. Unlike its neighbour, Southside is composed of finer sand and has a wide beach and surf zone, with rocks only outcropping toward the northern end of the beach. The beach is 1 km long, lying between Bells and Jarosite Headlands. It is backed by an amphitheatre of slumped sands and clays, that forms an eroding, 10 m high bluff and cobble storm beach along the back of the sand. It faces the south-east and receives waves averaging 1.5 m. Combined with the fine sand, these produce a wide surf zone, usually containing a permanent rip against each headland and one to two rips toward the centre.

The road to Bells Beach runs past the northern end of the beach and there is a cliff-top car park on Bells Headland, with a walking track down to the beach. The beach is also an official Optional Dress (nude) Beach.

Swimming: This is a potentially hazardous beach, with permanent rips and some rocks in the surf. Stay inshore on the bar and well clear of the rocks and headlands.

Surfing: The best known breaks are at the headlands, with a left called *Southside* off Bells Headland, and *Jarraside* out from the southern end of the beach.

Fishing: There are deep rip gutters off the headlands, as well as beach holes and gutters.

Summary: An energetic and potentially hazardous beach, fine for sunbathing but be careful if swimming.

304 ADDISCOT

Unpatrolled
Single bar: LTT/TBR
Beach Hazard Rating: **6**
Length: 1 800 m

Addiscot Beach is a 1.8 km long, curving, south-east facing beach, bordered and rimmed by red, slumping cliffs composed of unconsolidated sands and clays. The cliffs reach 80 m high toward the eastern Jarosite Headland. A road from the Great Ocean Road runs out to the southern Point Addis, where there is a car park and a track down to the 20 m high bluffs to the southern end of the beach. The beach is an official Optional Dress (nude) Beach.

The beach receives waves averaging 1 to 1.5 m, that increase in height toward Jarosite Headland. The waves and fine sand produce a low beach with a continuous bar, which is increasingly cut by rips to the north.

Swimming: The southern corner is the safest, as it has lower waves and is usually free of rips. Be very careful up the beach, as both the rips and cliffs are hazardous.

Surfing: There are beach breaks right along the beach, that increase toward Jarosite Headland.

Fishing: Best off the rocks at Point Addis. However, watch the waves that wash over the rocks at high tide.

Summary: An interesting beach and view, with the southern corner being the most protected.

305 POINT ADDIS

Unpatrolled
No bar: R
Beach Hazard Rating: **6**
Length: 80 m

At the base of 20 m high Point Addis is a narrow, 80 m long, sand beach, which is awash at high tide and fronted by rocks and reef flats at low tide. It can only be reached with difficulty around the rocks. It is not recommended for bathing.

Swimming: Not recommended; dominated by rocks, reef and a permanent rip.

Surfing: There is a right over the reef, which works in a low to moderate swell at high tide.

Fishing: There are reef holes and gullies off the rocks, however it is a hazardous location to fish.

Summary: A small, hazardous beach.

306, 307, 308 BLACK ROCKS, EUMERALLA, DEMONS BLUFF

Unpatrolled			
	Rating	Single bar	Length
306 Black Rocks	7	TBR/RBB	1 000 m
307 Eumeralla	7	TBR	2 200 m
308 Demons Bluff	8	TBR/RBB	1 500 m
		Total length	4 700 m

Between Point Addis and Anglesea is a spectacular, but hazardous, 5 km section of cliffed coast and energetic beaches. The cliffs are composed of weathered sandstone that continually slumps and, in places, falls onto the beaches. The three beaches face south-south-east and receive waves averaging 1.5 m which, beside producing wide, energetic surf, help to erode the base of the cliffs and bluffs. Access to these beaches is not recommended owing to rock falls.

The first beach (**Black Rocks Beach**) extends 1 km from Point Addis to Black Rocks. It can be reached via a walking track from the Point Addis car park. It usually has a single bar, cut by permanent rips against each headland, and two central rips. The backing bluffs are fairly subdued, reaching only 30 m. **Eumeralla Beach** is 2.2 km long and backed by massive slumps in its 90 m high bluffs. The beach sand is coarser, resulting in a surf zone dominated by rips every 300 m. **Demons Bluff**, as the name suggests, is a sheer, 30 to 50 m high, eroding cliff, fronted by a narrow, 1.5 km long beach that is awash at high tide. Rock falls commonly cover parts of the beach, which can only be reached on foot along the base of the cliffs. The beach has a wide surf zone, dominated by rips every 300 m. The Anglesea sewer works back the beach.

Swimming: These are three potentially hazardous beaches, dominated by moderate waves and persistent rips, with Black Rocks and particularly Demons Bluff having the added hazard of rock and cliff falls.

Surfing: All three have similar beach breaks, with the best access via Point Addis to Black Rocks. There is a point break at *Grinders* (Point), however access via the backing scout camp is steep and very difficult.

Fishing: The southern side of Point Addis is the best location, with both rock and beach fishing into deep rip holes.

Summary: This is a dynamic section of coast, both geologically and in the surf. Be wary of both, particularly if walking below the cliffs or in the surf.

309   ANGLESEA RIVER

Unpatrolled
Single bar: LTT/TBR
Beach Hazard Rating: **6**
Length: 700 m

The mouth of the **Anglesea River** protrudes from the coast in lee of an offshore reef. To either side of the river are the weakly consolidated bluffs of Demons Bluff and Point Roadknight. The river and its backing

estuary flow through a low point in the otherwise prominent bluffs. The 700 m long beach begins at the river mouth and runs in front of low bluffs, before the high Demons Bluff is encountered. On the bluffs behind the beach is a caravan park, and there is an access track from here down to the north side of the river mouth.

The beach faces east and, with protection from the reef and Point Roadknight, receives waves averaging about 1 m. These produce a continuous bar, only cut by rips during and following periods of higher waves.

Swimming: Moderately safe under low waves, however be careful if there is a surf, as currents run up the beach away from the river mouth. Stay clear of the river mouth when it is open.

Surfing: Usually low beach breaks.

Fishing: There is a greater chance of rip holes up the beach, while the river is a favourite spot, particularly when open.

Summary: A popular summer beach owing to its location below the caravan park.

310   ANGLESEA

Anglesea SLSC
Patrols: late November to Easter holidays
Surf Lifesaving Club: Saturday, Sunday and Christmas public holidays
Lifeguard: 10 am to 6 pm weekdays during Christmas holiday period
Single bar: LTT/TBR
Beach Hazard Rating: **5**
Length: 400 m
For map of beach see Figure 115

Anglesea Beach lies next to the mouth of the Anglesea River and fronts the town of Anglesea. The beach is 400 m long and curves in a south to south-east facing arc between the usually closed river mouth and the eroding rocks and cliffs in front of the bluff-top Anglesea Surf Life Saving Club. Access and parking are available at the river mouth, off the Great Ocean Road, and at the surf club.

The beach receives waves averaging 1 m. The larger ocean waves are reduced as they refract around Point Roadknight. They produce a wide, shallow, single bar, which is usually attached to the beach south of the surf lifesaving club. It is increasingly cut by rips toward the river mouth. The Anglesea Surf Life Saving Club was formed in 1952 and annually averages 12 rescues.

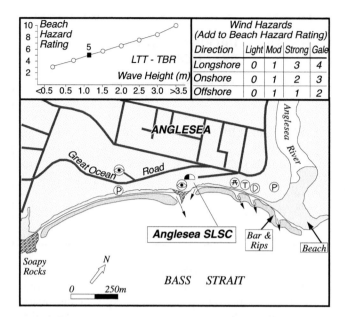

Figure 115. Anglesea Beach is located at the mouth of the Anglesea River, which it usually blocks. The surf lifesaving club is located on bluffs at the more protected and safer southern end of the beach. Stay at the southern end, as waves and rips increase toward the river mouth.

Swimming: A moderately safe beach under typical summer conditions, however avoid the rip against the southern rocks. Best at high tide as waves tend to dump at low tide. Stay on the bar and in the patrolled area.

Surfing: Popular with the less experienced surfers who use the wide, gently sloping surf zone.

Fishing: Beach fishing is best at the river mouth where rip holes are more prevalent.

Summary: A popular summer beach, offering good parking and access, and a moderately safe patrolled beach.

311   SOAPY ROCKS

Unpatrolled
Single bar: LTT/TBR
Beach Hazard Rating: **5**
Length: 750 m
For map of beach see Figure 115

Soapy Rocks refers to the eroding, red bluffs that back the beach. They also outcrop on the beach as slippery rocks (hence the name Soapy) and in the surf at either end of the 750 m long beach. The beach lies between Anglesea Surf Life Saving Club and Point Roadknight Beach and is protected by Point Roadknight and its easterly orientation. There is a patrol tower - lookout

above the centre of the beach. Access is from the western Soapy Rocks car park and along the beach from Anglesea or Point Roadknight. Beware of the unstable cliffs and slippery (soapy) rocks which make this a hazardous beach to access on foot.

The beach receives waves averaging less than 1 m. They break along a continuous bar, with wave height and the likelihood of rips increasing toward Anglesea.

Swimming: Safest at the Soapy Rocks end, where waves are lower and rips less likely.

Surfing: Usually a low beach break.

Fishing: Not recommended due to slippery rocks.

Summary: A narrow beach below steep, unstable bluffs, with access only over slippery rocks. Not recommended for access, swimming or fishing.

312   POINT ROADKNIGHT

Lifeguard
Patrols: Christmas holiday period
Lifeguard: 10 am to 6 pm all week
Single bar: LTT
Beach Hazard Rating: **4**
Length: 700 m

Point Roadknight is a narrow ridge of dune calcarenite that parallels the adjoining Urquhart Bluff Beach. The point and its reef protrude 500 m to the east and afford considerable protection to the beach. The beach is 700 m long and faces north-east. It lies between the slippery Soapy Rocks and the point. Beware of the slippery rocks which are a hazard to walk on. There is road access to the back of the beach, a large car park, a boat ramp and a yacht club.

Waves reaching the beach average less than 1 m, which results in a continuous, attached bar and usually no rips.

Swimming: This is the safest beach in the Anglesea region and is also patrolled daily by lifeguards during the Christmas holiday period.

Surfing: Usually too small to bother about.

Fishing: Better off the point than the beach.

Summary: A popular summer beach for those who want lower waves and the added safety of a patrolled beach.

313, 314   POINT ROADKNIGHT 'POINT', URQUHART BLUFF

Unpatrolled				
	Rating	Inner bar	Outer bar	Length
313 Point Roadknight 'Point'	7	TBR	-	100 m
314 Urquhart Bluff	7	TBR/RBB	RBB/LBT	5 000 m
			Total length	5 100 m

At Point Roadknight, the coast turns 180° and runs due west for 4 km, before finally swinging to the south in lee of Urquhart Bluff. In the vicinity of the point, calcarenite reefs dominate, and right on the point is the 100 m long, south facing **Point Beach**. Beyond this beach, the shoreline continues uninterrupted to **Urquhart Bluff**.

The Point Roadknight end has beach access tracks on the point and off Melba Parade, while the Great Ocean Road backs most of the remainder of the beach, with a large car park in lee of Urquhart Bluff.

Both beaches receive waves averaging 1.3 m, which combine with the fine sand to build a double bar system along the main beach. Rips cut the inner bar every 400 m and there are more widely spaced rips on the outer bar. The outer bar only breaks during bigger seas.

Swimming: Both beaches are hazardous. The more remote Point Beach has a strong, permanent rip against the reefs. Strong, persistent rips run the length of Urquhart Bluff Beach, with a strong, permanent rip running past the bluff car park. This rip has caused fatalities in the past.

Surfing: Toward the point the beach is called *Steps* (after the walk down the bluff) which has a number of breaks over the reefs. The remainder of the beach has numerous beach breaks, which are best in a low to moderate swell and northerly winds.

Fishing: The point and Steps area is a popular spot, with the reef providing permanent holes and gutters. There are many gutters along the beach, as well as the permanent rip at the bluff.

Summary: A long, energetic beach most suited to experienced surfers and beach fishers, with windsurfers making good use of the bluff area.

315 - 319   URQUHART BLUFF SOUTH, BEACHES 316 to 319

Unpatrolled			
	Rating	Single bar	Length
315 Urquhart Bluff South	7	TBR	200 m
316 Beach 316	5	R/LTT	100 m
317 Beach 317	6	R/LTT + reef	100 m
318 Beach 318	6	LTT	100 m
319 Beach 319	6	TBR + reef	500 m
		Total length	1 000 m

The 5 km of coast between Urquhart Bluff and Table Rock at Fairhaven is dominated by 20 to 50 m high, eroding bluffs composed of poorly consolidated limestone, tuffs, clays and silts. As they erode, they leave inter- and sub-tidal rock platforms and reefs. Running along the base of the bluffs are twelve small, exposed beaches, mostly dominated by the headlands, rocks and reefs.

The first five face south-east and extend from Urquhart Bluff south-west for 1 km. They can be reached at low tide around the rocks from Urquhart Bluff, or by climbing down some of the less steep bluffs. They are all exposed to waves averaging 1.3 m, but owing to the degree of protection or the presence of rocks and reefs, they have variable beaches and surf zones.

Urquhart Bluff South Beach (315) is 200 m long and has a wide, shallow surf zone with a permanent rip against the southern headland and reef. **Beach 316** is 100 m long and more protected by a large reef off the southern headland. It has a steep, narrow beach and narrow surf, however a permanent rip runs out by the reef. **Beach 317** is a narrow platform beach, backed by a steep, eroding bluff and fronted by a near continuous rock platform. **Beach 318** is a pocket of sand backed by steep bluffs, with prominent rock platforms at either end and a permanent rip. **Beach 319** is 500 m long and is backed by vegetated bluffs, with a narrow beach, and a reef and rip dominated surf zone. There is an access track leading from the houses behind the bluffs, and the locals use this beach.

Swimming: These are five hazardous beaches, owing to their relatively remote location and access ranging from difficult to dangerous. This is coupled with the presence of rocks, reefs and permanent rips.

Surfing: Moderate swells and a high tide produce right hand breaks over some of the reefs, most of which can be viewed from the top of Urquhart Bluff.

Fishing: Each of these beaches has good permanent rip holes and reef gutters. The biggest problem is access. Be very careful at high tide and in bigger seas, as the rocks and some of the beaches are awash.

Summary: Five beaches dominated by the bluffs, rocks and reefs.

320, 321   EAGLE NEST REEF NORTH & SOUTH

Unpatrolled			
	Rating	No bar	Length
320 Eagle Nest Reef North	6	R + reef	150 m
321 Eagle Nest Reef South	6	R + reef	100 m
		Total length	250 m

Eagle Nest Reef is a sandstone reef backed by 30 to 40 m high, red bluffs. Between the rock platform, reefs and the bluffs are two narrow beaches. On the north side, the beach is 150 m long, backed by a cobble storm beach at the base of the bluff and fronted by deep water and a permanent rip against the reefs. The south beach is hemmed in between two headlands and platforms. It has a single permanent rip channel straight off the beach. Both beaches are only accessible around the rocks at low tide.

A coastal walking track begins at the end of Boundary Road, just north of the northern Eagle Nest Beach, and follows the bluffs for 4 km to Fairhaven Surf Life Saving Club.

Swimming: These are two hazardous beaches, with deep water, permanent rips, rocks and reefs.

Surfing: Some waves over the reefs at high tide.

Fishing: Excellent deep holes off the beach and platforms. However, be careful as they are awash at high tide.

Summary: Two small beaches best suited for rock and beach fishing.

322, 323   AIREYS INLET, AIREYS INLET SOUTH

Unpatrolled			
	Rating	Single bar	Length
322 Aireys Inlet	6	TBR + rocks	500 m
323 Aireys Inlet South	6	LTT	50 m
		Total length	550 m

Aireys Inlet is a holiday settlement spread for 3 km along the back of the bluffs, on the north side of the actual Aireys Inlet. Below the bluffs is a series of small, south-east to south facing beaches. The only readily accessible beach is **Aireys Inlet Beach** (322), which is located at the mouth of a gully. It has a small car park, and steps down to the beach from the end of Eagle Rock Parade.

The beach is 500 m long, and is bordered by 20 m high headlands and rock platforms composed of red sandstone. Bluffs are eroding along the back of the beach, particularly the southern 300 m, which is a narrow strip of sand awash at high tide. The surf zone is 50 m wide, with a permanent rip against the northern rocks. **Aireys Inlet South Beach** (323) is a small, 50 m pocket of sand fronted by reefs. It is located below 30 m high bluffs and is essentially inaccessible.

Swimming: Aireys Inlet Beach has the best access in the area and is primarily used by the locals. However, it is a hazardous beach with a strong rip feeder current running along the beach and a rip running out past the headland. Take care if swimming here.

Surfing: There are beach breaks over the bars and southern reefs.

Fishing: The rock platform at the north end provides the best location to fish the permanent rip channel.

Summary: A popular beach with the locals, but one requiring caution.

324 - 327   BEACH 324, EAGLE ROCK, SPLIT ROCK, TABLE ROCK

Unpatrolled			
	Rating	Single bar	Length
324 Beach 324	6	LTT + reef	200 m
325 Eagle Rock	6	R + rocks & reef	50 m
326 Split Rock	8	R + rocks & reef	50 m
327 Table Rock	8	TBR + rocks & reef	200 m
		Total length	500 m

Between Eagle Rock and Table Rock at the mouth of Aireys Inlet, the coast is composed of 20 to 30 m high bluffs, fronted by rock platforms, reefs and sea stacks. In amongst the rocks are four small beaches, only the first of which (324) is accessible.

Beach 324 has a bluff-top car park and a steep track down to the 200 m long, east facing beach. Steep cliffs form the headlands and reefs lie offshore. **Eagle Rock** is a 20 m high sea stack backed by rocks, reefs and a narrow strip of sand at the base of the backing bluff. **Split Rock** is a lower rocky reef, with a narrow sand and cobble beach tucked into a 50 m long indentation in the backing bluffs. **Table Rock** forms the main headland and is the site of the lighthouse. On its southern side is a 200 m long, south facing beach that ends at the mouth of Aireys Inlet. Rocks outcrop on the beach and reefs lie offshore.

Swimming: These are all hazardous beaches, with only Beach 324 providing access and moderately safe conditions during low waves. The three Rock Beaches should be avoided, as they are difficult to access and are dominated by rocks, reefs and rips.

Surfing: There are a number of reef breaks along this section that are usually better at high tide.

Fishing: If you do not mind rock hopping, there are several permanent holes and gutters off the beach and in the rocks and reefs along the base of the bluffs. However be careful, as the rocks are awash at high tide and during larger swell.

Summary: Four beaches to view from the lighthouse and bluff-top walking track, but not recommended for anything other than fishing. The lighthouse was built in 1891 and stands 60 m above sea level. It features in the children's television series 'Around The Twist'.

328   FAIRHAVEN

Fairhaven SLSC
Patrols: late November to Easter holidays   **Surf Lifesaving Club:** Saturday, Sunday and Christmas public holidays   **Lifeguard:**   no lifeguard on duty or weekday patrols   Inner bar: TBR/RBB   Outer bar: RBB/LBT   Beach Hazard Rating: **7**   Length: 6 000 m   *For map of beach see Figure 117*

Six kilometre long **Fairhaven Beach** is the longest beach on the Great Ocean Road, from which it is readily accessible, as the road backs the entire beach. The beach runs due west from the mouth of Moggs Creek for 4 km, before slowly curving around to face east at the western Cinema Point.

Victorian Beaches - Region 7: West (Fairhaven to Cape Otway)

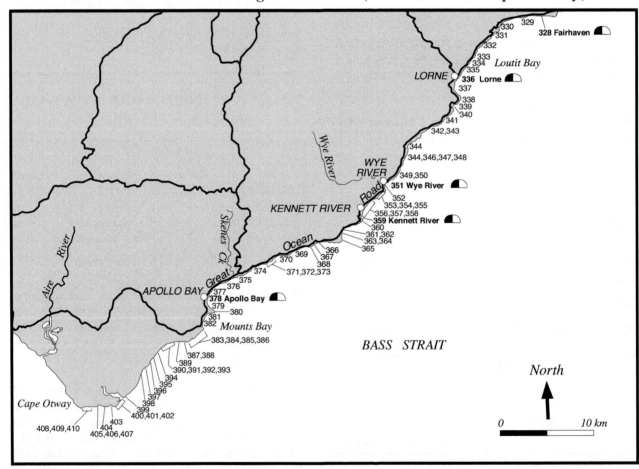

Figure 116.   Region 7: (western section) Fairhaven to Cape Otway (Beaches 328 - 410)

The southerly aspect exposes the beach to waves averaging 1.5 m, which combine with the fine to medium beach sand to produce a 200 m wide surf zone containing two bars. The inner bar is cut by rips every 300 m, resulting in up to 20 rips along the beach. The outer bar, which only breaks in higher waves, has more widely spaced rips, when it is active.

The Fairhaven Surf Life Saving Club, founded in 1957, is located toward the eastern end of the beach, and its members annually average 10 rescues.

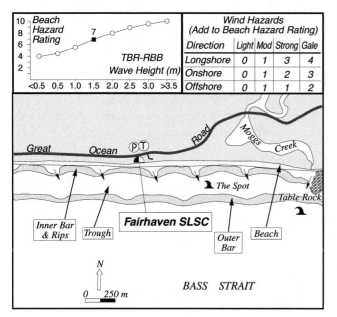

Figure 117. Fairhaven Surf Life Saving Club is located at the eastern end of 6 km long Fairhaven Beach. The beach has a double bar system, with rips cutting the inner bar every 300 m and the outer bar only active during larger swell.

Swimming: A potentially hazardous beach, with usually moderate waves and persistent and often strong rips. Westerly winds intensify longshore and rip currents. Stay in the patrolled area on the attached inner bar.

Surfing: The beach has numerous beach breaks and usually a good swell. However, it is exposed and works best with northerly winds. Some well-known spots along the beach include the mouth of *Moggs Creek,* where low summer lefts can be found; *The Spot,* a reef break just east of the surf lifesaving club; and further down at *Eastern View* and *Spouts Creek.*

Fishing: The good access and numerous rips and holes make this a popular, although usually uncrowded, spot for beach fishing. The mouths of Moggs and Spout Creeks are also popular, when they are flowing.

Summary: A long, natural beach more suited to experienced bathers and surfers, with the patrolled area in front of the surf club offering the safest bathing area.

Toward the western end of the beach is a Memorial Arch commemorating the construction of the Great Ocean Road during the depression years of the 1930s.

329, 330   CINEMA POINT NORTH, CINEMA POINT

Unpatrolled			
	Rating	Single bar	Length
329 Cinema Point North	4	R + reefs	50 m
330 Cinema Point	5	LTT	60 m
		Total length	110 m

At **Cinema Point**, the winding, cliff hanging section of the Great Ocean Road begins. Cinema Point is a 30 m high, grassy knoll, backed by the road and surrounded by sandstone rock platforms. There are two small beaches here, one on either side. The eastern one is below the car park and viewing area immediately behind the knoll. It is 50 m long, faces east, is backed by road fill and the knoll, and is dominated by platforms and reefs. On the western side of the point is the main beach, that occupies the deep gully carved by Grassy Creek. It consists of a veneer of sand over rock platforms.

Both beaches receive waves averaging about 1 m and have a surf zone entirely dominated by rocks and reefs.

Swimming: Be very careful if bathing here, as permanent rip currents drain out from both reefs, and rocks and reefs abound in the surf.

Surfing: The north side of Cinema Rocks is known as *Hunters* or *Shark Alley.* It has a moderate right hander during big swell.

Fishing: The extensive rock platforms at low tide provide good access to the rock gullies on either side of the point.

Summary: A favourite viewing site, with the beaches used by surfers and fishers, but unsuitable for safe bathing.

331   CARYARD

Unpatrolled
Single bar: TBR + rocks & reef
Beach Hazard Rating: **8**
Length: 800 m

One kilometre south of Cinema Point is a straight, south-east facing section of coast composed of steep, vegetated bluffs. At the base of the bluffs is 800 m long **Caryard Beach**, which is fronted by a near continuous rock platform, with a rip dominated surf zone further out. The name derives from the stolen cars that have been dumped over the cliff. The Great Ocean Road runs along the slopes 100 m above the beach, however parking is limited and access is via a small residential road and track, or along the rocks from Cinema Point.

Swimming: This is a dangerous beach, with safe bathing only possible under calm conditions in the tidal pools.

Surfing: Chance of a wave over the reef at high tide.

Fishing: Excellent rock platforms with numerous holes and gutters.

Summary: A difficult to find and access beach, dominated by the rock platform.

332   CATHEDRAL ROCKS

Unpatrolled
Single bar: TBR/RBB
Beach Hazard Rating: 7
Length: 300 m

Cathedral Rocks is another of the prime surfing spots along this section of coast. The Great Ocean Road hugs the bluffs along this section and you cannot miss seeing the beach and adjoining rock platforms and reefs. There are three small car parks off the road to park and view, or go surfing.

There is a 300 m long beach extending south from the mouth of the small creek. It has rock platforms at either end and a few rocks and reefs, particularly toward the creek. Permanent rips run out at each end.

Swimming: This is a dangerous beach dominated by rocks, reefs and strong rips.

Surfing: The rocks along the southern point provide the base of a long right hander.

Fishing: The easy access and rock platforms make this a popular location for rock and beach fishing.

Summary: A beach everyone sees while driving down the road and worth the stop to view the scenery, or the action if the surf is running.

333   BEACH 333

Unpatrolled
No bar: R + rocks & reefs
Beach Hazard Rating: **6**
Length: 150 m

Around the bend from the last car park at Cathedral Rocks is a similar beach. The Great Ocean Road runs along the slopes 20 m above the beach and access is a little more difficult. **Beach 333** is 150 m long and faces south-east. It has prominent rock platforms at either end and deep water off the northern end of the beach, where a permanent rip runs out.

Swimming: Not recommended owing to the rock, reef and, at times, a heavy shorebreak, with a strong, permanent rip off the beach.

Surfing: This is not as good as Cathedral Rocks, however there is a right hander over the southern rocks, and an easy paddle out in the rip.

Fishing: Good beach and rock fishing into deep, permanent holes and gutters.

Summary: There is a picnic area above the northern end of the beach, which is a nice spot to stop and take in the Great Ocean Road scenery.

334   REEDY CREEK

Unpatrolled
No bar: R + rocks
Beach Hazard Rating: **4**
Length: 80 m

Reedy Creek flows out under the Great Ocean Road 3 km north of Lorne. The road curves around the back of the beach and the road fill forms the backing boundary, with rock platforms to either side almost enclosing the small, 80 m long, south-east facing pocket beach. The platforms and reefs lower the waves to only about 50 cm at the beach, and the waters inside the small bay are usually calm.

Swimming: Relatively safe inside the reef, however watch the rocks and rip current flowing out between the two reefs.

Surfing: Only irregular breaks over the reefs at high tide.

Fishing: This is another very accessible spot, giving good access to beach and rock fishing.

Summary: The car park on the southern side is another popular viewing spot. One of the safer beaches along this section of coast.

335   NORTH LORNE

> **Unpatrolled**
>
> Single bar: TBR/LTT + rocks & reef
> Beach Hazard Rating: **6**
> Length: 1 300 m

South of Reedy Creek, the Great Ocean Road straightens out and heads for Lorne. At Stony Creek a narrow, crenulate platform beach begins, fronted by a near continuous rock platform and reefs. Rip dominated surf lies in and seaward of the reefs. The beach extends south-south-west for 1.3 km to the mouth of the Erskine River. The Great Ocean Road parallels the back of the beach, with two parking areas located immediately above the beach.

Swimming: There are some tidal pools along here that are relatively safe at mid to low tide, however be careful at high tide and seaward of the reefs, as rips and rocks dominate.

Surfing: Generally too much rock.

Fishing: A very popular spot to fish from the beach or rocks.

Summary: This is a popular summer beach for sunbathing and picnicking. It is reasonably safe to bathe in the tidal pools during low waves and at low tide.

336   LORNE

> **Lorne SLSC**
>
> **Patrols:** late November to Easter holidays
> **Surf Lifesaving Club:** Saturday, Sunday and public holidays
> **Lifeguard:** 10 am to 6 pm weekdays during Christmas holiday period
>
> Single bar: LTT
> Beach Hazard Rating: **5**
> Length: 1 200 m
>
> *For map of beach see Figure 118*

Lorne is one of Victoria's premier holiday destinations, with the popular beach paralleling the Great Ocean Road. A well-established foreshore reserve separates the road and main street from the beach. The reserve has extensive parking, parks and a pool. The Lorne Surf

Life Saving Club was formed in 1948 and occupies the southern corner. Its members average 50 rescues annually, attesting to its popularity, as well as the surf conditions.

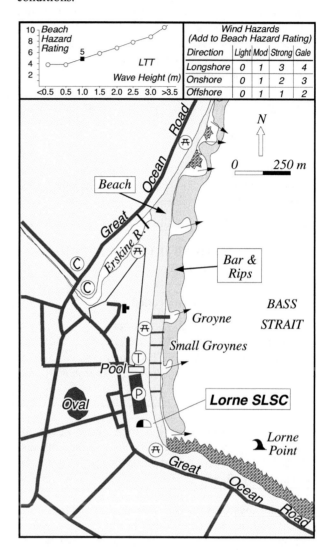

Figure 118. Lorne Beach is located between the mouth of the Erskine River and the protective rocks of Point Grey. The surf lifesaving club and patrol area are at the most protected southern end. Wooden and rock groynes cross the beach and rips intensify up the beach.

The beach is 1.2 km long, running almost due south from the usually closed mouth of the Erskine River and its adjacent rock platforms. The southern end is bordered by the long sandstone rock platform that extends 1 km out to Point Grey and the Lorne wharf. Point Grey affords the beach considerable protection, with waves averaging 1 m. These produce a low, wide, attached bar, with rips only occurring during and following higher waves. In addition, six small wooden groynes and one large rock groyne have been built across the beach, supposedly to prevent beach erosion.

Swimming: A relatively safe beach during average

summer conditions. Be careful of rips and groynes up the beach, with the safest bathing in the southern patrolled corner.

Surfing: Usually a low, wide beach break more suitable for beginners. During moderate to large swell, *Lorne Point* provides good right handers, while further out at *Point Grey* a fuller wave is preferred by windsurfers and waveskis.

Fishing: The best beach fishing is up at the Erskine River mouth and off the adjacent rocks, and the rock groyne at high tide.

Summary: One of Victoria's most popular beach and holiday destinations, offering a relatively safe beach backed by full resort facilities.

337, 338, 339   LORNE JETTY, POINT GREY, SHELLY

Unpatrolled			
	Rating	Single bar	Length
337 Lorne Jetty	3	R + rocks	50 m
338 Point Grey	6	TBR + reef	300 m
339 Shelly	8	TBR + rocks & reef	100 m
		Total length	450 m

A low rock platform runs out from Lorne Beach for 1 km to the jetty at Point Grey. Next to the jetty and crossed by a concrete boat ramp is the small, steep **Lorne Jetty Beach** (337). It receives low surging waves, but is not well suited for bathing. At the point itself, the coast and the Great Ocean Road swing to the south, and immediately south of the point are two beaches. There is a continuous car park behind the Point Grey Beach and a car park on the northern side of Shelly Beach.

The **Point Grey Beach** (338) extends south for 300 m from the extensive rock platform and reefs of the point to a protruding rock platform. This platform separates it from **Shelly Beach** (339), which is similar, but only 100 m long. Both beaches are dominated by the boundary platforms and numerous rocks and reefs off the beaches, and both have strong, permanent rips running out through the rocks.

Swimming: Be very careful if bathing here, with the southern end of Point Grey Beach offering the most protected location. However rocks, reefs and strong rips dominate.

Surfing: There is a good right hand reef break off the southern end of Point Grey Beach called *Vera Lynn*, with easy access via the beach rip. Out from Shelly

Beach is a second, shorter right called *Barrels*, and another heavier right hander a little further along called *Weeds*.

Fishing: These are three very accessible spots, with good beach and rock access to deeper holes and gutters. Boats are launched at the Point Grey boat ramp only during calm conditions, with a small beach on the south side of the jetty used when there is any swell.

Summary: Point Grey is a nice spot to get off the road for a picnic. It is not recommended for bathing, but is a popular fishing spot and occasional surfing spot.

340   ST GEORGE RIVER

Unpatrolled
Single bar: TBR
Beach Hazard Rating: **6**
Length: 80 m

The **St George River** reaches the coast via a deep valley 1 km south of Point Grey. The Great Ocean Road winds around and across the river, with a car park beside the small bridge. The river, which is more of a creek, is backed by a small lagoon and fronted by rock platforms and reefs, extending north from the southern bluffs that almost enclose the beach. The entire small bay is full of fine sand and has a surf zone that extends out to the rocks, with a permanent rip flowing out the narrow, 50 m wide entrance.

Swimming: Relatively safe inshore, however out toward the rocks the waves and rip intensify.

Surfing: When the bay is shallow, good right handers run off the south point and into the bay.

Fishing: The rip channel can be fished from the southern rock platform at low tide.

Summary: A picturesque little beach, creek and bay with a picnic area all right by the road.

ANGAHOOK - LORNE STATE PARK	
Area:	approximately 22 000 ha
Coast length:	25 km
Beaches:	340 - 365 (25 beaches)
Information:	Lorne (052) 989 1732

341   SHEOAK CREEK

Unpatrolled
Single bar: TBR + rocks & reef
Beach Hazard Rating: **8**
Length: 40 m

Sheoak Creek flows out of a deep, narrow valley under the small road bridge and over a 40 m long boulder and sand beach. The rock platforms to either side open out, permitting high waves to reach the beach and a strong permanent rip to run off the beach. There is a car park on the Great Ocean Road, however this is located so as to service the valley rather than the rocky beach. There are falls and caves a few hundred metres up the valley, with walking tracks leading to picnic and camping areas in the park.

Swimming: Not recommended, owing to the dominance of rocks and rips.

Surfing: Usually too rocky.

Fishing: The southern rocks provide the best location to fish into the small, open bay.

Summary: A small, non-recreational beach backed by a more useable valley with falls, caves and walking tracks.

342, 343   CUMBERLAND RIVER NORTH, CUMBERLAND RIVER

Unpatrolled			
	Rating	Single bar	Length
342 Cumberland River North	7	TBR	150 m
343 Cumberland River	7	TBR	250 m
		Total length	400 m

The **Cumberland River** flows through a steep-sided, 200 m wide valley containing a flat, riverside reserve. It reaches the coast in an open, south-east facing bay. The Great Ocean Road hugs the base of the bluff north of the river, then winds in to cross the river, before continuing south along the base of the bluffs. There is a 150 m long beach immediately north of the river mouth, with the road forming its rear boundary. The river mouth beach is 250 m long and is crossed by the creek and backed by a low, grassy area. There is a car park just north of the bridge and a caravan park on the west side of the road.

The two beaches face south-east and are exposed to waves averaging 1.5 m. The waves interact with the sand and rock platforms to produce an 80 m wide surf zone. This is dominated by one permanent rip to the north, as well as rips against each end of the river mouth beach.

Swimming: Be very careful if swimming here, as rip feeder currents run the length of both beaches, with strong rips at either end of both beaches.

Surfing: There are reasonable beach breaks on both beaches, that work in low to moderate swell.

Fishing: This is a popular location with the choice of creek, creek mouth, beach and rock fishing, plus a caravan park next door.

Summary: A picturesque valley and beach with good access, but a hazardous surf.

344   JAMIESON CREEK

Unpatrolled
Single bar: TBR + rocks & reef
Beach Hazard Rating: **8**
Length: 500 m

Jamieson Creek runs out of a small gap in the steep slopes to cross under the Great Ocean Road 1.5 km south of the Mount Defiance Lookout. The road runs right along the back of the beach, with car parks on both sides of the bridge. The 500 m long beach lies at the base of the 300 m high, forested slopes, with the creek flowing across the southern end. There are numerous scattered rocks on the beach and in the surf. Permanent rips exist at each end of the beach and one to two shifting rips occur along the beach.

Swimming: This is a hazardous beach, with waves averaging 1 to 1.5 m and a surf zone dominated by rocks and rips. Use extreme care if bathing.

Surfing: There are beach breaks amongst the rocks, which are better at high tide.

Fishing: You can fish the rocks and reefs from the beach, the creek (when flowing) and off the rock platform at the southern end.

Summary: An open and exposed beach that is unsafe for bathing, and is best suited to beach fishing.

345 - 348   ARTILLERY ROCKS, GODFREY CREEK, BEACH 347, BOGGALEY CREEK

Unpatrolled			
	Rating	Single bar	Length
345 Artillery Rocks	8	TBR + rocks & reef	50 m
346 Godfrey Creek	8	TBR + rocks & reef	100 m
347 Beach 347	8	TBR + rocks & reef	80 m
348 Boggaley Creek	8	TBR + rocks & reef	100 m
		Total length	330 m

South of Jamieson Creek, the Great Ocean Road continues to hug the lower parts of the 300 m high, forested slopes. The shore is predominantly resilient sandstone, which produces rock platforms along the base of the slopes and reefs further offshore. In amongst the 2 km of winding road south of **Artillery Rocks** are four pockets of sandy beaches. All four are less than 100 m long, face the east to south-east and are exposed to waves averaging 1 to 1.5 m. All four have sandy surf zones dominated by the rocks and reefs, with usually one strong, permanent rip draining each beach.

Swimming: These are four hazardous beaches and are unsuitable for safe bathing, apart from in the tidal pools at low tide.

Surfing: The waves tend to close out on the rocks along here. However at *Bog Alley*, just south of Boggaley Creek, there is a right hander that works in a big outside swell.

Fishing: There are many spots for rock and beach fishing along here. However, be very careful if rock fishing, as the rock platforms are awash at high tide or in big seas.

Summary: A picturesque section of coast, best suited for viewing.

349, 350   SEPARATION CREEK, SEPARATION CREEK SOUTH

Unpatrolled			
	Rating	Single bar	Length
349 Separation Creek	7	TBR	250 m
350 Separation Creek South	8	R + rocks & reef	100 m
		Total length	350 m
For map of beach see Figure 119			

Separation Creek is both a small creek and a settlement

of the same name located on relatively gentle slopes, right on the Great Ocean Road. A 250 m long beach lies on the east side of the road, with forested bluffs and rock platforms bordering each end. The creek drains across the southern end of the beach. Continuing around the southern rocks is a second 100 m long platform beach wedged in between the base of the road, a very small foredune and the continuous rock platform.

The main beach has permanent rips at each end, as well as boulders in the surf in front of the creek mouth.

Swimming: Both beaches are hazardous, with moderately safe bathing only possible at the main beach during low waves, and clear of the rocks, rips and reefs. On the south beach there are some relatively safe tidal pools at low tide.

Surfing: There are beach breaks at the main beach that work best in a low to moderate swell.

Fishing: The rock platforms provide good, if hazardous, access to deeper water, with safer fishing off the beach into the rip gutters at each end.

Summary: Two accessible beaches right on the road. They are suitable for a picnic or fishing, but are unsuitable for safe bathing.

351   WYE RIVER

Wye River SLSC

Patrols: late November to Easter holidays
Surf Lifesaving Club: Saturday, Sunday and Christmas public holidays
Lifeguard: no lifeguard on duty or weekday patrols

Single bar: TBR + rocks & reef
Beach Hazard Rating: **7**
Length: 200 m

For map of beach see Figure 119

The **Wye River** flows out through a narrow valley to deposit the sand that forms this 200 m long beach. A small settlement of the same name backs the beach. The beach is bordered by high valley slopes, with the Great Ocean Road winding behind, and sandstone rock platforms and reefs fringing each end. A caravan park, car park and the surf lifesaving club (founded in 1958) all lie between the road and the beach.

The beach faces south-east and receives waves averaging 1.4 m, that produce a wide, low gradient single bar, with strong permanent rips against the rocks at each end of the small beach.

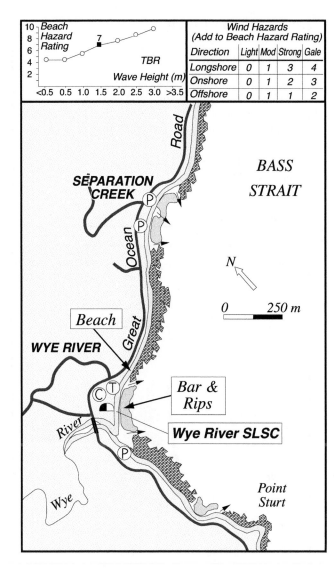

Figure 119. Wye River Beach is a small sand beach at the mouth of the Wye River. The surf lifesaving club is located in the centre and patrols the entire beach. This is a hazardous beach, with permanent rips against the rocks.

Swimming: A potentially hazardous beach, owing to the moderate waves, strong rips and reefs. Stay in the centre on the bar and between the flags, as an average of 26 people are rescued annually on this small beach.

Surfing: A popular and very visible spot. The beach breaks are popular under low swell with less experienced surfers, while bigger swell tends to close out across the bay. Best at high tide.

Fishing: A popular spot with good access to the river, river mouth and permanent rip holes against the rocks. As a result, there is a choice of river, beach or rock fishing.

Summary: A small, but highly visible, and relatively popular spot in summer, with the added attraction of the surf club patrols.

During major floods the Wye River sweeps the sandy beach out to sea, leaving a cobble-boulder base. Fortunately the sand always returns after a few months.

352   WYE RIVER SOUTH

Unpatrolled
Single bar: LTT + rocks & reefs
Beach Hazard Rating: **6**
Length: 200 m

On the southern side of the **Wye River** mouth, as the Great Ocean Road slowly climbs toward Point Sturt, there is a 200 m long, east facing platform beach. It is wedged in between rock platforms and reefs, with a 50 m opening to the surf. Waves are usually lower than at Wye River and the surf zone is shallow, however a permanent rip runs out between the platforms.

Swimming: A moderately safe beach when waves are low in the sand section. However, stay well clear of the rocks and rip current.

Surfing: There is a break off the southern point, known as *Baldy Rock*, that only works occasionally.

Fishing: There are excellent spots for beach and rock fishing out along the southern rocks.

Summary: Usually slightly lower waves than Wye River, but care is needed because of the rocks, reefs and rip.

353 - 358   POINT STURT SOUTH 1 & 2, VIEW POINT, MONASH GULLY, BEACH 357, HITCHCOCK GULLY

Unpatrolled			
	Rating	Single bar	Length
353 Point Sturt South 1	8	LTT + rocks & reef	100 m
354 Point Sturt South 2	8	LTT + rocks & reef	100 m
355 View Point	7	TBR + rocks & reef	250 m
356 Monash Gully	7	R + rocks & reef	80 m
357 Beach 357	7	LTT + rocks & reef	800 m
358 Hitchcock Gully	7	LTT + rocks & reef	50 m
		Total length	1 380 m

Between Point Sturt and Hitchcock Gully is 4 km of steep, forested slopes, with the Great Ocean Road winding along about 20 m above the bordering rock platforms. In amongst the rocks, and behind them in places, are seven generally small sand beaches, each dominated to some degree by the rock platforms and reefs. They all tend to face south-east, exposing them to waves averaging 1 to 1.5 m. Access to all the beaches requires a climb down the bluffs from the road, with parking limited to the side of the road.

Point Sturt South 1 is a 100 m long platform beach, with a gully cutting the centre of the platform, out of which flows a permanent rip.

Point Sturt South 2 is another platform beach, with a deeper gully against the southern rocks, which at low tide forms a tidal pool.

View Point has a 250 m long beach, partly located behind platforms and toward the centre in lee of deeper reefs, with a permanent rip running out the centre.

Monash Gully is a creek mouth with an 80 m long pocket of sand, bordered by rock platforms and fronted by continuous reef. At low tide it becomes a large tidal pool.

Beach 357 is an 800 m long, sinuous platform beach that winds along between the rocks, the platforms and reefs. For the most part it has continuous platforms, that are exposed at low tide, with reefs and tidal pools in a few spots.

Hitchcock Gully is a creek mouth with a 50 m pocket of sand wedged in between two rock platforms. It is fronted by a wide surf zone and permanent rip.

Swimming: All these beaches are dominated by moderate to high waves, rocks, reefs and rips and are unsuitable for bathing. Only at low tide is it safe to swim in some of the protected tidal pools.

Surfing: The only break of note along this section is *Sawmills*, located adjacent to an old sawmill site. It is a right hand reef break that can handle big swell.

Fishing: There are many good rock and beach fishing sites along here. However, care is required on the rock platforms, which are awash at high tide and during large waves.

Summary: Most people drive along here unaware of the small beaches, which are best left to the knowledgable locals.

359   KENNETT RIVER

Kennett River SLSC

Patrols: late November to Easter holidays
Surf Lifesaving Club: Saturday, Sunday and Christmas public holidays
Lifeguard: no lifeguard on duty or weekday patrols

Single bar: TBR + rocks & reef
Beach Hazard Rating: **7**
Length: 200 m

For map of beach see Figure 120

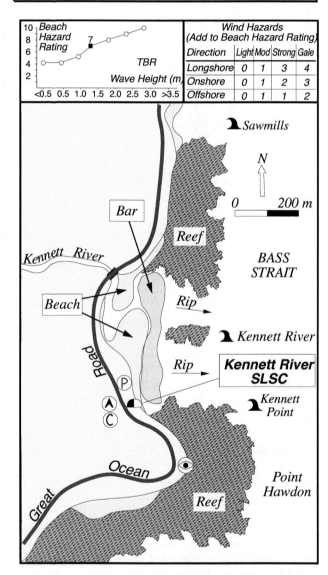

Figure 120. Kennett River Beach is located at the mouth of the river, and the small sand beach is surrounded by the road, rocks and reefs. This is a hazardous beach with permanent rips against the rocks, so stay between the flags if bathing.

Kennett River winds out through a 200 m wide, flat valley to enter the sea across **Kennett River Beach**.

The beach is 200 m long, faces east and is moderately protected by Point Hawdon, which protrudes 500 m to the south-east. The beach is hemmed in by steep, vegetated valley sides, sandstone rock platforms and extensive intertidal and submerged reefs. The Great Ocean Road backs the beach, with a large caravan park across the road. The Kennett River Surf Life Saving Club and car park are wedged in between the road and the beach. The surf club averages 5 rescues each year.

Waves average 1.3 m and interact with the sand, rocks, reefs and river mouth to form a narrow surf zone dominated by strong, permanent rips against the rocks and reefs.

Swimming: Moderately safe when patrolled under low summer conditions, however a very hazardous beach when any swell is running, as strong rips flow out of the bay.

Surfing: A very popular spot offering a range of breaks, depending on wave height. The *Rivermouth* breaks in moderate swell, with bigger waves breaking off *Kennett Point*, while really big swell breaks over the *Sawmill* reef.

Fishing: A very popular fishing spot with good access to river, river mouth, beach and rock fishing.

Summary: A small, but very accessible and popular beach with holiday makers, when waves are low. During higher waves the surfers arrive.

---

360 - 365   **POINT HAWDON, ADDIS BAY 1 & 2, GREY RIVER, SHRAPNEL GULLY, CAPE PATTON**

Unpatrolled			
	Rating	No bar	Length
360 Point Hawdon	8	R + platform	150 m
361 Addis Bay 1	7	R + rocks & reef	50 m
362 Addis Bay 2	7	R + rocks & reef	50 m
363 Grey River	8	R + rocks & reef	50 m
364 Shrapnel Gully	8	R + rocks & reef	300 m
365 Cape Patton	8	R + rocks & reef	300 m
		Total length	900 m

Point Hawdon is a prominent sandstone headland on the south side of Kennett River. From the point, the coast sweeps south-west for 4 km to the next major headland at Cape Patton. In between, steep, forested slopes with rock platforms at their base, dominate the coast. The Great Ocean Road winds along up to 50 m above the sea. Below the road are six sand beaches. They all face south-east to east and receive waves averaging between 1 and 1.5 m. However, except for the Cape Patton

Beach, they are fronted by continuous platforms and reefs, which dominate the surf and currents.

The **Point Hawdon Beach** lies on the south side of the point, with a steep track leading to the beach from the point car park. The beach is fronted by a continuous rock platform, that is exposed at low tide.

In the apex of open **Addis Bay** are two 50 m pockets of sand. Steep tracks lead down from the road to each of these. The first has a narrow opening to the sea, while the second has a continuous platform.

Grey River Beach is the only one of the six that has public access, with a car park and shady picnic area next to the beach. Unfortunately, the beach is often more boulder than sand, and is usually unsuitable for bathing.

Shrapnel Gully lies 300 m south of Grey River. It is a near continuous strip of sand fronted by a continuous platform, that links it with Grey River Beach.

Finally, out on the north side of **Cape Patton** is a 300 m long, east facing beach that runs from the point to a small gap in the platform. This beach is backed by private property with no public access.

Swimming: All these beaches are unsuitable for safe bathing, apart from in the protected tidal pools at high tide.

Surfing: There is generally too much rock and not enough reef along this section for rideable surf.

Fishing: Most of the tracks to these beaches are made by fishers, who fish the extensive rock platforms.

Summary: A spectacular section of coast, best left for viewing and rock fishing.

---

366 - 369   **CARISBROOK CREEK, SAN JOSE, SUGARLOAF, WHITE CREST**

Unpatrolled			
	Rating	Single bar	Length
366 Carisbrook Creek	9	TBR + rock platform & reef	150 m
367 San Jose	7	TBR + rock platform & reef	600 m
368 Sugarloaf	7	TBR + rock platform & reef	1 200 m
369 White Crest	8	R + rock platform & reef	100 m
		Total length	2 050 m

At Cape Patton most visitors stop to take in the spectacular view west along the coast toward Apollo Bay. The most noticeable feature is the steep, vegetated slopes with a rock platform running along the base. The rocks dominate the coast for 3 km until the mouth of Carisbrook Creek. From here, near continuous platform beaches run along the back of the rock platforms past San Jose and Sugarloaf Hill. Then there is a gap before the final pocket of sand at White Crest. The Great Ocean Road parallels the back of the beaches and there is good vehicle access to all four beaches. All the beaches face south-east and receive waves averaging 1.5 m.

Carisbrook Creek Beach extends for 150 m from the boulder-strewn mouth of the creek. It is fronted by a continuous rock platform and is unsuitable for bathing.

San Jose Beach is a 600 m long section of predominantly platform beach, with 50 m of open surf, which is used by surfers to reach a reef break called *Boneyards*.

Sugarloaf Hill is a prominent, 30 m high knoll, around which a platform beach runs for 1.3 km. There is then a rocky gap of 1 km, before the small, 100 m long **White Crest Beach**, which consists of a high tide sand and boulder beach fronted by a continuous low tide platform.

Swimming: Of these beaches, only San Jose provides some access to the water and even here there is a permanent rip. The remainder are fronted by platforms, with only tidal pools or rocks at low tide.

Surfing: There are two recognised breaks along this section: *Boneyards*, out from San Jose Beach; and *Juniors*, out off Sugarloaf Hill.

Fishing: This is a popular section of coast for rock fishers, with easy access, generally low waves and safe platforms to fish from.

Summary: There are several parking areas along this section which people use for picnics and fishing, however be very careful if you intend getting wet. The tidal pools at low tide are safest.

---

370   SMYTHE CREEK

Unpatrolled
Single bar: TBR + rocks & reefs
Beach Hazard Rating: **7**
Length: 60 m

Smythe Creek flows through a steep, V-shaped valley to reach the coast in a small, south facing break in the

steep slopes. The Great Ocean Road winds around the small bay, with a car park located in a disused quarry. Below the road is a small, 60 m long beach composed of sand, but with rocks and boulders spread around. Waves average 1.5 m and the beach is unsuitable for safe bathing, but does provide surfers access to a right hand break out off the southern point.

Swimming: Unsuitable owing to numerous rocks and boulders, plus a permanent rip.

Surfing: A short right hander off the south point in moderate swell.

Fishing: Good rock platform to either side of the beach, plus a rip channel right off the beach.

Summary: Mainly used by fishers, and surfers when the point is breaking.

371, 372, 373   BEACH 371, BROWNS CREEK, BEACH 373

Unpatrolled			
	Rating	No bar	Length
371 Beach 371	7	R + rock platform	1 000 m
372 Browns Creek	7	R + rock platform	1 000 m
373 Beach 373	7	R + rock platform	200 m
		Total length	2 200 m

Past Smythe Creek, the coast continues to the south-west. The generally rocky shore is backed by the lower, largely cleared slopes and the Great Ocean Road. Rock platforms and reefs dominate the shore, however a strip of sand runs along the back of the platforms, and forms three beach areas, with waves averaging 1.5 m off the platforms. All three beaches are readily accessible from the road.

Beach 371 is 1 km long and begins just past Smythe Creek. It is backed in places by a low foredune and fronted by a broad intertidal rock platform, which truncates jagged beds of sandstone and shale. The platform is exposed at low tide. On either side of **Browns Creek** is a 1 km strip of sand and boulders also fronted by broad, irregular intertidal rock platforms, with occasional gullies in the rocks. **Beach 373** is a shorter, 200 m strip of sand and boulders right beside the road. It is fronted by a sloping rock platform.

Swimming: All three beaches are dominated by the rock platforms which, coupled with boulders and breaking waves, make this a very hazardous section of coast. Bathing is not recommended outside of some protected tidal pools.

Surfing: There are no recognised breaks along this section.

Fishing: The very accessible rock platforms are low along here and can only be safely fished at low tide.

Summary: An accessible, rock dominated section of coast, with enough sand to sunbathe and picnic on, but not suitable for swimming.

374   BEACH 374

Unpatrolled
Single bar: RBB
Beach Hazard Rating: **7**
Length: 500 m

Beach 374 is a 500 m long, open sand beach that faces south-east and is backed by the Great Ocean Road. A small creek crosses the middle of the beach, where it has deposited a bed of cobbles. It is bordered by two headlands, the northern one being more a collection of boulders, while there is a wide rock platform off the southern end. There are also reefs off the beach. Waves average 1.5 m and maintain a strong permanent rip against the northern end and a second rip down the beach.

Swimming: Deep troughs, reefs and strong rips dominate the surf, so stay close inshore, and preferably toward the southern end.

Surfing: There are breaks over the reefs at high tide.

Fishing: Both the beach and southern rock platform give good access to deeper rip holes and gutters.

Summary: This is the first open, predominantly sand beach since Kennett River, however reefs still dominate the surf.

375   SKENES CREEK

Unpatrolled
Single bar: RBB
Beach Hazard Rating: **7**
Length: 200 m

Skenes Creek is the first ocean beach free of platforms and reefs since Lorne. The Great Ocean Road runs along the back of the beach and there are car parks to either side of the small bridge. A camping area is located behind the northern end of the beach. The creek flows through a small, narrow lagoon and crosses the centre of the beach. The beach faces south-east and

receives waves averaging 1.5 m. These produce a moderately wide beach and surf zone, with strong permanent rips against the rock platforms at each end.

Swimming: Stay close inshore on the attached part of the bar and well clear of the rips and headlands.

Surfing: Skenes Creek offers some beach breaks, that are best on a low to moderate swell and mid to high tide.

Fishing: There are permanent rip channels that can be fished from either end of the beach or the rock platforms at low tide.

Summary: A popular summer camping spot for holiday makers, surfers and fishers.

376, 377   BEACH 376, SEAFARER

Unpatrolled			
	Rating	Single bar	Length
376 Beach 376	**7**	TBR + rocks	150 m
377 Seafarer	**6**	TBR	700 m
		Total length	850 m

Five hundred metres south of Skenes Creek is **Beach 376**. It lies right beside the Great Ocean Road, and is a 150 m long sand beach bordered by rock and shore platforms, with a few rocks on the beach and in the surf. Around the southern rocks is **Seafarer Beach**: a 700 m long, east facing beach, which has rocks and reefs scattered along the beach and in the surf. It is backed by a low foredune, then the road.

Both beaches receive waves averaging between 1 and 1.5 m and have 100 m wide surf zones, with permanent rips against the boundary rocks and the larger reefs.

Swimming: Be careful of the rocks, reefs and rips on these two beaches. If swimming, stay close inshore and on the attached part of the bar.

Surfing: There are usually low to moderate beach breaks amongst the reefs and rocks.

Fishing: There are several good holes and gutters formed by both the rips and reefs that can be fished from the beaches, or the rocks at low tide.

Summary: Two sandy beaches next to the road, but with enough rock and reef to require care if bathing or surfing here.

378   APOLLO BAY

Apollo Bay SLSC

Patrols: late November to Easter holidays
Surf Lifesaving Club: Saturday, Sunday and
Christmas public holidays
Lifeguard: no lifeguard on duty or weekday patrols

Single bar: LTT/TBR
Beach Hazard Rating: **6** (north)   **5** (south)
Length: 3 000 m

For map of beach see Figure 121

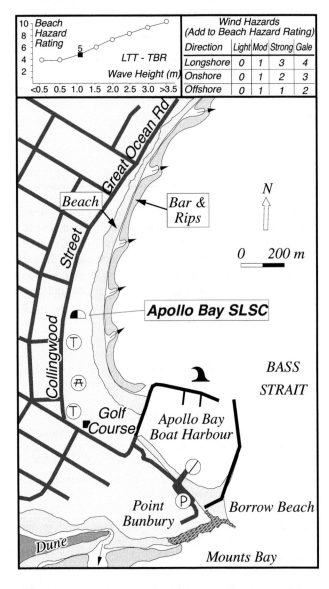

Beach Hazard Rating	Wind Hazards (Add to Beach Hazard Rating)			
	Direction	Light	Mod	Strong	Gale
LTT - TBR	Longshore	0	1	3	4
Wave Height (m)	Onshore	0	1	2	3
	Offshore	0	1	1	2

Figure 121. Apollo Bay Surf Life Saving Club is located at the southern end of the 3 km long beach, with the harbour wall forming the southern boundary. It is relatively protected by Point Bunbury and the harbour, and is moderately safe in the patrolled area, with waves and rips increasing up the beach.

Apollo Bay is the southernmost town on the Great Ocean Road. It is fronted by a 3 km long, east facing, relatively safe beach, which is very popular during the summer holidays.   The beach is protected by its orientation, Point Bunbury and the Apollo Bay Boat Harbour seawalls. As a result of this protection, the southern end of the beach has built out tens of metres along the northern harbour wall.   Shoaling of the harbour has been a continual problem since it was constructed in the 1950s, with dredging often taking place.

The beach receives waves averaging 1 m at the southern end, which slowly increase in height up the beach. The southern end is safest, with a usually continuous, attached, shallow bar and few rips.   Rip size and intensity increase up the beach. The Apollo Bay Surf Life Saving Club was founded in 1952 and averages 8 rescues annually.

Swimming: Relatively safe along the southern end and in front of the surf lifesaving club, so it's best to stay here and between the flags. Be careful of higher waves and rips further up the beach.

Surfing: Usually low to moderate beach breaks of variable quality. Westerly winds blow offshore. During big swell a small right hander runs down the western *Harbour Wall*.

Fishing: Best beach fishing is up the beach where the rips are more persistent.

Summary: An attractive town and beach, offering relatively safe bathing and a beach sheltered from the westerlies.

379   APOLLO BAY BOAT HARBOUR

Unpatrolled
No bar: R
Beach Hazard Rating: **2**
Length: 300 m

Apollo Bay Boat Harbour was constructed in the 1950s, replacing a jetty that ran out from the southern end of the beach. The harbour, however, continues to fill with sand and requires constant dredging. The sand that accumulates in the harbour forms a 300 m long beach along the southern side, with a boat ramp and jetty in the centre. There is also a large car park behind, and the harbour walls to either end.

Swimming: This is usually a quiet and relatively safe beach in against the shore, however there are dredged holes off the beach and boat traffic in the harbour.

Surfing: None in the harbour, however there is a right hand break along the western *Harbour wall* in a big swell.

Fishing: The harbour walls are very popular locations to fish the deeper outside water.

Summary: This is an artificial beach created by the harbour construction and modified by dredging. It offers good access and quieter waters, with boating activity in the harbour.

380, 381   BORROW, POINT BUNBURY

	Unpatrolled		
	Rating	Single bar	Length
380 Borrow	6	LTT + rocks	100 m
381 Point Bunbury	8	R + rock platform	500 m
		Total length	600 m

Point Bunbury is a low sandstone headland covered by dune sand and now occupied by Apollo Bay Golf Course. The harbour lies on the north side of the point, with two beaches along the south-east side. The first is wedged in between the eastern harbour wall and a rock groyne on the point. Sand moving around the point collects here in a 100 m long gap. It is called **Borrow Beach**, as sand is mined from the beach to both help reduce the harbour fill and for use in aggregate. The beach faces north-east and receives waves averaging 1 m. It usually has a shallow bar with a rip off the beach.

On the south side of the groyne is **Point Bunbury Beach**. This is a 500 m long strip of sand backed by a low dune and the golf course, and fronted by a continuous rock platform and reefs. The beach is awash at high tide while the platform is exposed at low tide. The beach ends at the mouth of the Barham River in Mounts Bay.

Swimming: Bathing is only possible at Borrow Beach, however be careful here as there are rocks and a permanent rip present.

Surfing: No rideable waves.

Fishing: Both top fishing spots, however be careful on the point as the rocks are awash at high tide and in big seas.

Summary: Two very accessible but rock dominated beaches, mainly used for fishing.

382   MOUNTS BAY

Unpatrolled
Beach Hazard Rating: **6**
Single bar: LTT/TBR
Length: 1 800 m

Mounts Bay is an exposed, south-east facing bay on the south side of Point Bunbury. It extends for 1.8 km to the southern Hayley Point or Cape Marengo. Little Henty Reef, 500 m off Hayley Point, provides additional protection to the southern half of the beach. The beach is part of a sand barrier and is backed by a low dune and the floodplains of the Barham River, which exits at the northern end of the beach. The Great Ocean Road runs along the back of the beach, providing good access once across the bridge. There are several car parks behind the beach and at the point.

The beach receives waves averaging 1 m in the south and increasing northward. The beach has a narrow, attached bar in lee of the reef, with rips usually beginning halfway up the beach and intensifying to the north.

Swimming: The southern corner in lee of the reef is the safest location. Avoid the rips up the beach.

Surfing: The best beach breaks are toward the centre and the northern end where the waves are usually larger and the surf wider.

Fishing: The best beach fishing is in the rip holes up the beach and at the river mouth. On Hayley Point there is a wide, intertidal rock platform to fish the deeper water between the reef and the point.

Summary: A very accessible and relatively natural beach, with a more protected southern end, and higher surf, the river mouth and rock platforms to the north.

383 - 386   MARENGO, SWELL POINT, STORM POINT NORTH & SOUTH

	Unpatrolled		
	Rating	No bar	Length
383 Marengo	9	R + sewer outlet	150 m
384 Swell Point	9	R + sewer outlet	50 m
385 Storm Point North	7	R + rock platform & reefs	40 m
386 Storm Point South	7	R + rock platform	100 m
		Total length	340 m

At Hayley Point, the coast swings to the south-west and is dominated by a series of protruding sandstone rock platforms and reefs. These are divided by small beaches that extend 2.5 km from the point to The Blowhole. They are only accessible via Marengo, with a walk along the rocks required to reach the southern beaches.

The first beach is called **Marengo** after the name of the settlement on Hayley Point, and is accessible via a road that runs around the point to the rear of the beach. The beach faces east, is 150 m long and is tucked in between two rock platforms with some reefs off the beach. It receives waves averaging 0.5 m and has a steep, barless beach with deep water close inshore. However, it is also the site of a sewer outlet.

Swell Point is the name of the next platform and is a smaller version of Marengo, with 50 m of access to the water between them. It too is next to the sewer outlet.

Storm Point is a more continuous stretch of platform, with a small, 40 m long beach on the north side and a 100 m long, south facing platform beach on the southern side. Both beaches are only awash at high tide, with exposed rocks at low tide.

Swimming: These four beaches are all sheltered from large waves. Marengo and Swell Point are relatively safe in a physical sense, but they are polluted by the sewer and are not suitable for bathing. The Storm Point beaches are dominated by rocks and only offer tidal pools at low tide.

Surfing: The reefs at Marengo, called *Infinities*, provide left and right handers during moderate to high swells. They are the only rideable waves along this rocky section of coast.

Fishing: This is a very popular stretch of coast for fishing, with both the beaches and rock platform giving good access to holes and gullies.

Summary: This is a nice section of coast for a walk or fishing, but is unsuitable for bathing or surfing.

387, 388, 389   THE BLOWHOLE SOUTH 1 & 2, BEACH 389

Unpatrolled			
	Rating	No bar	Length
387 The Blowhole South 1	7	R + rock platform	50 m
388 The Blowhole South 2	7	R + rock platform	100 m
389 Beach 389	7	R + rock platform & reef	300 m
		Total length	450 m

The Blowhole is a narrow gap in the sandstone rock platforms 2.5 km south of Marengo. On the south side of the point are two small, south-west facing beaches backed by grassy slopes. Just beyond the second beach the coast swings to the south-west. A 300 m long strip of sand runs between the slopes and an irregular rock platform, that contains two more open tidal pool areas. These beaches are backed by private property and can only be reached along the rock platforms from Marengo.

All three beaches are fronted by rock platforms, with surf averaging 1.5 m outside. Only a narrow gap at the western end of the second beach and two gaps in Beach 389 provide access to the sea.

Swimming: All three beaches are unsuitable for bathing, as they are dominated by the platforms that are exposed at low tide.

Surfing: There is a sand and reef break out off the Blowhole beaches that works in a low to moderate swell.

Fishing: The Blowhole itself and the rock platforms at low tide have several good gullies.

Summary: This is a scenic section of coast, best suited for a coastal walk or rock fishing.

OTWAY NATIONAL PARK			
Area:	12 750 ha	Camping Areas:	Blanket Bay
Coast length:	56 km		Aire River
Beaches:	390 to 409, 411 to 432 (30 beaches)		Johanna Beach
Picnic Area:	Shelly Beach	Park Information:	Colac: (052) 9325 799
			Lavers Hill (052) 9373 243
			Apollo Bay (052) 9376 889

390 - 394   SHELLY NORTH, SHELLY, GEARY RIVER, GEARY RIVER SOUTH, BEACH 394

Unpatrolled				
	Rating	No bar		Length
390 Shelly North	7	R + rock platform		300 m
391 Shelly	7	R + rock platform		300 m
392 Geary River	7	R + rock platform		60 m
393 Geary River South	7	R + rock platform		250 m
394 Beach 394	7	R		40 m
		Total length		950 m

Shelly Beach Picnic Area is located 1 km off the Great Ocean Road, 6 km south of Apollo Bay. The picnic area is set in forests with a 1 km walking track to Shelly Beach. This track and the rock platforms provide the only access to Shelly Beach and the beaches on either side.

The coast consists of steep, forested slopes, with low cliffs fronted by extensive sandstone rock platforms. All the beaches are platform beaches lying at the base of the cliffs and toward the rear of the platforms, and are only awash at high tide. Waves average 1.5 m seaward of the platforms.

Both **Shelly Beach North** and **Shelly Beach** are 300 m long strips of irregular sand fronted by the near continuous platforms, that have occasional tidal pools and gullies through them.

The **Geary River** is a creek that reaches the coast 600 m south-west of Shelly Beach. It can be reached along the platforms and has a boulder filled creek, with a small platform beach on its northern side and a 250 m long platform beach to the south.

Beach 394 lies 1 km past Geary River and is remote and difficult to access. It is a 40 m long pocket of sand lying at a small creek mouth.

Swimming: None of these beaches are suitable for bathing, except in protected rock pools and gullies at low tide.

Surfing: There are no recognised surfing spots along here, with more rock than surf.

Fishing: This is the domain of rock fishers, who will find numerous holes and gullies in the rock platforms. However be careful, as the platforms are awash at high tide and in large waves.

Summary: A natural, forested section of coast well worth visiting for the scenery, but only suitable for walking and rock fishing.

395, 396   BEACHES 395 & 396

Unpatrolled			
	Rating	Single bar	Length
395 Beach 395	7	TBR	50 m
396 Beach 396	5	LTT/TBR	60 m
		Total length	110 m

Beaches 395 and 396 are two isolated pockets of sand and surf located 1.5 and 0.5 km north of Blanket Bay, respectively. They are only accessible by boat and neither are suitable for land-based recreational pursuits. The Stony Creek walking track reaches the coast near Beach 395.

397, 398   BLANKET BAY NORTH, BLANKET BAY

Unpatrolled			
	Rating	Single bar	Length
397 Blanket Bay North	6	R + rock platform & reef	50 m
398 Blanket Bay	4	R/LTT + rock platform & reef	200 m
		Total length	250 m

Blanket Bay comprises a small collection of fishing and holiday shacks, and a camping area, located at the end of the circuitous Blanket Bay vehicle track, that runs off the Cape Otway Lighthouse Road. The bay is part of the Otway National Park. It faces east and is bordered by forested slopes to the west and rimmed by rocks and reefs to the east. **Blanket Bay Beach** is protected by the reefs and is used by the fishers to launch their boats. The beach is 200 m long. The smaller northern beach is located just north of the bay. Both beaches are fronted by a rock platform and reefs. Waves average 1.5 m outside the reefs, but are usually less than 1 m at the beach.

Swimming: The bay is the safest place to swim, particularly in the calmer areas in lee of the reefs. Beware of the rocks, reefs and the currents flowing out through the reefs off the beach.

Surfing: There are only irregular breaks over the outer reefs at high tide.

Fishing: This is a fishing centre, with most of the fishing done from boats, but there are also good holes off the beach and adjoining rock platforms.

Summary: An out of the way spot used by the locals, shack owners and visiting campers.

---

399   PARKER RIVER

Unpatrolled
Single bar: TBR
Beach Hazard Rating: **6**
Length: 100 m

The **Parker River** flows through steep, forested slopes to enter the coast in a 200 m deep bay, with just a 50 m wide entrance between two 50 m high headlands. The river and waves have built a 100 m long beach inside the bay, with a small, low, sand barrier and estuary behind the beach. The bay is shallow, with a bar and permanent rip occupying the area between the beach and the entrance. Rock platforms line either side of the entrance. There is a parking area on the southern slopes above the bay, with a walking track down to the partially cleared flats behind the beach.

Swimming: It is moderately safe to swim inshore at the beach, however the surf and a rip draining the bay intensify toward the entrance.

Surfing: There is a break over the sand and reef at the entrance.

Fishing: The rock platforms to either side can be used to fish the permanent rip channel flowing out of the bay.

Summary: A picturesque little bay.

400, 401, 402   ERIC THE RED, BEACH 401, POINT FRANKLIN EAST

Unpatrolled			
	Rating	Single bar	Length
400 Eric the Red	**8**	R + platform	100 m
401 Beach 401	**6**	R/LTT + rocks & reef	150 m
402 Point Franklin East	**8**	R + rocks & reef	100 m
		Total length	350 m

Between the Parker River and **Point Franklin** is 1 km of coast, with the backing slope dropping from the 80 m high Parker Hill to the low, dune covered rocks of Point Franklin. A vehicle track runs out to the point, however it is on private property and public access is restricted. The shoreline is dominated by rock platforms and reefs, with the three beaches tucked in between the base of the steep bluffs and the platforms. The three beaches face south-east and receive waves averaging 1 to 1.5 m.

However, the waves break over the reefs and platform and the beaches are only awash at high tide.

Swimming: These are three hazardous beaches, with moderately safe bathing only possible in the more protected rock pools at low tide.

Surfing: The surf is dominated by rocks and reefs and is generally unsuitable for surfing.

Fishing: The reefs can be fished from the beaches at high tide, while the rock holes and gutters can be fished from the platforms at low tide.

Summary: A winding, rock dominated section of coast with the narrow beaches providing access for fishing.

The ship *Eric the Red* was wrecked at this site in 1880, with the loss of 4 lives.

403, 404   POINT FRANKLIN WEST, SEAL POINT

Unpatrolled			
	Rating	No bar	Length
403 Point Franklin West	5	R + reefs	150 m
404 Seal Point	8	R + platform	200 m
		Total length	350 m

At Point Franklin the coast turns and heads almost due west to Cape Otway, the southernmost point in western Victoria. Along this 3.5 km section of rocky coast are seven small beaches and bays. The first two lie immediately west of Point Franklin, with a vehicle track off the Cape Otway Lighthouse Road providing access to Seal Point, where there used to be a few shacks above the bay.

Point Franklin West Beach faces south-west and, although exposed to strong winds, it is protected from the waves by offshore reefs. This results in a steep, barless beach, with rock platforms at either end. Active dunes cross over the backing point, to spill onto the eastern beach.

Seal Bay curves around to **Seal Point** and faces south. It is backed by 20 m high, grassy bluffs and the beach is a strip of sand between the base of the bluffs and a continuous, but irregular, rock platform.

Swimming: Point Franklin is the only spot where swimming is possible, however you need to watch the shorebreak, which can be strong at times, and a permanent rip running out against the rocks.

Surfing: Waves do break over the reefs at high tides, however there are no recognised breaks here.

Fishing: This is a good location for both beach and rock fishing into the deep water off the point and in Seal Bay.

Summary: An exposed, rocky section of coast only suitable for beach and rock fishing.

405, 406, 407   CRAYFISH BAY, CRAYFISH BAY WEST, CRAYFISH POINT

Unpatrolled			
	Rating	Single bar	Length
405 Crayfish Bay	4	R + reefs	150 m
406 Crayfish Bay West	5	R + rocks & reefs	70 m
407 Crayfish Point	7	R/LTT + rock platform	50 m
		Total length	270 m

Crayfish Bay and the adjoining point used to house a few fishing shacks. These have been removed now that the area is part of Otway National Park. A vehicle track from the Cape Otway Lighthouse Road runs down to the bluffs above the bay. The coast here consists of 20 to 30 m high, vegetated bluffs fronted by rock platforms and reefs.

Crayfish Bay is 200 m long and contains two beaches, both protected by offshore reefs, with Seal and Long Points forming the boundaries. The main beach is 150 m long, faces south and is backed by a strip of flat land below the bluffs, where the shacks used to be located.

The beach is steep, with no bar and deep water lying between the beach and the reefs 50 m offshore. At the western end of the bay, next to the point, is a second 70 m long beach with bluffs running down to the back of the beach and continuous rocks and reef along the front. On the western side of **Crayfish Point** is the third small beach, a 50 m strip of protruding sand surrounded by rock platforms and backed by the point.

Swimming: The main bay beach is moderately safe close inshore at low tide. There are, however, reefs and currents further offshore. The West and Point Beaches are both platform beaches and are unsuitable for bathing, with the platforms becoming exposed at low tide.

Surfing: There are breaks over the reefs at high tide.

Fishing: This has long been a popular fishing area, originally with the shack dwellers and now with the National Park visitors and campers.

Summary: An attractive and relatively calm bay, surrounded by rocks and reefs.

408, 409, 410   KELP COVE, SHELLY, OTWAY COVE

Unpatrolled			
	Rating	Single bar	Length
408 Kelp Cove	8	R + rocks & reef	40 m
409 Shelly	7	TBR	50 m
410 Otway Cove	7	RBB	50 m
		Total length	140 m

At Cape Otway, the calcarenite capped sandstone cape reaches 70 m in height and sheer cliffs form the coastline. The lighthouse was constructed in 1848 and stands right on the point, 91 m above sea level. The cliffs slope gradually away to the east. In the first 2 km there are three pockets of sand at the base of the cliffs. The Cape Otway Lighthouse Road runs along the top of the bluffs, however there is no safe access down the bluffs to the beaches. All three face south and are exposed to waves averaging 1 to 1.5 m. Rocks and reefs dominate each beach, however there is also sufficient sand to produce a small surf zone and permanent rips.

Kelp Cove is just 40 m long, with a 100 m wide surf zone containing a permanent rip against the eastern Red Point. **Shelly Beach** is similar, but with more reef in the surf. **Otway Cove** is 600 m east of Cape Otway. It is a 50 m long beach, which has a large cavern cut into the backing calcarenite. The 100 m wide surf zone contains both rocks and a permanent rip.

Swimming: These are three hazardous beaches, in terms of both access and bathing.

Surfing: There are beach and reef breaks off each of the three beaches.

Fishing: Both the beaches and the adjoining platforms provide access to deep rip gutters.

Summary: Three beaches to view from Cape Otway, but unsuitable for recreational activities.

8. CAPE OTWAY TO CHILDERS COVE

Cape Otway to Childers Cove

OTWAY NATIONAL PARK
PORT CAMPBELL NATIONAL PARK
BAY OF ISLANDS COASTAL PARK

Beaches 411 - 498
Coast length 118 km

For maps of region see Figures 122 & 125

Cape Otway marks the western boundary of Bass Strait. Its lighthouse was often the first indication of land to be sighted by sailing ships travelling from Europe to Melbourne. The coast west of Cape Otway, all the way to the South Australian border, is an exposed, high wave energy region, with the prevailing westerly winds blowing directly on shore. Consequently, the coast has been worked over by the wind and waves and today contains some of Australia's highest energy beaches and biggest surf, backed by large dune systems andor eroding cliffs.

In this book the west coast is divided into two regions. The first (Region 8), from Cape Otway to Childers Cove, is dominated by limestone and calcarenite and contains generally short, narrow, rock and reef dominated beaches (Figure 123). This section contains 118 km of coast and 88 beaches, many of them small and located at the base of cliffs. The longest beach, at Johanna, is only 3.6 km long.

The second area (Region 9), from Warrnambool to iscovery Bay, has long, sweeping beaches and dunes contained in larger, bedrock controlled bays. This section is nearly twice as long as the first, at 211 km, but contains only 62 beaches. The longest of these is a 29 km long section of the 44 km long iscovery Bay.

Region 8 contains some of Australia's most spectacular coast. For this reason, all the coast between Cape Otway and Peterborough is contained in the Otway and Port Campbell National Parks. West of Peterborough, the Bay of Islands Coastal Park takes in another 33 km of coast. In all, 100 km of the 118 km of coastline are contained within National Park land.

Victorian Beaches - Region 8: East (Cape Otway to Port Campbell)

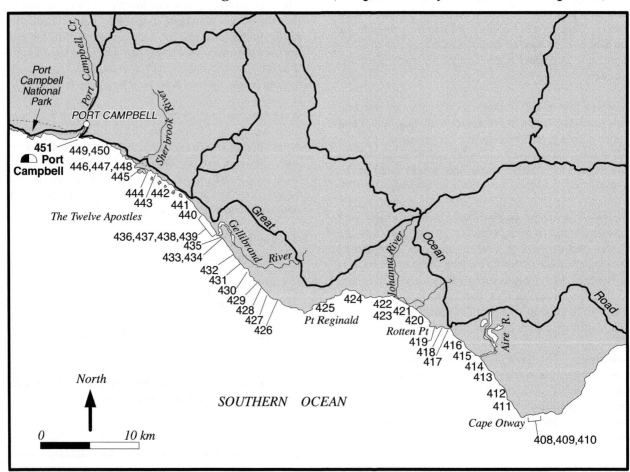

Figure 122.   Region 8 (eastern section): Cape Otway to Port Campbell (Beaches 411 - 451)

Figure 123.　Cape Otway - Port Campbell coast is characterised by prominent limestone cliffs and rocks, with beaches tucked into any available location.　Here at the Bay of Martyrs, the low limestone bluffs form a series of headlands, stacks and reefs, with low energy beaches in between.

411　POINT FLINDERS

Unpatrolled
No bar R rock platform reef
Beach Hazard Rating **7**
Length 100 m

At Cape Otway the coast turns and runs north-west for 2.5 km to **Point Flinders**. The coast consists of 50 m high cliffs and rock platforms.　Just before Point Flinders is a 100 m long, south facing beach, backed by a low foredune below 60 m high bluffs.　It is fronted by a rock and reef dominated surf zone. The beach is only active at high tide, with the rocks exposed at low tide.

Swimming:　This is a remote and hazardous beach and is unsuitable for bathing.

Surfing:　No rideable waves.

Fishing:　There are platforms right along this section, however they are exposed to high waves and are very dangerous to fish from.

Summary:　A small, rock dominated beach, with a sea arch on the point just west of the beach.

412　STATION

Unpatrolled
Single bar RBB
Beach Hazard Rating **7**
Length 2 800 m

Station Beach is the first of the high wave and wind energy, west coast beaches.　It is 2.8 km long and faces south-west, directly into the wind and waves.　There is no direct access to the beach and it can only be reached on foot along the coast from Glenaire Beach.

The beach receives waves averaging over 1.5 m, which produce a 250 m wide surf zone containing a detached bar and longshore trough.　Large, strong rips occur every 350 m, including permanent rips against Point Flinders and the central and western rocks.　The strong winds have blown dunes up to 4 km inland, and today bare calcarenite bluffs with occasional dune ramps back the beach. This beach once extended for 8 km all the way to Castle Cove. However, as sand was lost from the beach to the dunes it narrowed, and is now split by the rocks into seven beaches. These do still however share a common surf zone and during bigger swell, waves break continuously on the outer bar all the way from Point Flinders to Castle Cove.

Swimming: This is a remote and isolated beach and is very hazardous for bathing, with high waves and strong rips, plus a few rocks and reefs dominating the surf.

Surfing: There are beach breaks along the length of the beach, which work best in lower swell and northerly winds.

Fishing: Persistent beach and permanent reef holes and gutters are a feature of this beach.

Summary: A wild and remote beach.

413, 414   GLENAIRE EAST, GLENAIRE

Unpatrolled			
	Rating	Single bar	Length
413 Glenaire East	7	RBB	200 m
414 Glenaire	7	RBB	750 m
		Total length	950 m

The Aire River enters the sea at Glenaire Beach. It flows through a 500 m wide gap in the dune calcarenite, that has been piled up on either side by the strong westerly winds. The now lithified dunes date back several hundred thousand years, with the most recent active sand being deposited during the past 6 000 years. An elongate lagoon forms the lower stretch of the river, which is usually blocked at the beach. Access to these beaches is via the road to the National Park's riverside camping area. A 2 km walking track leads from the camping area along the river to **Glenaire Beach**.

Glenaire and its eastern beach face south-west and receive waves averaging over 1.5 m. These produce a 200 m wide surf zone dominated by detached bars and strong rips every 300 m, as well as against the larger rocks. **Glenaire East Beach** usually has one massive rip, while Glenaire usually has two large rips, that begin to flow seaward immediately off the beach.

Swimming: These are two hazardous beaches and swimmers should use the lagoon. Only venture into the surf if you are very experienced, and even then it is not advisable.

Surfing: There are usually beach breaks along here, with best conditions during low swell and northerly winds.

Fishing: This area offers lagoon, beach and rock fishing, with deep permanent holes and gutters along the beach.

Summary: The Glenaire region offers a range of natural environments, including the river, lagoon, beach, dunes and rocky coast. The coast is hazardous, so take care.

415, 416   SENTINEL ROCKS, EAGLE NEST ROCK

Unpatrolled			
	Rating	Single bar	Length
415 Sentinel Rocks	8	RBB platform	500 m
416 Eagle Nest Rock	8	RBB rocks reefs	1 000 m
		Total length	1 500 m

Sentinel and Eagle Nest Rocks lie at either end of a major rock fall, below steep, 70 m high, vegetated bluffs. To the east of **Sentinel Rocks** is a 500 m long strip of sand lying below the bluffs and fronted by a continuous rock platform and reef, with bars further offshore. A similar beach runs for 1 km to the west of **Eagle Nest Rock**. In addition to the rocks and reefs, there are large rips in the outer surf zone that link up with Glenaire Beach to the east and Castle Cove to the west.

There is no direct access to either beach, with Sentinel Rocks the more accessible from the Aire camping area via Glenaire Beach.

Swimming: Two very hazardous beaches that are unsuitable for bathing.

Surfing: While there are breaks in the outer surf, the barricade of rocks and reefs make this section unsuitable for safe surfing.

Fishing: There are numerous rock holes and gutters, plus the rip holes that are accessible from the rocks at low tide in calm seas. However be very careful, as this is a dangerous coast from which to rock fish.

Summary: Two rock dominated, high energy beaches.

417, 418   CASTLE COVE EAST, CASTLE COVE

Unpatrolled			
	Rating	Single bar	Length
417 Castle Cove East	**8**	RBB rocks reef	100 m
418 Castle Cove	**8**	RBB reef	300 m
		Total length	400 m

At **Castle Cove**, the Great Ocean Road reaches the coast for the first time since Mounts Bay. From the roadside car park 30 m above the beach, you can view the surf and the large rip that dominates the spot.

The two beaches face south-west and receive waves averaging over 1.5 m. These interact with the sand and reef to produce one massive rip straight out from below the car park and a second rip to the east, that runs out past the east beach. The main beach consists of 300 m of sand below the steep, vegetated bluffs, while the east beach is a narrow strip of sand backed by the bluffs and fronted by patchy rocks and reef.

Swimming: These are two hazardous beaches dominated by high waves, reefs, an often heavy shorebreak and a very strong permanent rip. They are unsuitable for safe bathing.

Surfing: *Castle Cove* is a popular surfing spot and works best during low to moderate swell. The strong rip maintains a channel, with lefts and rights to either side.

Fishing: You can fish the deep rip hole right off the beach.

Summary: Most people view this beach from the road, with only experienced surfers and fishers venturing down to the beach.

419   DINOSAUR COVE

Unpatrolled
Single bar RBB
Beach Hazard Rating **8**
Length 80 m

Dinosaur Cove is located below 40 m high bluffs on the eastern side of Rotten Point. There is a track to the top of the bluffs with a steep climb required to reach the cove. The beach consists of an 80 m long boulder beach with 50 m of sand, fronted by a bar with scattered rocks and reef. A permanent rip runs out of the 200 m wide mouth of the cove.

Swimming: nsuitable for safe bathing owing to boulders, rocks, reef and rips.

Surfing: Waves break over the bar, however rocks dominate.

Fishing: There are rock platforms to either side of the beach, with deep water also off the beach.

Summary: A hazardous and difficult to access location.

420   JOHANNA

Unpatrolled
Single bar RBB
Beach Hazard Rating **8**
Length 3 600 m

Johanna Beach is the best known surfing location west of Cape Otway. It was the site of the World Surfing Championships in 1970, and its famous left and right breaks are a mecca for surfers. The beach can be reached on a loop road from the Great Ocean Road. There is a National Park camping and picnic area on the grassy slopes behind the western end of the 3.6 km long beach. Two parking and viewing areas overlook the beach. The beach faces south-west and receives waves averaging over 1.5 m. The Johanna River crosses the centre of the beach, with dunes climbing the 30 to 100 m high bluffs behind the eastern half of the beach. Rotten Point forms the eastern boundary and Slippery Point the western.

The most noticeable features of the surf zone are the distinct bars and deep rip channels. The bar is 250 m wide and the rips are spaced every 350 m, forming eight bar-rip systems, each capable of holding waves from 1 to 3 m. These transverse bars and rips are a product of the medium sand and persistent high waves. Their spacing is determined by a phenomenon called edge waves (see page 22). The result is the production of bars with deep channels to either side, forming left and right hand breaks off each bar. Because the channels are deep, the beach can hold relatively high, surfable waves before closing out.

Swimming: Each of the embayments along the beach is the start of a large rip. So if you want to swim, stay on the bars and well clear of the rips.

Surfing: *Johanna* is one of the best beach breaks in the world, holding swell up to 3 m and more.

Fishing: Always deep rip holes along the beach.

Summary: A beautiful farmland setting with good access, camping and picnic facilities, great surf and good beach fishing.

421, 422, 423   SLIPPERY POINT, SUTHERLAND, DEEP CREEK

Unpatrolled			
	Rating	Single bar	Length
421 Slippery Point	8	RBB	100 m
422 Sutherland	8	RBB	600 m
423 eep Creek	8	RBB rocks	70 m
		Total length	770 m

Slippery Point and Sutherland Beaches are a western extension of Johanna Beach. **Slippery Point Beach** is 100 m long and bordered by 30 m high, short headlands and rock platforms. Six hundred metre long **Sutherland Beach** is backed by higher slopes and has cliffs to the west. The two beaches are backed by steep, 30 to 50 m high, vegetated bluffs and the easiest access is around the rocks at low tide from neighbouring Johanna Beach. A steep bluff at the end of Sutherland Beach separates it from the 70 m long, pocket sand beach at the mouth of **Deep Creek**. This beach shares the outer surf zone with Sutherland Beach, but has an inner surf zone dominated by rock reefs.

The three beaches face south-west and receive waves averaging over 1.5 m, which produce distinct bars and rips, like at Johanna Beach. The added impact of the headlands also produces permanent rips.

Swimming: These are two hazardous beaches with usually high waves and strong, permanent rips off the beach.

Surfing: There are beach breaks here almost as good as Johanna and usually not as crowded. Like Johanna, they are best in a low to moderate swell.

Fishing: eep, permanent rips run off the beach and can also be fished from the rocks at low tide. However be careful if rock fishing, as waves wash over the platforms.

Summary: Two beaches to head for if Johanna is too popular.

424   MELANESIA

Unpatrolled
Single bar TBR reefs
Beach Hazard Rating   7
Length   1 800 m

Melanesia Beach is located in an open, 2 km wide bay between 100 m high Lion Headland and Bowker Point. The beach is 1.8 km long, faces south-south-west and for the most part is fronted by patchy reef. The reefs lower the waves at the beach, but also form six permanent rips. The beach is backed by steep, forested slopes rising to 200 m that are cut by two valleys. One is called Nettle Pass and drains Melanesia and Running Creeks. The creek sometimes forms a small lagoon behind the beach, while the deeper water off the creek mouth provides the only area clear of rocks and reef. The beach is backed by private farm land, with one vehicle track and one foot track leading to the beach. The foot track is accessible from the end of Hornes Road.

Swimming: The reefs provide some protection for the beach and there are some quieter tidal pools and holes, particularly at low tide. However, the beach is remote and isolated, and off the beach are reefs and strong rips, so take care.

Surfing: There are breaks over the reefs, which are best at high tide in a low swell.

Fishing: There are permanent rip holes against the reefs that can be fished from the beach, or the rocks at low tide.

Summary: An isolated, crenulate beach, backed by steep, forested slopes. Partially protected by a near continuous, patchy reef.

425   CAPE VOLNEY

Unpatrolled
No bar R rocks reef
Beach Hazard Rating   7
Length   50 m

Cape Volney and adjacent Point Reginald are prominent headlands, backed by steep slopes that rise to 190 m. One kilometre north of the cape is a small valley, at the base of which is a 50 m long, south-east facing sand and boulder beach, which is fronted by continuous reefs. The beach is essentially inaccessible, except by boat in calm conditions at high tide.

Swimming: nsuitable for safe bathing owing to rocks, reef and its isolated location.

Surfing: None, too rocky.

Fishing: There are high rock platforms on the southern head, with reef lying off the beach.

Summary: A remote, isolated and hazardous beach.

426   THE GABLE

Unpatrolled
No bar R rocks
Beach Hazard Rating **7**
Length 200 m

The Gable is a prominent rock feature located 1 km west of Moonlight Head. Tertiary limestone rises over 100 m at Moonlight Head and then slopes down toward The Gable, where 10 to 50 m high cliffs back a 200 m long beach. The beach can be seen from the eastern Moonlight Head car park. It can be reached from the spur off the Moonlight Head Road, with a 1 km walk down the steep, grassy slopes to the beach.

The crenulate beach contains three pockets of sand, each separated by intertidal rock platforms and fronted by deeper reefs. Rips flow out between the platforms. The beach faces south-west and receives waves averaging 1.5 m, which break over the reefs and produce a steep, barless beach face.

Swimming: The reefs lower waves at the beach, however three permanent rips flow seaward of each sandy section, so take care if bathing here.

Surfing: There is surf over the reefs at high tide.

Fishing: You can fish deep water right off the beach in the rips, or from the inner reefs at low tide.

Summary: A remote but accessible beach, which is only frequented by adventurous surfers and fishers.

427, 428   WRECK (MARIE GABRIELLE), MOONLIGHT

Unpatrolled			
	Rating	Single bar	Length
427 Wreck	**8**	RBB	1 200 m
428 Moonlight	**7**	R rocks reef	400 m
		Total length	1 600 m

Wreck Beach is named after the wreck of the ship Marie Gabrielle in 1869 and sometimes also goes by that name. It is located below the 80 m high Moonlight Head car park and immediately west of Cat Reef.

Wreck Beach is 1.2 km long and consists of a continuous strip of sand arranged in six semi-circular bays, with rock platforms forming the boundary of each bay and reefs offshore. **Moonlight Beach** continues on past a small headland for another 400 m. The two beaches face south-west and are backed by steep, 20 to 80 m high bluffs, with a large rotational slump behind Moonlight Beach. Patchy reef occupies the surf off Wreck Beach and becomes more continuous along Moonlight Beach.

The waves average over 1.5 m and combine with the reefs and sand to produce four large, permanent rips along Wreck Beach, and one off Moonlight Beach. All the rips flow out through gaps in the reefs.

Swimming: These are two hazardous beaches with a rip and reef dominated surf. There are some quieter rock pools exposed at low tide, which are the safest places to swim. Be very careful of the deep, calm channels between the reefs, as these contain rip currents.

Surfing: There are breaks over the reefs at high tide, with the rip channels giving easy access to the outer breaks.

Fishing: There are several good, deep, permanent rip holes along the beach, plus the reef off the beach.

Summary: Two accessible beaches that are most popular with keen surfers and fishers.

429, 430   OLIVER HILL, PEBBLE POINT

Unpatrolled			
	Rating	No bar	Length
429 Oliver Hill	7	R platform	100 m
430 Pebble Point	7	R platform	100 m
		Total length	200 m

Oliver Hill is a 130 m high bluff just west of Moonlight Beach. At the base of the bluffs is a narrow, 100 m long, south-west facing beach, with sheer bluffs behind and on the sides, and reefs to eit her side. Around its western head is a similar beach below 60 m high bluffs, with **Pebble Point** forming its western boundary. A walking track from Moonlight Head runs along the back of the bluffs, however there is no access to the beaches.

Both beaches are exposed to waves averaging over 1.5 m which are, however, lowered by the reefs and points. The result is two steep, narrow beaches, with reefs to either end and a deep, reef controlled rip channel running out the centre.

Swimming: These are two isolated and hazardous beaches, dominated by the reefs and rips.

Surfing: There are breaks over the reefs at high tide, with a long left off the Pebble Point Beach. However, access is a problem.

Fishing: The rip channels provide permanent, deep holes right off the beaches.

Summary: These beaches are only for the adventurous surfers or fishers.

431 - 435   GELLIBRAND EAST 1,2, 3 & 4; GELLIBRAND

Unpatrolled			
	Rating	Single bar	Length
431 Gellibrand East 1	7	TBR reefs	200 m
432 Gellibrand East 2	7	RBB	1 000 m
433 Gellibrand East 3	8	RBB rocks reefs	100 m
434 Gellibrand East 4	7	RBB	100 m
435 Gellibrand (Princetown)	7	RBB	600 m
		Total length	2 000 m

The Gellibrand River flows out beside Point Ronald a 50 m high, limestone cliff capped by dune calcarenite. A continuous surf zone extends for 3 km to the east of Point Ronald, backed by five separate beaches, with calcarenite and limestone bluffs and platforms separating the beaches.

Access to the main **Gellibrand Beach** is by the Princetown Reserve Road, which ends at the National Park camping area located behind the main beach, also known as Princetown Beach. A 500 m long walking track runs along the river bank to the beach. The two easternmost beaches can be reached by a vehicle track along the National Park boundary, which leads to an access track to the second beach (432), 2 km east of the camping area. To reach the middle two beaches (433, 434) requires a walk along the bluffs or around the headland.

The five beaches face south-west and receive waves averaging over 1.5 m, which interact with the coarse sand to produce alternating bars and deep rip channels every 300 m. The easternmost beach (431) is 200 m long, with protruding reefs at each end and 60 m high bluffs behind. It usually has one large rip draining out from the middle of the beach. The second beach (432)

is 1 km long, has reefs at each end and usually two to three large rips along the beach. There are 20 to 40 m high, vegetated bluffs behind the beach. The middle beaches (433, 434) are each 100 m long pockets of sand backed by steep bluffs. The main beach (435) has the small Gellibrand River mouth in the western corner and a vegetated, 1 km wide dune field behind. One large rip usually runs out through the surf.

Swimming: These are three energetic and rip dominated beaches, with deep rip channels against the shore. se extreme care if swimming here and stay in lee of the bars.

Surfing: There are excellent beach breaks along these three beaches, with best conditions in low to moderate swell and northerly winds.

Fishing: The deep rip channels provide good deep holes to fish from the beaches.

Summary: Three natural, rip dominated beaches, with access via the camping area. Popular with more experienced surfers and beach fishers.

PORT CAMPBELL NATIONAL PARK

Area	1 750 ha
Coast length	32 km
Beaches	436 to 457 (21 beaches)

Camping Port Campbell

Park Information Centre
          Port Campbell   (055) 998 6382

The Port Campbell National Park contains 32 km of some of the most spectacular coastal scenery in Australia. Much of it is steep limestone cliffs and sea stacks, with a few exposed beaches that usually have high waves and a wide surf. The Great Ocean Road parallels the coast, providing many excellent access and viewing points, together with picnic facilities. The small town of Port Campbell is nestled in a protected bay in the centre of the National Park and provides camping, accommodation and most other facilities. Pick up a Park Guide or visit the Information Centre at Port Campbell before driving along this coast.

436 - 440   POINT RONALD, BEACHES 437 & 438, BROWN HILL, GIBSON

Unpatrolled			
	Rating	Single bar	Length
436 Point Ronald	8	RBB	150 m
437 Beach 437	9	RBB reef	200 m
438 Beach 438	8	RBB	50 m
439 Brown Hill	8	RBB	800 m
440 Gibson	8	RBB	1 500 m
		Total length	2 700 m

At Point Ronald, the coast runs due north-west for 6 km to the Twelve Apostles. The first 4 km consist of dune capped, 50 to 60 m high calcarenite bluffs. These sit on a limestone base with rock platforms and reefs scattered along the shore. A continuous, 200 m wide surf zone is dominated by large rips every 250 m. There is no direct access to any of these beaches.

In amongst the rocks and reefs are five narrow beaches, most lying between large slumps in the bluffs that now divide the once continuous beach. The first beach lies 300 m west of **Point Ronald**. It is backed by sheer and slumped bluffs and has a central bar, with rips to either side, along with reefs. **Beach 437** consists of three pockets of sand that are joined at low tide, with scattered rocks in the surf and two large rips. **Beach 438** is a 50 m pocket of sand, with a bar in front and rips to either side. The beach below **Brown Hill** is 800 m long, with 50 to 70 m high bluffs and slumps behind. It also has scattered rocks on the beach, plus reefs in the wide, rip dominated surf zone. **Gibson Beach** is the longest at 1.5 km. The bluffs decrease in height to 30 m at the western end and the beach has fewer rocks and reefs. Alternating bars and rips continue along the beach.

Swimming: These are five isolated and largely inaccessible beaches with rip, rock and reef dominated, high energy surf. They are unsuitable for safe swimming.

Surfing: There are many breaks out on the bar that are free of rocks and can provide good surf in low to moderate swell. The best known is called *Princetown Peak*, which is located on Brown Hill Beach and is accessible by way of a vehicle track that runs just south of the bridge.

Fishing: The rips and reefs provide permanent holes and gutters along the length of this section of coast.

Summary: Some of these beaches can be viewed from the bluffs, however access is difficult and dangerous and the beaches and surf are hazardous.

441   GIBSON STEPS

Unpatrolled
Single bar RBB
Beach Hazard Rating **8**
Length 1 100 m

Gibson Steps provide one of the few access points to a beach along this cliff dominated section of coast. The beach is backed by a 20 to 30 m high, sheer limestone cliff. Next to the cliff-top car park, steps were cut by an early fisherman into the soft limestone to provide access to the beach.

The beach is 1.1 km long and faces south-west, receiving waves averaging over 1.5 m. The eastern half of the beach has a normal, 200 m wide surf zone usually with two large rips. The western half toward the Twelve Apostles not only narrows, as it winds below the increasingly high cliffs, but is fronted by more continuous reefs. These reefs lower waves at the beach.

Swimming: This is an accessible but hazardous beach. The only safe places to swim are in lee of some of the reefs at low tide, when waves are low. on't even think of going down the stairs when waves are high, as they wash right up to the cliffs.

Surfing: This is a favourite surfing spot, because of both the access and the eastern breaks over the sand and reefs.

Fishing: The steps were cut by a fisherman and today it remains a popular spot to fish the deep rip holes, gutters and adjoining reefs.

Summary: Many visitors stop here to view the beach, however only venture down the steps if it's low tide and calm, or you are an experienced surfer or fisher.

442   TWELVE APOSTLES

Unpatrolled
Single bar RBB reef
Beach Hazard Rating **8**
Length 800 m

The **Twelve Apostles** are a series of 50 to 60 m high, limestone sea stacks, standing off the cliffs of the same height. They are one of the most photographed spots on the Australian coast. The Great Ocean Road runs along the top of the cliffs, with a good viewing spot located to the eastern end of the stacks.

At the base of the cliffs is a narrow, crenulate, inaccessible, sandy beach. It is backed by the cliffs and fronted by a 200 to 300 m wide sand and reef surf zone containing six of the Apostles. High seas run right up to the cliffs in places, and there are only two small areas of foredune.

Swimming: This beach is inaccessible, hazardous and not suitable or used for bathing.

Surfing: There are reef and sand breaks off this beach, but access is a major problem. A paddle from Gibson Beach would be required.

Fishing: There is no access for fishing.

Summary: Many people see this beach when viewing the Twelve Apostles, but very few have ever been on it.

443, 444   BEACH 443, BRIDGE ISLAND

Unpatrolled			
	Rating	Single bar	Length
443 Beach 443	**8**	TBR rocks reefs	100 m
444 Bridge Island	**9**	TBR rocks reefs	100 m
		Total length	200 m

These are two narrow, inaccessible beaches located 1 km and 100 m east of Loch Ard Gorge, respectively. They lie at the base of 50 m high cliffs. They are awash at high tide and fronted by rock and reef dominated surf. A permanent rip runs out from each beach. They are unsuitable for any recreational activity.

445   LOCH ARD GORGE

Unpatrolled
No bar R
Beach Hazard Rating **4**
Length 80 m

Loch Ard Gorge is named after the famous 1878 shipwreck on nearby Mutton Bird Island. Fifty-two people were lost from the Loch Ard and only two teenage survivors were fortunate enough to be washed into the only safe gorge on the coast. Other people on the ship were washed into adjacent gorges and perished. Loch Ard Gorge is unusual for this section of coast, in that it has a relatively calm interior. It has a low energy beach and moderate slopes that can be climbed, as they were by wreck survivor Tom Pearce when he went to seek help. Today the Great Ocean Road backs the gorge, together with a car park, a small cemetery (where

the four recovered wreck victims are buried) and steps down to the beach.

The beach is 80 m long and is located 300 m deep inside the narrow gorge. Even during high seas, waves are low at the beach, while a surf may break across the entrance.

Swimming: Relatively safe at the beach. However, do not venture out into the gorge as a rip current runs out the entrance.

Surfing: None.

Fishing: The water is moderately deep off the beach.

Summary: Another very popular tourist stop with one of the few calm beaches on this section of coast.

446, 447, 448   BROKEN HEAD, SHERBROOK CREEK, RUTLEDGE CREEK

Unpatrolled			
	Rating	Single bar	Length
446 Broken Head	**5**	R	150 m
447 Sherbrook Creek	**7**	TBR	200 m
448 Rutledge Creek	**7**	TBR	100 m
		Total length	450 m

Broken Head is the most seaward protruding of the limestone headlands along this section of coast. It is a good place to view the coast and surf. On the western side of the head are a car park and toilets. **Broken Head Beach** is located immediately below the car park. Sherbrook Creek flows out on the western side of the car park, with the beach usually blocking the creek. **Rutledge Creek Beach** is located just around the next bluff and is accessible directly from the Great Ocean Road. A parking area is located above the western end of the beach.

The three beaches all face south-west and are backed by 10 to 20 m high cliffs. In addition, there is a small dune area behind **Sherbrook Creek Beach**. The beaches are composed of medium sand and are consequently steep. Broken Head Beach is 150 m long and slightly more protected, with waves averaging less than 1 m. This usually results in a steep beach with no bar, a heavy shorebreak and deep water off the beach. Sherbrook Creek Beach is more exposed and has a steep beach slope, with a single, often detached bar and a permanent rip off the beach. Rutledge Creek Beach is similar, with a single bar and permanent rip.

Swimming: Sherbrook and Rutledge Creek Beaches are dominated by higher waves and rips and should be avoided. Broken Head Beach is the most accessible and has lower waves, however there can be a heavy shorebreak and currents off the beach. Take care if bathing on any of these beaches.

Surfing: There are beach breaks over the bars and reefs off Sherbrook and Rutledge.

Fishing: There is deep water off all three beaches as well as rip holes at Sherbrook and Rutledge that can be fished from the beach. In addition, Broken Head is used for cliff-top fishing.

Summary: These are three reasonably accessible beaches, with parking and toilets at Broken Head.

449, 450   BEACH 449, SENTINEL ROCK

Unpatrolled			
	Rating	Single bar	Length
449 Beach 449	8	TBR	50 m
450 Sentinel Rock	8	TBR	100 m
		Total length	150 m

Beach 449 is located in a 200 m long gorge at the base of 50 m high cliffs. Next door is a 100 m long beach at the base of 60 m high cliffs, that is fronted by a solitary sea stack, with **Sentinel Rock** lying at the western end of the beach. Both beaches are dominated by high waves and rips and neither is accessible or suitable for recreation.

451   PORT CAMPBELL

Port Campbell SLSC

Patrols: late November to Easter holidays
Surf Lifesaving Club: Saturday, Sunday and Christmas public holidays
Lifeguard: no lifeguard on duty or weekday patrols

Single bar RLTT
Beach Hazard Rating **4**
Length 150 m

For map of beach see Figure 124

Port Campbell is the only town along this section of coast, and its beach is one of the few sheltered and relatively safe bathing spots on a notorious stretch of exposed coast. The town and beach occupy a 200 m wide, partially infilled valley, where the small town spreads over the eastern slopes. The 150 m long beach is bounded by the valley sides on the east and the entrance to Port Campbell Creek in the west.

A reserve, car park and caravan park, together with the Port Campbell Surf Life Saving Club, back the beach. The narrow port entrance reduces the waves to a height averaging 0.5 m at the beach. These produce a moderately steep beach fronted by a continuous, narrow bar. Rips only occur during big seas. These big seas often erode the beach, exposing a rocky substrate. The occasional erosion also required a low seawall to be built along the back of the beach. The surf lifesaving club, which was formed in 1963, averages 4 rescues each year.

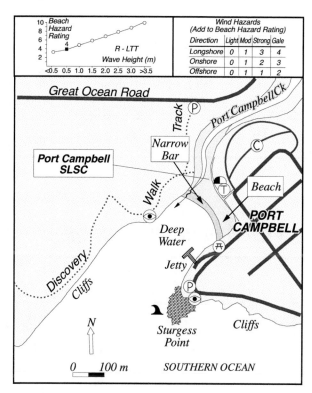

Figure 124. Port Campbell occupies a 200 m wide gap in the seacliffs. The small bay contains the jetty, a 150 m long, protected beach and the town of the same name. The beach is relatively safe and surrounded by a good park, including picnic facilities, with the added safety of the surf lifesaving club.

Swimming: A relatively safe beach with usually low waves, however the water is deep close inshore, so stay between the flags and away from the rocks to either end. Bigger seas can produce a heavy shorebreak and rips.

Surfing: sually a low swash at the beach. Only in big swell does a reef off the eastern point produce a rideable left.

Fishing: Professional fishers launch their boats from the Port Campbell Jetty. The beach offers fairly placid waters, with the creek and creek mouth the most popular areas. The rocks are exposed and very hazardous and should only be fished in low waves, and if you are very experienced.

Summary: Port Campbell is right on the Great Ocean Road and is a major stop-over for sightseers.   The reserve between the road and beach offers a lovely spot for a picnic, while the beach is the safest and only patrolled one on this dangerous section of coast. It is the best place to bathe in summer.   There is also a iscovery Walk from the first car park on the western side of the bay, to the Two Mile Bay Road.

452, 453, 454   SHELLY, TWO MILE BAY, TWO MILE BAY WEST

Unpatrolled			
	Rating	Single bar	Length
452 Shelly	7	R platform	300 m
453 Two Mile Bay	7	LTT	100 m
454 Two Mile Bay West	8	TBR rocks reef	150 m
		Total length	550 m

Two Mile Bay is a very open, 3 km long bay that extends west of Port Campbell. nlike the adjacent coast, the bluffs here are protected by an ancient raised platform, capped by the remnants of a beach, dune and swamp. The vegetated, 60 m high bluffs show what the entire coast would have looked like before the sea level rose (about 6 000 years ago) and reactivated the cliffs.

There is a road out to the bluffs with a car park, and a track down to Two Mile Bay Beach. **Shelly Beach** lies immediately to the east. The beaches face south and, while exposed to high waves, are partially protected by reefs extending 200 to 300 m offshore. Shelly Beach is 300 m long and fronted by a continuous calcarenite platform that is exposed at low tide, with reefs further offshore.

The main **Two Mile Bay Beach** is 100 m long and backed initially by low calcarenite spurs, then by bluffs rising to 60 m. It has direct access to the sea with usually a heavy shorebreak, but extensive reefs off the beach.

Victorian Beaches - Region 8: West (Port Campbell to Warrnambool)

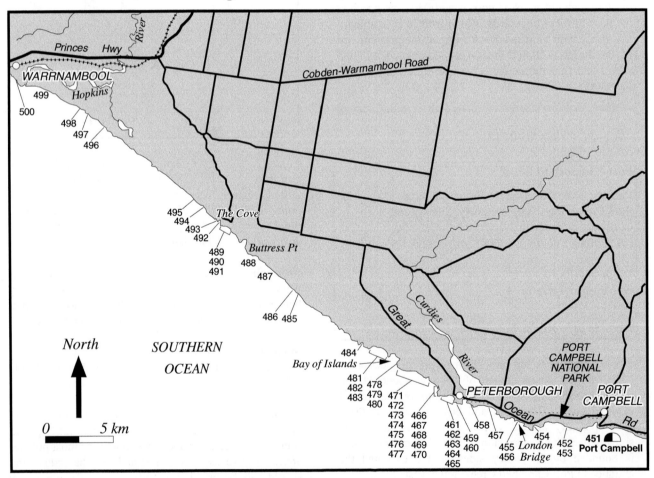

Figure 125.   Region 8 (western section): Port Campbell to Warrnambool (Beaches 452 - 498)

Further west, the bluffs continue and a narrow, 150 m long beach lies at their base, before the bluffs turn south to The Arch. At this point the Great Ocean Road lies immediately above this inaccessible and hazardous beach.

Swimming: Rocks and reef dominate these three beaches and none are suitable for safe bathing.

Surfing: Experienced big wave riders surf the outer reefs off the eastern end of Two Mile Bay, called the *Rifle Range*.

Fishing: At low tide you can fish off the platforms into the deep water inside the reefs.

Summary: A geologically interesting area, used by big wave riders and fishers.

455, 456   LONDON BRIDGE, LONDON BRIDGE WEST

Unpatrolled			
	Rating	Single bar	Length
455 London Bridge	5	RLTT	250 m
456 London Bridge West	6	R	40 m
		Total length	290 m

London Bridge is another of the more famous stops on the Great Ocean Road. nfortunately, the Bridge collapsed in 1990, and visitors can no longer enjoy the thrill of walking out over the crashing waves below. The remaining seaward part of the bridge does now, however, resemble an outline of Australia, when viewed from the air.

On the western side of the Bridge is a 250 m long, south-west facing beach. It is partially protected by the Bridge stack and reefs and usually has waves averaging 1 m, that surge heavily on the steep beach face. There are steps leading down the 20 m high bluffs to the beach, however these are blocked by a locked gate as the beach, backing dunes and cavern are a penguin colony. Just around the western end of the beach is a second 40 m long pocket of sand, nearly surrounded by 20 m high cliffs and inaccessible to the public.

Swimming: Both beaches have heavy shorebreaks, with deeper water off the beaches.

Surfing: There is surf during moderate swell on the reefs off the beach, however access is a problem.

Fishing: The best way to fish here is from the cliffs. Watch the waves during moderate to high swell.

Summary: The beaches are off limits, however people still come to see the remains of London Bridge.

457, 458   NEWFIELD BAY, PETERBOROUGH

Unpatrolled			
	Rating	Single bar	Length
457 Newfield Bay	6	LTTTBR	700 m
458 Peterborough	5	RLTT	800 m
		Total length	1 500 m

The town of Peterborough lies at the mouth of Curdies Inlet a moderately large, open lagoon. The inlet is blocked by a 1.5 km long, 300 to 400 m wide sand barrier. This consists of 10 to 20 m high dunes and a cuspate foreland in lee of Schomberg Reef, containing two beaches. The Great Ocean Road runs along the back of the barrier and crosses the inlet to the small town of Peterborough.

The two beaches are **Newfield Bay Beach**, on the eastern side of the sandy foreland, and **Peterborough Beach** to the west. Newfield Bay is accessible from the road, with a parking area at the eastern end. The beach is 700 m long, faces south and is partially protected by the offshore reefs. As a result, the beach is steep and barless toward the reef. Closer to the car park, the waves are slightly higher and there is a continuous attached bar with an increasing chance of rips.

Peterborough Beach is 800 m long, faces south-south-west and is also protected by the offshore reefs and those near the inlet. It usually has a steep beach with surging waves. There are parking and picnic areas just before the bridge and, when the inlet is closed, the beach is also accessible from the Peterborough foreshore reserve.

Swimming: These are two moderately safe beaches. The Peterborough inlet area is the safest, when it is closed. Higher waves will produce a heavy shorebreak and rips on Newfield Bay Beach, with a permanent rip at the eastern end below the car park.

Surfing: There are usually beach breaks in front of the Newfield Bay car park, and shorebreaks along the rest of the two beaches.

Fishing: There is excellent fishing in the lagoon, the inlet (when open) and off the beach and rocks around Peterborough.

Summary: These are two of the few accessible and moderately safe beaches on the west coast, with all the facilities of Peterborough next door.

Newfield Bay is named after the ship *Newfield*, which ran aground here in 1892 with the loss of 8 lives.

459, 460   JAMES IRVINE, WATER TOWER

Unpatrolled			
	Rating	No bar	Length
459 James Irvine	6	R platform	60 m
460 Water Tower	6	R rocks reef	100 m
		Total length	160 m

James Irvine and Water Tower Beaches are two small pockets of sand located out on the Peterborough headland. There is a car park on the low headland and a grassy reserve on the 10 m high bluffs behind the two adjoining beaches. The water tower stands on the bluffs above the western end of the beach that is named for it. **James Irvine Beach** is fronted by a continuous, calcarenite, intertidal platform that is exposed at low tide. **Water Tower Beach** is similar, but with some deeper water close inshore.

Swimming: Only possible at high tide, however stay close inshore as there are rocks and reefs off the beaches.

Surfing: None.

Fishing: The shallow reefs off the beaches can be fished at high tide.

Summary: Two very accessible, but small, pocket beaches surrounded by calcarenite bluffs and platforms.

461 - 465   WILD DOG COVE 1 to 5

Unpatrolled			
	Rating	No bar	Length
461 Wild og Cove 1	4	R	70 m
462 Wild og Cove 2	4	R	40 m
463 Wild og Cove 3	4	R	50 m
464 Wild og Cove 4	4	R	60 m
465 Wild og Cove 5	4	R	500 m
		Total length	720 m

Wild Dog Cove is a 500 m wide bay on the western side of Peterborough. It is a circular, south facing bay, backed by 10 m high red limestone cliffs, with a shallow bay floor dominated by limestone reefs. The bay has 1 km of shoreline, with four small pocket beaches and one longer beach adjacent to the Great Ocean Road. All the beaches are located at the base of the cliffs and receive low to often calm waves, particularly at low tide.

A foreshore reserve backs all the beaches. The four small beaches (461 to 464) lie in a row on the eastern side of the bay. Each is backed by the red bluffs and is well protected by the shallow bay reefs. Access to each is by steps or a climb down the bluffs. The 500 m long main beach (465) parallels the road and has vegetated bluffs, a car park and steps in the centre of the beach.

Swimming: All these beaches are relatively safe, with usually low waves or calms. Bathing is best at high tide, as many of the reefs are exposed at low tide.

Surfing: Only on the outer reefs.

Fishing: The bay is very shallow, with the best spots being on the outer rocks.

Summary: A very accessible and well-protected bay. The main beach is located right by the Great Ocean Road, is suitable for picnics and offers relatively safe bathing.

466   BEACH 466

Unpatrolled
No bar R platform
Beach Hazard Rating **5**
Length 70 m

Beach 466 is located 200 m east of the car park at the Falls of Halladale wreck site. It is located below 10 m high, red bluffs and is fronted by an intertidal, limestone platform. Waves are usually low to calm at the beach. It is also protected by reefs further offshore.

Swimming: A relatively safe beach when waves are low. Best at high tide.

Surfing: None at this beach.

Fishing: There are numerous spots along the adjoining bluffs and rocks.

Summary: A quiet, small beach best suited for sunbathing.

467, 468   FALLS OF HALLADALE 1 & 2

Unpatrolled			
	Rating	No bar	Length
467 Falls of Halladale 1	4	R	80 m
468 Falls of Halladale 2	4	R	40 m
		Total length	120 m

The ship Falls Of Halladale was wrecked on reefs off these two beaches in 1908. Today a vehicle track and car park are located on the bluffs above the site. The two adjoining beaches are located on the western side of the 15 m high, limestone headland. Both beaches are pockets of sand backed by the bluffs and fronted by sea stacks and reefs off the beaches.

Swimming: Relatively safe close inshore, with deeper water and reef out from both beaches.

Surfing: There is a surfing break directly in front of the main car park called *The Well* or *Flyhole*. It is a heavy right breaking over the reef. It requires a paddle out through a narrow gap in the reefs called the Flyhole. A second break on the west side of the point is a left called *Boneys*.

Fishing: There is relatively deep water off the beaches, plus plenty of low bluffs and rocks to fish from.

Summary: Two small, accessible and relatively safe beaches.

469   WORM BAY

Unpatrolled
No bar R
Beach Hazard Rating **4**
Length 100 m

Worm Bay is a small gap in the limestone bluffs located on the eastern side of the Bay of Martyrs, and 300 m west of the Falls of Halladale car park. It lies just off the Great Ocean Road and there are car parks at both ends of the 100 m long beach. Steps lead down to the beach from the 10 m high limestone bluffs.

The beach faces west and is partially protected by two headlands and the numerous reefs in the Bay of Martyrs. Waves tend to be low in the bay and surge up the beach face.

Swimming: A relatively safe beach under normal conditions. The shorebreak intensifies as wave height increases.

Surfing: None in Worm Bay, however there are several breaks on the reefs out in the Bay of Martyrs.

Fishing: There is relatively deep water right off the beach, otherwise the bluffs and headlands inside the bay provide good access to deep water.

Summary: A very accessible and protected beach, with a little bit of swell. Relatively safe for bathing.

470, 471   BAY OF MARTYRS, MASSACRE HILL

Unpatrolled			
	Rating	Single bar	Length
470 Bay of Martyrs	4	R	500 m
471 Massacre Hill	4	LTT	100 m
		Total length	600 m

The Bay of Martyrs is an open, 2.5 km long, south-west facing bay containing numerous reefs and sea stacks. The shoreline is composed predominantly of 10 m high, red limestone bluffs. Within the bay are several smaller bays and beaches, two of which are named Massacre Bay and Crofts Bay. The main **Bay of Martyrs Beach** lies at the eastern end of the bay. It is 500 m long, faces south-west and is backed by a mixture of irregular bluffs and dunes, with the Great Ocean Road just behind. There is a car park and beach access at the eastern end.

Massacre Hill Beach lies immediately to the west. It is a narrower beach, backed by steep bluffs, with access at low tide from the Bay of Martyrs Beach. Both beaches are fronted by extensive reefs, with a large sea stack just off the Bay Beach. Waves are reduced by the reefs to about 0.5 m and the beaches are usually steep, with little or no bar and no rips.

Swimming: The Bay of Martyrs Beach is the most accessible and safest location to swim. It is better at high tide when the reefs are covered.

Surfing: There is usually no surf at the beaches, while there are breaks on the reefs offshore.

Fishing: You can fish the shallow reefs from the beaches at high tide.

Summary: These are two relatively long beaches, next to the road and relatively safe for bathing.

472, 473, 474   MASSACRE BAY 1, 2 & 3

Unpatrolled			
	Rating	Single bar	Length
472 Massacre Bay 1	4	RLTT	100 m
473 Massacre Bay 2	4	RLTT	60 m
474 Massacre Bay 3	4	LTT	50 m
		Total length	210 m

Massacre Bay is a 250 m wide bay located in the centre of the larger Bay of Martyrs. The Great Ocean Road skirts the back of the bay, with a car park on the bluffs just to the west. The bay consists of 20 m high, red limestone bluffs, with these three beaches lying below them and separated by protruding sections.

The first beach is 100 m long and has some vegetated dunes that have climbed the bluffs, crossed the road and now extend up to 300 m inland. The two smaller beaches are only accessible at low tide from the main beach. All three receive low waves averaging 0.5 m, that surge across a narrow, continuous bar.

Swimming: These are three relatively safe beaches, with the main beach having the best access.

Surfing: There is no surf at the beaches or in the bay.

Fishing: The bay is relatively shallow and the bluffs are dangerous to fish from.

Summary: Most travellers miss these three beaches. The main beach is reasonably wide, usually has a low surge at the beach face and is relatively safe for bathing.

475, 476   BEACH 475, BEACH 476

Unpatrolled			
	Rating	No bar	Length
475 Beach 475	4	R	50 m
476 Beach 476	4	R	60 m
		Total length	110 m

Between Massacre Bay and Crofts Bay and just off the Great Ocean Road, are 500 m of irregular bluffs, within which are two gorges containing two small pockets of sand. The first lies immediately west of the headland car park. The first beach, **Beach 475**, is 50 m long and surrounded by 10 m high bluffs, while the second, **Beach 476**, is 60 m long with similar bluffs. Both are fronted by extensive inter- and sub-tidal reefs, and receive low waves at the shore.

Swimming: Both are relatively safe, with exposed rocks and reefs at low tide and usually calm conditions at high tide.

Surfing: None at these beaches.

Fishing: Only at high tide from the beaches, however the adjoining bluffs provide access to deep water.

Summary: Two accessible pockets of sand, if you know where to look.

477   CROFTS BAY

Unpatrolled
No bar R
Beach Hazard Rating **4**
Length 1 200 m

Crofts Bay is the westernmost of the smaller bays within the larger Bay of Martyrs. It lies 4 km west of Peterborough. The Great Ocean Road runs behind the low bluffs and dunes that back the 1.2 km long bay. There are car parks off the road in the centre and at the western end of the bay. The western car park has steps down the bluff to the beach.

The beach faces south and is protected by extensive reefs extending up to 1 km offshore. These reduce the waves to an average of 0.5 m at the beach, resulting in a steep, barless beach, with a low, surging shorebreak and usually no bar or rips. There are a few rocks along the beach and reefs that are exposed at low tide.

Swimming: This is a relatively safe beach, with typically low waves and no rips. Water is deep off the beach, particularly at high tide.

Surfing: No surf at the beach, and it's a long paddle to the outer reef breaks.

Fishing: There are reefs off the beach that can be fished at high tide. A few locals store their fishing tinnies in caves under the bluff at the western end.

Summary: An accessible, more quiescent, long, sandy beach, fine for picnics and bathing, but no surf.

478, 479   BEACH 478, BEACH 479

Unpatrolled			
	Rating	No bar	Length
478 Beach 478	4	R	100 m
479 Beach 479	4	R	50 m
		Total length	150 m

Three hundred metres west of the western Crofts Bay car park are two small pocket beaches. These are located at the base of the bluffs that form the low headland between the Bay of Martyrs and the Bay of Islands. The first, **Beach 478**, is 100 m long and bordered by small cliffs and a sea stack, with reefs extending 200 to 300 m offshore. The second, **Beach 479**, is in a small gorge and is 50 m long. There are channels off the beach, and platforms, reefs and sea stacks to either side and offshore. The reefs reduce the waves, which average less than 0.5 m at the shore.

Swimming: These are two relatively safe beaches, with the longer one fronted by reefs, while the second, smaller beach has deep water off the beach.

Surfing: None at these beaches.

Fishing: You can fish the reefs from the beaches at high tide or off the adjoining bluffs.

Summary: Two small beaches just off the Great Ocean Road, with the western one having the better access down the bluffs.

480   BAY OF ISLANDS BOAT RAMP

Unpatrolled
No bar R
Beach Hazard Rating **2**
Length 70 m

The **Bay of Islands** is an irregular, semi-circular bay that faces south-west and has several large sea stacks or islands dotted about the bay, together with rocks and reefs. Most of the bay shore is made up of steep, 10 to 20 m high limestone bluffs. However, tucked in the eastern corner of the bay and right next to the bend in the Great Ocean Road, is a gorge containing a 70 m long beach.

The beach is used for boat launching and there is a steep ramp and steps descending from the bluffs to the beach. The beach itself is narrow, with deep water offshore, particularly at high tide. The reefs filter out most waves, with usually calm conditions at the beach.

Swimming: A relatively calm location, only watch for boats.

Surfing: None at this beach.

Fishing: Most fishers go out from here in boats, however there is deep water off the adjoining bluffs.

Summary: A very accessible beach, but should be left to the boat users when busy.

BAY OF ISLANDS COASTAL PARK

Coast length   20 km
Beaches   481 to 493

This 20 km long park includes several kilometres of sculptured limestone bluffs; the energetic Flaxman Beach and its steep, climbing foredunes; and the more intricate, small bays around Childers Cove.

Park Information   Port Campbell   (055) 98 6382

481, 482, 483   **BEACHES 481, 482 & 483**

Unpatrolled			
	Rating	No bar	Length
481 Beach 481	**8**	R	150 m
482 Beach 482	**8**	R	50 m
483 Beach 483	**8**	R	100 m
		Total length	300 m

At the Bay of Islands, the Great Ocean Road turns north and leaves the coast. The road at this point forms the eastern boundary of the Bay of Islands Coastal Park, a narrow, 20 km long park that occupies the former foreshore reserve. Most of the park shoreline is composed of limestone and calcarenite bluffs, reefs and sea stacks, with only a few beaches. The first three beaches lie 1 to 1.5 km west of the bend in the road, and the bluffs above the beaches are accessible along the northern park boundary.

The three beaches total 300 m in length and all lie at the foot of sheer, 30 m high cliffs, with no access down to the sand. They are narrow, covered in parts by cliff debris, and awash at high tide and in big seas. A large part of the cliff fell on **Beach 483** in 1994, burying much of the beach beneath the rubble. They are fronted by scattered reefs and rocks, but still receive waves averaging 1.5 m, resulting in permanent rips next to the reefs.

Swimming: These are three inaccessible and hazardous beaches dominated by high cliffs, high waves, reefs and permanent rips.

Surfing: There is surf over the reefs off the beach, however access is a problem.

Fishing: Again, there are good rip channels off the beaches, but no access.

Summary: Three beaches and cliffs worth viewing but difficult to set foot on.

484   RADFORDS

Unpatrolled
No bar LTT rocks
Beach Hazard Rating **8**
Length 50 m

Five hundred metres east of the end of Radfords Road is a south-east facing pocket of sand lying at the base of 40 m high cliffs. The beach is inaccessible, backed by sheer cliffs and fronted by a rock and reef dominated surf zone. It is unsuitable for swimming or surfing.

485 - 488   BEACH 485, FLAXMAN HILL, FLAXMAN, BUTTRESS POINT

Unpatrolled			
	Rating	Single bar	Length
485 Beach 485	7	TBR	150 m
486 Flaxman Hill	8	RBB	1 300 m
487 Flaxman	8	RBB	3 400 m
488 Buttress Point	7	RBB	250 m
		Total length	5 100 m

To the west of the Bay of Islands and Radfords Beach, the next 5 km of coast consist of 50 to 60 m high, crenulate, limestone cliffs, with well-developed rock platforms at their base. A few farm roads run down to the Coastal Park boundary, otherwise the coast has limited access. Beaches begin again just east of Flaxman Hill. The hill is an 80 m high sand dune sitting on top of 50 m high cliffs. Whites Road runs down to the park boundary 1 km west of the hill, where there is a car park behind the 50 m high bluffs, but no facilities. Vehicle tracks follow the park boundary to the bluffs above the other beaches.

There are four exposed, south-west facing beaches along this 5.5 km section of coast. The first, **Beach 485**, lies just east of Flaxman Hill and is 150 m long, backed by vegetated bluffs, with headlands to each end. It is fronted by a bar and two permanent rips, one against each headland.

The second beach lies immediately below **Flaxman Hill**. It is 1.3 km long, with 60 m high, dune-capped bluffs backing the eastern half, and a climbing sand dune running up a wide gully in the western half. This dune illustrates how cliff-top dunes were emplaced, when there were more beaches and sand, and cliff-climbing sand ramps along the coast. This beach has a headland at the eastern end, while a 50 m high bluff forms the western boundary.

Just past this bluff is 3.4 km long **Flaxman Beach**. This energetic beach has a few reefs in front of the car park, but for the most part it has a 300 m wide surf zone, with large, strong rips every 300 m. These alternate with a detached bar. The winds have blown the sand from the beach to form a series of grassy sand ramps that have attempted (but so far failed) to reach the top of the backing 40 to 50 m high bluffs.

The fourth beach lies just around an eroding bluff, that forms the western end of Flaxman Beach. It is 250 m long and bounded to the west by **Buttress Point**. It has a continuous reef extending 200 m off the beach, which lowers the waves at the shore and forms a permanent rip between the beach and the reef.

Swimming: These are four very hazardous and little used beaches. se extreme caution if you plan to swim, as a continuous deep trough runs along the beach and leads to strong, persistent rips.

Surfing: There are good beach breaks for experienced surfers. They work best in a low swell with north-east winds. Locals know this as *Terrys* Beach.

Fishing: The rips produce deep holes and gutters right along the beaches. In addition, there are some rock platforms at the base of some of the points.

Summary: There is an excellent view of Flaxman Beach from the car park, and the view is worth the drive. However, do not enter the water unless you are very experienced.

489 - 493   STANHOPE BAY, DOG TRAP BAY, BUCKLEY CREEK, MURANE BAY, CHILDERS COVE

Unpatrolled			
	Rating	Single bar	Length
489 Stanhope Bay	3	LTT	250 m
490 og Trap Bay	7	TBR	250 m
491 Buckley Creek	4	R	500 m
492 Murane Bay	5	LTT	80 m
493 Childers Cove	6	TBR	100 m
		Total length	1 180 m

Childers Cove is a well-known, small cove located about 20 km east of Warrnambool, with a road out to the cove called Childers Cove Road. This sealed road leads out to the cove and two adjoining beaches, with a vehicle track heading east to og Trap and Stanhope Bays. Five beaches lie in protected bays and coves within 3 km of coast, that mostly consists of eroded, 20 to 30 m high, limestone bluffs, cliffs, headlands, sea stacks and reefs. They all face south-west, however the five beaches vary considerably in size and nature, depending on the impact of the rocks, reefs and waves.

Stanhope Bay is an almost circular bay, with a 200 m wide opening to the sea and a curving, 250 m long beach inside. The narrow beach lies at the base of the cliffs and is fronted by a shallow sand and reef bay floor. Waves are low in the bay, with the water often calm at low tide.

Dog Trap Bay is a more open and exposed, 250 m long bay, with bright red limestone headlands. There are some dunes climbing the bluffs behind the beach. Waves break over the patchy reefs and rocks in the surf zone. There is a permanent rip on the eastern side of a small sea stack, and a cobble shoreline at the rear of the beach.

Buckley Creek drains into the eastern end of a 500 m long beach that bulges out to the west in lee of some of the largest sea stacks after the Twelve Apostles. The three stacks lie right off the western end of the beach and can be viewed at ground level. The stacks and reefs lower waves at the beach to about 0.5 m, which results in a steep, barless beach with a strong, surging shorebreak. Higher waves break heavily at the base of the beach.

Murane Bay is a more deeply embayed, 80 m long beach, that still receives waves averaging 1 m. There are toilets and a car park above the beach, with steps down to the beach. The beach is low and flat and awash at high tide. It has a shallow surf including a few reefs and rocks. A permanent rip runs out of the bay and there is a sea arch in the making on the eastern headland.

Childers Cove itself is another small, embayed beach. It is 100 m long with a bluff crossing the western end of the beach. It is backed by 30 m high, grassy bluffs, and 40 m high, red headlands guarding the cove. The beach is low and flat, with a shallow bay floor. It has a few reefs and one narrow sea stack just off the beach. A permanent rip drains out of the cove.

Swimming: Stanhope Bay is the safest beach with the lowest waves, but is the most difficult to access. Of the other beaches, the inner, shallower reaches of Murane Bay and Childers Cove are moderately safe, so long as you do not venture out into the bays. og Trap Bay is dominated by reefs and rips, and Buckley Creek, while it often has low waves, has deep water close inshore and at times a heavy shorebreak.

Surfing: There are no breaks on these beaches, with the larger waves breaking on the outer reefs.

Fishing: This is a very popular spot for beach and rock fishing into the numerous rock holes and gutters.

However be careful, as the rock platforms are awash at high tide and in big seas.

Summary: A reasonably popular destination for tourists and day trippers, with interesting coastal scenery and some reasonably safe beaches.

> Childers Cove is named after the wreck of the ship *Children*. It grounded here in 1839 with the loss of 11 lives.

494, 495   BEACH 494, BEACH 495

Unpatrolled			
	Rating	Single bar	Length
494 Beach 494	7	RBB	200 m
495 Beach 495	8	RBB	80 m
		Total length	280 m

One and two kilometres west of Childers Cove are two small, remote beaches. They are backed by the Coastal Park and farm land, with no direct public access, other than a walk along the bluffs. Both lie below 70 m high cliffs, with rock falls to each end and rocks and reefs offshore. In addition, there is a sand bar with large, strong rips further out.

Swimming: These are two almost inaccessible beaches with very hazardous surf zones, and are unsuitable for bathing.

Surfing: There is surf off the beaches, if you can find a way down.

Fishing: There are permanent, deep rip channels off both beaches, but no safe access.

Summary: Two essentially inaccessible and hazardous beaches, that are there for viewing only.

496, 497, 498   BEACHES 496, 497 & 498

Unpatrolled			
	Rating	Single bar	Length
496 Beach 496	8	RBB	100 m
497 Beach 497	8	TBR	50 m
498 Beach 498	8	TBR	500 m
		Total length	650 m

Port Campbell Limestone dominates the 60 km of coastline between the Gellibrand River and the eastern reaches of Warrnambool. From Childers Cove the limestone cliffs continue along the coast for 13 km, initially as 70 m high cliffs that gradually decrease in height to 20 m, before turning inland and being replaced by a lower coast and sandy beaches. Along the final 2 km of limestone are three exposed, high energy beaches lying below the cliffs. There is no formal access to these beaches, that are backed by the Coastal Park and farmland.

The first, **Beach 496**, is 100 m long, backed by 30 m high cliffs, with rocks and reef in the surf and a permanent rip running offshore. The second, **Beach 497**, is just a 50 m pocket of sand at the base of a hollow in the 30 m high cliffs, with rocks and reef in the surf. The final beach, **Beach 498**, is longer, but narrow and irregular, lying at the base of the last section of cliffs. It has rocks and reef in the surf and permanent rips.

Swimming: These are three difficult to find and access beaches, that are unsuitable for bathing.

Surfing: The only recognised surfing spot along this section of coast is *Backyards*, a left hand reef break reached by the Lake Gillear Road, and then a walk across the farm land.

Fishing: This is a hazardous coast to fish, with high waves breaking over the irregular rocks along the base of the cliffs.

Summary: This coast offers exciting coastal scenery but should only be used by very experienced surfers and rock fishers.

9. WARRNAMBOOL TO NELSON

> **Warrnambool to Nelson**
> **(Hopkins River to South Australia)**
>
> Beaches 499 to 560
> Coast length 211 km
>
> *For maps of region see Figures 126, 130 & 131*

At the Hopkins River mouth, the nature of the coast changes dramatically. The rugged terrain and high cliffs and bluffs that have dominated the coast since Torquay, give way to a wide, low coastal plain. The occasional headlands tend to be low and basaltic, with dune calcarenite sometimes capping the basalt, while elsewhere the calcarenite forms jagged cliffs and headlands. In between these is a series of low, curving bays and beaches, beginning with Lady Bay at Warrnambool. This is followed by Armstrong Bay, Port Fairy, then the long Portland Bay, Nelson and Bridgewater Bays (Figure 35) and finally, the long and exposed iscovery Bay (Figure 127) which runs on into

South Australia. These bays and the anchorages they provide also form the sites of the western coastal ports and towns of Warrnambool, Port Fairy and Portland.

	Coast length	Beaches	No.
Figure 126:			
Warrnambool (Lady) Bay	10 km	499 - 501	3
Armstrong Bay	15 km	502 - 507	6
Port Fairy Bay	11 km	508 - 513	6
Griffiths Island - Cape Reamur	11 km	514 - 522	9
Figure 130:			
Portland Bay	68 km	523 - 543	21
Grant	16 km	544 - 547	4
Nelson Bays			
Figure 131:			
Bridgewater Bay	20 km	548 - 555	8
iscovery Bay	61 km	556 - 560	5

Victorian Beaches - Region 9: East (Warrnambool to Yambuk)

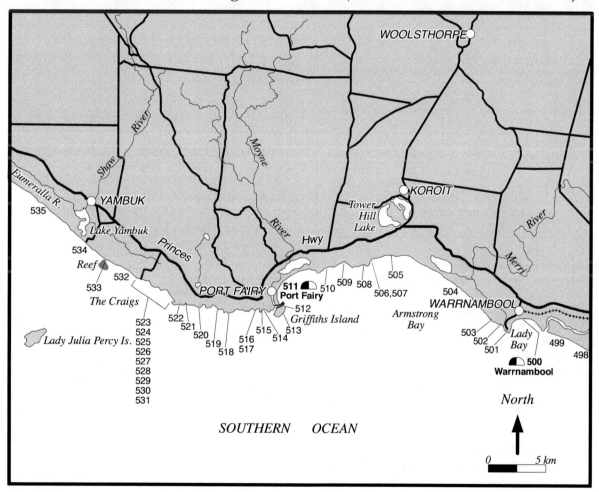

Figure 126.   Region 9 (eastern section): Warrnambool to Yambuk (Beaches 499 - 534)

Figure 127.   A view from the high foredune at Nelson in the state's far west.   This western section of Discovery Bay typifies the high energy western Victorian coast.   The fine sand, low gradient beach is fronted by a 400 m wide surf zone, containing three bars.   Rips dominate the first two bars.

499   LOGANS

Unpatrolled
Single bar TBRRBB
Beach Hazard Rating **7**
Length 4 300 m

Logans Beach forms the geological boundary of the limestone coast of Port Campbell and the calcarenite and basalt coast to the west. It begins as the limestone bluffs, that have dominated the coast since the Gellibrand River, are replaced by Holocene dunes capping older Pleistocene dune calcarenite. The Hopkins River Road runs a few hundred metres behind the beach, with access to the beach by way of two car parks. One of these is on the banks of the river, the other is in the dunes a few hundred metres east of the river mouth.

The beach is 4.3 km long, faces south-west and runs from the end of the 20 m high limestone cliffs to the mouth of the Hopkins River. There are 10 m high calcarenite bluffs at the river mouth, called Point

Ritchie. The beach receives waves averaging 1.5 m, which produce a rip dominated surf. Strong, persistent rips are spaced every 300 m, often with detached bars in between. In addition, patchy calcarenite reef occurs in the surf, particularly off the river mouth. This further intensifies some of the rips.

Swimming: This is a hazardous, rip dominated beach. Council signs warn against swimming here.

Surfing: There are numerous beach breaks up the beach and more consistent breaks over the reefs just north of the river mouth. This area is called *Japs*, and works best in low to moderate swell and northerly winds.

Fishing: There are permanent rip holes in lee of the river mouth reefs, and shifting rips up the beach.

Summary: This beach is a popular site for whale watching, with a platform erected by the council at the northern car park.

500   WARRNAMBOOL

Warrnambool SLSC

Patrols: late November to Easter holidays
Surf Lifesaving Club: Saturday, Sunday and
Christmas public holidays
Lifeguard: 10 am to 6 pm weekdays during Christmas
holiday period

Single bar LTTTBR
Beach Hazard Rating **6**
Length 3 500 m

For map of beach see Figure 128

Warrnambool is the largest coastal town in western
Victoria. The city has a population of 25 000 and is also
a very popular holiday destination. The entire western
foreshore is given over to recreation, with extensive
parks, lakes, caravan parks and additional facilities.

The beach fronts the foreshore reserve and forms the
shoreline of Lady Bay. It is 3.5 km long and faces
south, but swings around to face east in lee of Middle
Island and the harbour breakwater. Since breakwater
construction began in the 1850s, the western end of the
beach has built out a few hundred metres. To the east of
the surf lifesaving club, the beach is backed by grass
covered sand dunes.

Today the beach extends from Point Ritchie in the east
to the western breakwater. ue to the change in
orientation and degree of protection, wave height is low
at the breakwater averaging less than 0.5 m. This
increases to about 1.3 m at the Warrnambool Surf Life
Saving Club and over 1.5 m at Point Ritchie. Likewise
the beach adjusts to this change in wave height, with a
narrow, continuous bar near the breakwater. The first
rips occur near the surf club, and these increase in size
and intensity toward Point Ritchie.

The Warrnambool Surf Life Saving Club, one of the
oldest in Victoria, was formed in 1930, and averages 10
rescues a year.

Swimming: A moderately safe beach at the surf club
and toward the breakwater, where waves are usually
lower, rips less common and the bar wide and shallow.
o not bathe further up the beach as waves and rip
currents intensify. Stay in the patrolled area between the
flags. Best in the morning, as afternoon westerly winds
can intensify the waves and rips.

Surfing: Best up the beach where the waves are more
consistent, the most popular spot being *The Flume*,
where some rock and reef in the surf generate a more
reliable break. There is a car park here and a short walk

across the dunes that provides good access. Toward
Point Ritchie is a reef break called *Granny's*, while in
front of the caravan park is a beach break called
Maginines. In addition, there are several beach breaks
along the beach.

Fishing: The breakwater is very popular, with a boat
ramp for offshore fishers. The best beach fishing is up
the beach where the rip holes and gutters are more
prevalent.

Summary: Warrnambool and its beach offer
everything for the holiday maker, bather, surfer and
fisher, with a range in waves, beach type and fishing
locations. As a result, this is a very popular summer
holiday destination.

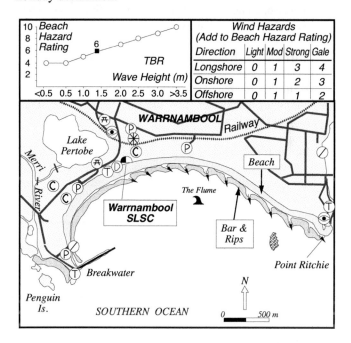

*Figure 128. Warrnambool Surf Life Saving Club is
located toward the western end of Lady Bay. Extensive
car parks, parks and facilities are located nearby.
Bathing is relatively safe in the patrolled area, with
waves decreasing toward the harbour and breakwater,
while they increase up the beach, producing a wider
surf and stronger rips.*

501   MERRI

Unpatrolled

Single bar LTT
Beach Hazard Rating **4**
Length 150 m

Merri Beach is actually a low sand bar at the mouth of
the Merri River. The sand is wedged in between the
long breakwater that forms Warrnambool Harbour on
the east and the 10 to 15 m high dune calcarenite of

Middle and Merri Islands and Pickering Point on the west. There is good access from the breakwater road, with a footbridge across the rear of the beach that provides access to the beach and the coast walking track around Thunder Point.

The beach lies in lee of Middle Island, with low waves approaching from both sides around the island. This results in a low, cuspate sand bar and beach in lee of the island, that is usually awash at high tide. A low, shallow bar runs out to the island, which can be reached on foot at low tide.

Swimming: A relatively safe beach when waves are low, with protection provided by the islands. However be careful out toward the islands as waves and currents increase. This also happens during falling tides when water flows out of the river mouth.

Surfing: sually only a wide, low beach break.

Fishing: A popular spot with fishing from the adjoining breakwater, into the river and off the rocks around the island.

Summary: This is a popular kiddies' beach and a place to explore the tidal pools at low tide. Be careful on the island however, as the exposed side is very hazardous.

502, 503   **BEACH 502, SHELLY**

Unpatrolled			
	Rating	Single bar	Length
502 Beach 502	7	TBR	500 m
503 Shelly	7	RBB	600 m
		Total length	1 100 m

West of the Merri River mouth is an exposed, high energy section of coast called Armstrong Bay. It extends for 15 km from the river mouth to the low basalt outcrops of Sisters Point. The first 5 km are dominated by dune calcarenite, capped by 20 to 30 m high cliff-top dunes. The Thunder Point walking track provides foot access to the two beaches. It also has some good views of the coast, as well as a display describing the area and its evolution.

Three kilometres west of the river mouth are two adjoining, exposed, south-west facing beaches, both bordered by calcarenite headlands and reefs. The first, **Beach 502**, is 500 m long with reefs dominating either end, and patchy reef off the beach, as well as some calcarenite along the beach. The result is a reef dominated surf with a strong, permanent rip running out the western end of the beach. Next door is 600 m long

Shelly Beach that is similar, though with less reef in the surf. It has a wide surf with strong, permanent rips running out along each point.

Swimming: These are two relatively isolated beaches dominated by high waves, strong rips, rocks and reefs. They are unsuitable for safe swimming.

Surfing: There is a combination of sand and reef breaks off both beaches, with best conditions in a low swell and north-easterly winds.

Fishing: There are good rip holes and gutters that can be fished from the beach or the rocks at low tide.

Summary: Two more isolated beaches, dominated by the waves and reefs and backed by active dunes.

---

504   **ARMSTRONG BAY-LEVY POINT**

Unpatrolled	
Inner bar TBRRBB	Outer bar RBBLBT
Beach Hazard Rating **8**	
Length 7 600 m	

Th **Armstrong Bay-Levy Point Beach** is a 7.6 km long, south-west facing beach that extends from the calcarenite cliffs west of Thunder Point to the mouth of Kelly Swamp at The Cutting. The beach is backed by 20 to 30 m high active dunes, with the low Kelly Swamp behind. The swamp drains both to the east through the Merri River, and to the west by way of The Cutting. There is access to the beach in the east by a car park and a 300 m long walk over the dunes, and in the west at The Cutting car park. For the most part it is only accessible along the beach on foot.

The beach receives waves averaging over 1.5 m, which combine with the finer sand to break over a 400 m wide surf zone. This contains an inner detached bar with strong rips every 500 m, and an outer bar with more widely spaced rips.

Swimming: A hazardous beach and surf dominated by usually high waves and a deep longshore trough containing strong longshore and rip currents.

Surfing: Surfers use the Levy Point track to surf *Levy Beach*. The beach picks up any swell and is best in low swell with north-easterly winds.

Fishing: There are usually good rip holes and gutters that can be fished from the beach.

Summary: An exposed, high energy beach and active dune field that is only suitable for experienced surfers and beach fishers.

MAHOGANY SHIP

The famous missing Mahogany Ship is located in the dunes of Armstrong Bay. It was seen a number of times in the 19th century, the last time being in 1880, before being buried by the dunes. It is suspected to be a 16th century Portuguese ship which, if correct, would rewrite the history of the European discovery of southern Australia.

TOWER HILL VOLCANIC ERUPTION

The low basalt rocks, points and reefs between The Basin and Reef Point are remnants of Australia's last lava flows. These occurred some time between 6 000 and 10 000 years ago. The eruption formed the tuffaceous hills around the volcano, with a lava flow spreading south and crossing the coast between the Basin and Reef Point, including Sisters Point. The lava continued out to sea where it now forms the many black reefs off the beach and points.

505　BELFAST

Unpatrolled

Inner bar LTTTBR　　Outer bar TBRRBB
Beach Hazard Rating　**6**
Length　2 000 m

Belfast Beach lies on the west side of The Cutting and forms the western end of the longer Armstrong Bay-Levy Beach. It faces south and runs for 2 km from The Cutting to the first low, basalt point. Grassy dunes back the beach. There is vehicle access from the Princes Highway via Cormans Road, which runs due south to the Belfast Coastal Reserve. A car park is located at the end of the road and there is also one at The Cutting.

The beach receives high waves at The Cutting, where there is a wide, rip dominated, double bar surf zone. Offshore reefs and Sisters Point cause waves to gradually decrease to the west. The result is a single, rip dominated bar at the first car park and a wide, low, continuous, attached bar with a few rips west of the car park, toward the low point.

Swimming: If you intend swimming, head west of the car park and stay inshore on the shallow bar. Avoid the rips at, and to the east of, the car park and at The Cutting.

Surfing: Surfers head for *The Cutting*, where a good left at times peels off the outer bar. Further out to sea is a right hand reef break called *Pelicans*, that is occasionally surfed in big swell.

Fishing: There are usually good rip holes at The Cutting and the backing creek, which is usually closed.

Summary: A very accessible beach, offering both higher waves for surfers to the east and a more protected beach to the west.

506, 507　THE BASIN EAST & WEST

Unpatrolled

	Rating	Single bar	Length
506 The Basin East	6	LTTTBR	600 m
507 The Basin West	3	LTT	100 m
		Total length	700 m

The Basin is a south-east facing, protected section of beach in lee of Sisters Point. The low basalt points and offshore reefs reduce the waves, and form two beaches bordered by the low points and backed by grassy dunes. The Basin vehicle track runs through the dunes from Killarney Beach to a car park and a beach boat launching area at The Basin.

The 600 m long eastern beach receives waves averaging about 1 m, which break across a wide bar. A few rips exist along the beach and there is a permanent rip against the basalt rocks below the car park. The smaller, 100 m long west beach has lower waves and usually a continuous, shallow bar with no rips.

Swimming: The smaller west beach is the safer of the two, and is usually free of rips.

Surfing: The west beach has the larger and better surf breaking over the wide bar. This is a place to check in offshore westerly winds.

Fishing: The west beach has the more persistent rip holes, along with the permanent gutter against the rocks. At low tide, deep water can be fished off the moderately protected basalt points.

Summary: Two accessible and moderately protected beaches, suitable for relatively safe bathing and with lower surf.

508, 509, 510   SISTERS POINT, KILLARNEY, REEF POINT

Unpatrolled			
	Rating	Single bar	Length
508 Sisters Point	5	LTTTBR	1 700 m
509 Killarney	4	RLTT	1 000 m
510 Reef Point	5	LTTTBR	1 700 m
		Total length	4 400 m

Between The Basin and Reef Point is 5 km of shoreline dominated by the low basalt point, with extensive basalt reefs offshore. Between the points are three crenulate, south facing beaches, backed by marram covered dunes. They are fronted by variable surf zones, depending on the presence and influence of the reefs.

The first beach is 1.7 km long and lies between The Basin and **Sisters Point**. It is accessible off The Basin track, with two car parks and tracks over the dunes to the beach. It has continuous reefs off the eastern half, and offshore reef at the western half. The beach has an attached bar, that is interrupted by the reefs. There are rips to the west, with a permanent rip flowing out in lee of the offshore reef.

Killarney Beach is located at the end of Beach Road and is just over 1 km from the Princes Highway. There are camping and picnic areas and a car park, as well as a beach boat launching area at the second car park. The 1 km long beach is protected by continuous offshore reefs and waves are usually low to calm at the beach. The bar is shallow and continuous with usually no rips.

Reef Point forms the western boundary of the third beach. The beach is 1.7 km long and is fronted by continuous offshore reefs, with additional rocks and reef on the beach. There is a continuous, shallow bar along the beach with rips forming against some of the rocks.

Swimming: These are three moderately safe beaches, with Killarney being the safest. The other two beaches have some rips, inshore rocks and reefs.

Surfing: There are low beach breaks that are bigger and better at high tide, when they break over some of the reefs. The western end of Killarney is one of the more consistent spots.

Fishing: The beaches tend to be shallow, except near the rips, while the rocks and reef provide a number of good, moderately protected spots to reach deep water.

Summary: Killarney Beach has long been a popular summer camping and holiday spot. There are limited facilities, but good access to a relatively safe beach.

511   PORT FAIRY (EAST)

Port Fairy SLSC

Patrols: late November to Easter holidays
Surf Lifesaving Club: Saturday, Sunday and Christmas public holidays
Lifeguard: on duty weekdays during Christmas holiday period

Inner bar LTTTBR   Outer bar RBB
Beach Hazard Rating **5**
Length 5 800 m

For map of beach see Figure 129

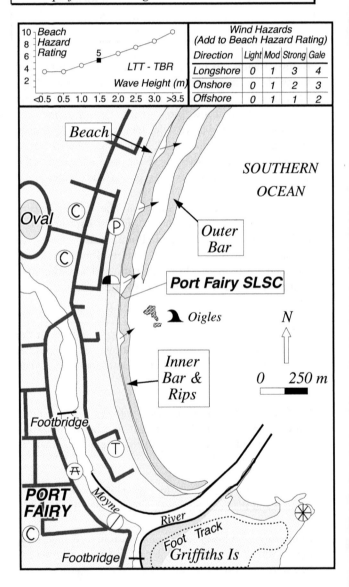

Figure 129. Port Fairy Surf Life Saving Club is located 1 km east of the Moyne River entrance walls. Waves decrease toward the walls, but increase to the north of the beach where there is a wide, rip dominated and more hazardous surf.

Port Fairy is an attractive fishing and holiday town located in lee of Griffiths Island. A fishing fleet operates out of the Moyne River, that enters the sea through the two entrance walls constructed in the 1870s. The town has all holiday facilities, with a large caravan park located behind the beach. Houses back the southern end of the beach, while dunes and the golf course back the centre and northern end.

The beach is 5.8 km long, extending in a broad, curving arc from Reef Point in the east, where it faces south, to the North Mole or harbour entrance wall in the south, where it faces east. Erosion along the southern end has resulted in the construction of a rough seawall and several wooden groynes. The Port Fairy Surf Life Saving Club, founded in 1950, is located on the foredune 1 km north of the north entrance wall. Its members average 10 rescues each year.

The beach is composed of fine, white sand. It receives waves that average less than 0.5 m in the south, about 1 m at the surf lifesaving club (where it is also called East Beach) and up to 1.7 m by Reef Point. In response to the changing waves, the beach is fronted by a single, continuous, attached bar in the southern corner, with rips rarely present. However, by the surf club the surf is over 150 m wide with two bars. The inner bar is cut by rips every 250 m, while the outer bar has more widely spaced rips. Further up the beach, the waves and rips intensify. uring and following high seas, a 300 to 400 m wide surf zone and a third outer bar can form.

Swimming: Summer conditions are moderately safe, when waves tend to be lower and rips less frequent and intense. However, always bathe in the patrolled area between the flags, and avoid the northern end of the beach where rips can be very intense. Be careful if on a bodyboard or surfboard as westerly winds will blow you out to sea.

Surfing: There are numerous beach breaks in the wide, shallow surf zone, with shape depending on waves, bars and wind. A popular spot called *Oigles*, just south of the surf club, works in a big swell, as it breaks over an old shipwreck.

Fishing: Port Fairy usually has a wide and shallow surf, so look for rip holes and gutters north of the surf club. There is also good fishing off the harbour walls, into either the river or ocean.

Summary: Port Fairy is a very popular summer holiday destination offering all facilities. It has a moderately safe patrolled beach at the southern end, with a more energetic beach toward the north.

512   SOUTH MOLE

Unpatrolled
Single bar LTT
Beach Hazard Rating **3**
Length 200 m

South Mole Beach has formed during the past century, since the construction of the harbour moles or entrance walls at the mouth of the Moyne River. The beach lies on the north side of Griffiths Island, between the low dune calcarenite that forms the island and the wall. It can be reached on foot from the car park via a footbridge to the island.

The beach is 200 m long and faces north-east. It is protected by the island and waves average about 0.5 m at the beach. This results in a low, continuous bar and usually no rips.

Swimming: A relatively safe beach with usually low waves and a shallow bar.

Surfing: sually a very low beach break.

Fishing: The harbour moles or breakwaters are the most popular spot to fish the channel.

Summary: A nice beach backed by low, marram covered dunes, with protection from waves and westerly winds.

513   GRIFFITHS ISLAND

Unpatrolled
No bar R
Beach Hazard Rating **3**
Length 100 m

Griffiths Island was attached to the mainland with the construction of the Moyne River harbour moles and a second breakwater, to prevent water flowing around the back of the island and into the harbour. A footbridge now provides access from the car park to the island.

On the south-east side of the island is a small, protruding beach lying in lee of a reef. Most waves break over the reef and the beach is often fronted by a calm lagoon, particularly at low tide.

Swimming: The small lagoon in lee of the reef is the safest spot, particularly at low tide.

Surfing: None at the beach, however on the outer reef is a big right hander called *Green Island*. It is a long paddle out and only surfed in big swells.

Fishing: There are numerous spots around the island to fish from the rocks, into either the tidal gutters and pools, the Back Passage or the sea.

Summary: There is a walk around this small island which, besides the beaches, features a low lighthouse built in 1859. The island is a wildlife reserve and houses a shearwater (mutton bird) colony.

514, 515   PEA SOUP/SOUTH, BEACH 515

Unpatrolled			
	Rating	Single bar	Length
514 Pea SoupSouth	3	LTT	500 m
515 Beach 515	3	LTT	200 m
		Total length	700 m

On the south side of Port Fairy, Ocean rive runs past the caravan park and west along the shore for 2 km. There are three car parks along the road that give direct access to the shore. The shore is fringed by continuous basalt reefs lying 100 to 200 m offshore.

In lee of the reef are two beaches, both bordered by low, basalt rocks. The first is **South** or **Pea Soup Beach**. It is a 500 m long, south facing beach. It adjoins a smaller, 200 m long beach that is backed by scarped dunes and, in places, a protective basalt seawall. The reefs completely protect the wide, low beaches at low tide. Shallow lagoons lie between the beaches and the reefs. At high tide, small waves reach the beaches and wash over continuous, shallow bars with no rips.

Swimming: Both beaches are relatively safe, particularly at low tide. However at high tide, be careful of the submerged rocks and holes.

Surfing: None at the beach. However, at the eastern end of Pea Soup is a short, right hand point break called *The Passage*, which lies right in front of the car park. Off the second beach are two reef breaks a left called *Garbos* and a right called *Gooloos*.

Fishing: The Back Passage is good at high tide, while the beaches and inner rocks are generally too shallow at low tide.

Summary: Two very accessible and relatively safe beaches.

516 - 522   BEACHES 516 to 521, CAPE REAMUR

Unpatrolled			
	Rating	Single bar	Length
516 Beach 516	4	R	800 m
517 Beach 517	4	R	800 m
518 Beach 518	4	R	100 m
519 Beach 519	4	RLTT	600 m
520 Beach 520	4	LTT	500 m
521 Beach 521	4	LTT	100 m
522 Cape Reamur	5	R	500 m
		Total length	3 400 m

West of Port Fairy, the coast runs due west for 8 km to Cape Reamur. Low basalt points and offshore reefs dominate the shore. In amongst the rocks are a number of embayments, containing seven crenulate, sandy beaches. All these are dominated by the rocks and reefs and are backed by marram covered foredunes. At low tide, the reefs stop most waves from reaching the shore. A shallow lagoon lies between the beaches and the outer reefs, with patches of rocks and reefs in the lagoons. The Princes Highway parallels the coast 1 km inland. However, most of the land between the highway and the coast is private property and access is limited to walking along the shore.

Beach 516 comprises a series of three arcing beaches with basalt dominating the low tide area. A few boats are anchored at this protected spot. **Beach 517** lies in lee of a 1 km long reef and protrudes from the run of the coast. **Beach 518** is a 100 m pocket of sand lying in lee of an outer reef. This reef continues on past the 600 m long **Beach 519**, which is a relatively straight beach with an outer reef and a reef-filled lagoon. **Beach 520** is a curving, 500 m long beach with a lagoon. **Beach 521** is a very protected, 100 m long pocket of sand deep inside extensive shallow reefs. **Cape Reamur Beach** is a narrow, high tide beach fronted by basalt boulders, a lagoon and outer reefs at low tide. It ends at Cape Reamur a low, basalt point capped by dunes moving in from the west.

Swimming: All these beaches are relatively safe, so long as you stay close inshore and clear of the many rocks and reefs. Bathing is better at high tide, with a shallow tidal pool - lagoon environment at low tide.

Surfing: None at the beaches, however on the outer reefs are a variety of breaks that can be surfed at high tide.

Fishing: Fishing is better at high tide from the beaches, with no really deep water until you get to the outer reefs.

Summary: A presently isolated series of basalt dominated, low energy, sandy beaches and lagoons.

523   MCKECHNIE CRAIG

Unpatrolled
Single bar LTTTBR
Beach Hazard Rating **6**
Length 1 700 m

At Cape Reamur, the coast swings to the north-west and begins the 50 km long, sweeping arc that becomes Portland Bay. The first 10 km between the cape and Yambuk Inlet are dominated by calcarenite rocks and reefs, interspersed with 11 beaches. The first of these beaches begins right at the cape and runs for 1.7 km to the north-west. The nearest vehicle access is at The Craigs, 2.5 km to the west.

McKechnie Craig Beach is exposed to the westerly winds and waves, however basalt reefs in the east and calcarenite reefs in the west lower the waves toward each end of the beach. As a result, only the central few hundred metres receive high waves. The beach consists of a low energy eastern arc with reefs offshore, and a high energy central-western arc with reefs to either side and a surf zone with two to three rips. Active dunes back the beach and extend up to 1.5 km inland.

Swimming: This is a variable and, in places, hazardous beach. The more protected eastern end is the safest location, with rips dominating the centre and toward the west. In addition, reefs dominate the outer surf.

Surfing: There are reef breaks to either end and beach breaks in the centre.

Fishing: This beach has a variety of locations, with deep water off the protected eastern beach and reefs offshore, rips in the centre and rock platforms to either end.

Summary: A relatively isolated beach, with a section of safer water for bathing, plus beach and reef surf.

524 - 531   BEACHES 524 to 528, THE CRAIGS EAST, THE CRAIGS WEST 1 & 2

Unpatrolled			
	Rating	Single bar	Length
524 Beach 524	4	R	100 m
525 Beach 525	5	R	100 m
526 Beach 526	5	R	250 m
527 Beach 527	5	LTT	100 m
528 Beach 528	6	RLTT	80 m
529 The Craigs East	6	R	300 m
530 The Craigs West 1	4	R	60 m
531 The Craigs West 2	5	R	80 m
	Total length		1 070 m

The Craigs is a well-known coastal access spot located at the end of 3 km long Craigs Road. The road ends at the crest of 20 m high calcarenite bluffs, from where there are good views of the jagged coast. A few small, south-west facing beaches lie at the base of the bluffs, for the most part fronted by shallow calcarenite reefs. There are six beaches to the east of the car park beach and two to the west.

All the beaches have narrow, high tide strips of sand, with variable degrees of fronting reef. All receive low to no waves at low tide and lowered waves that have broken on the outer reefs at high tide.

Beaches 524 and **525** are both 100 m long and occupy the eastern and western halves of a small bay, with exposed reef flats fronting the beaches at low tide and outer reefs offshore. **Beach 526** is 250 m long and has a reef lying just off the beach, which forms a small lagoon and channel at low tide, with deeper reefs further offshore. **Beach 527** is bordered by calcarenite reefs, with a deep central channel running out from the beach between the reefs. The channel is occupied by a permanent rip.

Beach 528 is an 80 m pocket of sand, backed and bordered by low calcarenite bluffs with reef flats off the beach. Amongst these flats are tidal pools at low tide. **The Craigs East Beach** is a 300 m long, crenulate, rock-studded beach with low bluffs behind. Reefs lie to either end and in the centre, and two permanent rips run out between the reefs. The first **Craigs West Beach** lies at the end of The Craigs car park. It is a 60 m strip of sand at the base of the bluffs, fronted by shallow reefs and tidal pools. The second western beach is another narrow, 80 m long sand strip partly covered by rock falls. It is fronted by a continuous reef flat with tidal pools exposed at low tide.

Swimming: Most of these beaches have low waves at the shore, owing to the extensive inner and outer reefs. However, be very careful if swimming here as the reefs are irregular and full of deep holes. On the more open beaches, permanent rips run out through the gaps in the reefs, while the outer reefs are extremely treacherous.

Surfing: There are no recognised breaks along this section, however there may be rideable surf on the many outer reefs.

Fishing: This coast is best suited for rock fishing, with numerous spots along the beaches and points, including deep gullies and gutters in the inner reefs.

Summary: Most people drive out to The Craigs to view the rocky bluffs and reefs. It is best suited for fishing and is only safe for swimming in the protected inner tidal pools.

532, 533   BEACH 532, HUMMOCK

	Unpatrolled			
	Rating	Inner bar	Outer bar	Length
532 Beach 532	9	RBB	-	500 m
533 Hummock	8	TBR	RBB	4 200 m
		Total length		4 700 m

The calcarenite bluffs, rocks and reefs that dominate The Craigs give way to more open beaches 1 km west of the car park. The first is a 500 m long, south-west facing beach with some reefs along its eastern half and offshore reefs at either end. Two permanent rips drain the beach and flow out past the bordering reefs. Past the western reef is 4.2 km long **Hummock Beach**, which is exposed to waves averaging over 1.5 m. It has a 400 m wide surf zone, with an inner bar dominated by rips spaced every 400 m, and an outer bar with more widely spaced rips. Active dunes extend up to 1 km inland of the two beaches. Access to these beaches is limited to walking in from The Craigs car park.

Swimming: These are two exposed, rip dominated, high energy beaches and are unsuitable for safe swimming.

Surfing: There are reef and beach breaks along both beaches, however access is a problem.

Fishing: The first beach has permanent rip channels at each end, while large, persistent rips and a continuous longshore trough characterise the longer Hummock Beach.

Summary: Two isolated, high energy and hazardous beaches, best suited for beach fishing.

534   YAMBUK

Unpatrolled
Inner bar TBR   Outer bar RBBLBT
Beach Hazard Rating **7**
Length 1 500 m

Yambuk Beach lies at the mouth of the Eumeralla River and Lake Yambuk. Yambuk Road runs out 4 km from the highway to the beach, picnic area and caravan park. The beach faces south-west and is 1.5 km long. It is located between two calcarenite reefs, with the river flowing out behind the western reef. The car park is behind the dunes in the centre of the beach, with high, active dunes to the east and lower dunes between the river and the beach to the west.

Swimming: This is a very hazardous beach, dominated by strong rips every 500 m and permanent rips against the reefs. Be very careful if swimming here and stay close inshore in the swash or on the inner bar if attached, and well clear of the rips and reefs. The safest place to swim is in the lake.

Surfing: This is a relatively popular spot due to the good access, with best beach breaks on a low swell and northerly winds.

Fishing: There is both lake and surf fishing here, together with a few holes around the reefs.

Summary: A popular summer holiday campsite, with a quiescent lake, but a hazardous beach.

535   EUMERALLA

Unpatrolled
Inner bar TBR   Outer bar RBB
Beach Hazard Rating **7**
Length 19 000 m

The mouth of the Eumeralla River at Yambuk marks the beginning of a 43 km long stretch of sand, that extends all the way to the rocks on the eastern side of Portland. The beach consists of three sections, that are separated by river mouths. The first runs 19 km from the Eumeralla River to the Fitzroy River. The second extends 13 km from the Fitzroy River to the Surry River mouth, and the final section runs 11 km from the Surry River to the town of Portland. At the Eumeralla River mouth, the beach faces south-west and receives high waves. The beach then gradually swings to face the south and finally south-east, by which time waves are lowered to less than 1 m.

Victorian Beaches - Region 9: Central (Yambuk to Cape Nelson)

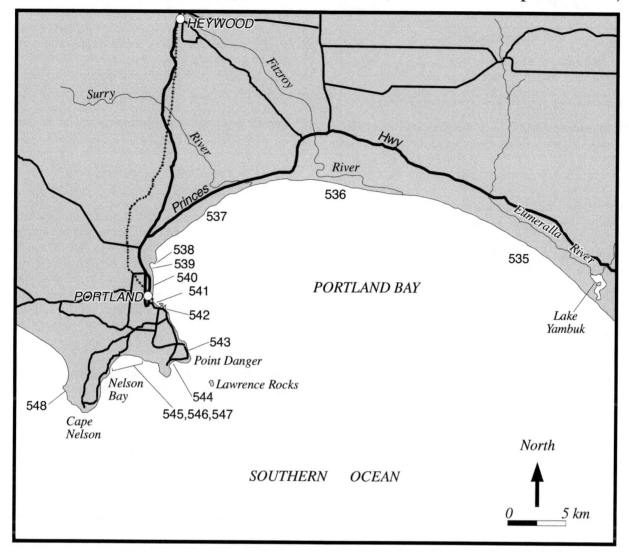

Figure 130.   Region 9 (central section): Yambuk to Cape Nelson (Beaches 535 - 548)

Eumeralla Beach begins at the rocks that lie next to the river mouth and runs in nearly a straight line toward the Fitzroy River. It faces south-south-west and is exposed to westerly winds and waves averaging over 1.5 m. The waves interact with the generally fine beach sand to maintain a 250 m wide surf zone containing two bars. The inner bar consists of alternating bars and rips, with rips spaced every 500 m, while the outer bar has more widely spaced rips. A deep trough lies between the two bars. The beach is backed for the most part by a 10 to 20 m high, narrow series of foredunes, then a long, narrow swamp and a second series of foredunes behind. Farms back the beach and public access is limited to the Fitzroy River Mouth Coastal Reserve, where a 500 m long walking track leads from the car park to the beach.

Swimming: This is a long and relatively isolated beach containing approximately 40 rips. Be very careful if swimming here and stay inshore on the shallow, attached parts of the bars. Stay clear of the many rips, and river mouths when open.

Surfing: There are many beach breaks along here on the inner and outer bars, with best conditions occurring during a low swell and northerly winds.

Fishing: The two river mouths and backing lagoons are the most popular locations, with good rip channels also along the beach.

Summary: A long, isolated beach with limited access.

536   FITZROY RIVER

Unpatrolled

Inner bar LTTTBR   Outer bar LBT
Beach Hazard Rating **6**
Length 13 000 m

Fitzroy River Beach starts at the Fitzroy River mouth and runs for 13 km almost due west to the Surry River mouth, at the small settlement of Narrawong. The beach is only accessible at either end, through the Fitzroy River Mouth Coastal Reserve in the east and at the large riverfront car park and caravan park at Narrawong on the Surry River mouth. The remainder of the beach is backed by a narrow strip of 10 to 20 m high foredunes and a backing long, narrow swamp, all of which is on farm land.

The beach receives waves averaging about 1.5 m. These have been slightly reduced by the extensive Julia Reef, which is part of a lava flow that extends more than 30 km seaward of the coast. In addition, Cape Nelson begins to shield the western end of Portland Bay. The surf zone averages about 200 m wide, with an inner bar that alternates between attached bars and rips spaced about every 500 m. Higher waves break on an outer bar.

Swimming: This is a long, relatively isolated beach exposed to moderate to high waves and with more than 20 rips. Take care if swimming here, stay on the inner sections of the attached bars and clear of the rips and river mouth, when open.

Surfing: There are always beach breaks along here, with the best conditions during low to moderate swell and northerly winds.

Fishing: There are river mouths and lagoons at each end of the beach, with rip channels along the beach, particularly following higher waves.

Summary: A long, low beach with good access at each end, but little in between.

537   SURF

Unpatrolled

Inner bar LTTTBR   Outer bar LBT
Beach Hazard Rating **5**
Length 11 000 m

Surf Beach is the first surfing beach east of Portland. It begins at the mouth of the Surry River and runs for 11 km south-west to Portland. The shoreline meanders due to a series of 1 to 2 km long shoreline protrusions.

Toward the west at utton Way, erosion has resulted in loss of the western end of the beach back to the seacliffs. To combat the beach erosion, a 3 km long basalt boulder seawall and several groynes have been constructed since 1960. The seawall follows the shoreline protrusions and for the most part has replaced the beach. Most of the beach is backed by a series of low foredunes, in places 1 km wide.

The beach is accessible at Narrawong and in the west along utton Way, which parallels the beach and seawall. The seawall area is dangerous for bathing. There is one small pocket beach in the seawall next to the caravan park.

Wave height is reduced westward along the beach by refraction around the large Cape Nelson and by Minerva Reef. It averages 1.5 m at Narrawong, but reduces to 1 m along utton Way. The surf zone is about 100 m wide, with two bars in the east and one to the west. The inner bar is usually low and attached, with rips more common following high seas. The rips are spaced every 500 m. High tide waves crash against the seawall and often over the road.

Swimming: A moderately safe beach under normal waves and clear of the seawall. However, high waves will cut rips and make the seawall area particularly dangerous.

Surfing: The best chance is east of the seawall, as the waves gradually increase in size. Best conditions are during a moderate outside swell and northerly winds.

Fishing: The lagoon and river mouth at Narrawong offer the best chance. The beach is often shallow inshore, with less persistent rips.

Summary: A long and accessible, but heavily modified beach. The beach erosion is attributed to the Portland port construction. The resulting seawall has ruined the eastern end of the beach.

538, 539   ANDERSON POINT, WHALERS POINT

Unpatrolled

		Rating	Single bar	Length
538	Anderson Point	4	LTT	100 m
539	Whalers Point	5	LTT seawall	250 m
			Total length	350 m

Anderson and Whalers Points are two adjoining, 30 m high bluffs, located 1 km north of Portland Harbour. Whalers Point Lighthouse stands 41 m above sea level and was built in 1859. Along the base of the bluffs are two narrow, low beaches. The beaches are often awash at high tide with waves reaching the bluffs. Along Whalers Bluff, a seawall has been built on the beach to protect the bluff from erosion. Both beaches receive waves averaging about 1 m, which break over a shallow, attached bar. Some rocks and reefs are present off the beaches.

There is a reserve and car park on **Anderson Point**, and houses on **Whalers Point**. It is a steep descent to both beaches. Whalers Point Beach is also accessible around the rocks from Nunns Beach.

Swimming: Moderately safe during normal waves. However, the seawall on Whalers Point Beach now makes it unsuitable for bathing.

Surfing: sually low, sloppy beach breaks over the bar and reefs.

Fishing: sually shallow off the beaches.

Summary: Two narrow beaches that are difficult to access and are generally unsuitable for bathing.

540   NUNNS

Unpatrolled
Single bar LTT
Beach Hazard Rating **3**
Length 300 m

Nunns Beach is a heavily modified and now artificial beach. The beach lies immediately north of Portland Harbour and is backed by a car park. A long beach originally ran below the white limestone cliffs on either side of Portland Harbour. The harbour construction resulted in erosion of most of this beach and exposure of the backing cliffs. Some sand accumulated on the northern side of the harbour breakwater to form Nunns Beach. In addition, a groyne has been built at the northern end of the 300 m long, east facing beach. In 1990, sand from Pivot Beach was dumped here to replenish the beach.

The beach is well protected by Cape Nelson and the harbour walls. Waves average less than 0.5 m which, with the fine beach sand, ensures a low, shallow beach and attached bar.

Swimming: A relatively safe beach that gradually shelves out into the bay.

Surfing: None.

Fishing: Better off the breakwater and groyne than the beach.

Summary: An accessible, quiet beach suitable for safe swimming.

541   HENTY PARK

Unpatrolled
No bar R
Beach Hazard Rating **2**
Length 500 m

Henty Park Beach is located inside Portland Harbour and adjacent to downtown Portland. This quiet sand beach was originally part of a more open beach that ran from here to Nunns Beach and below Whalers Point. With harbour construction, the beach has been surrounded by breakwaters and jetties, with a road, park, playground and picnic area behind the beach. Fairy penguins are occasionally seen on this beach at dusk, and covered sanctuaries have been constructed for them in the park as protection from attack by dogs. The beach itself receives very low waves and is usually calm, with a moderate slope into the harbour.

Swimming: This is a relatively safe beach with no waves, however water is deep off the beach.

Surfing: None.

Fishing: The adjoining jetties and breakwaters offer better fishing spots than the beach.

Summary: A quiescent harbour beach with a nice park behind.

542   PIVOT

Unpatrolled
Single bar LTT
Beach Hazard Rating **5**
Length 80 m

Pivot Beach was purposefully created following the Portland Harbour construction, to trap sand that was infilling the harbour. It is located inside a south facing seawall that traps the sand between the wall and the limestone bluffs of Observatory Point. When the beach is full, sand is trucked from here to Nunns Beach on the other side of the harbour.

The beach itself is 80 m long, faces south-east and receives waves averaging 0.5 to 1 m. These break over a shallow, attached bar. A permanent rip runs out against the seawall and there are rocks off either side of the beach. A road runs down the bluffs to the seawall where there is an informal car park and beach boat launching area.

Swimming: A moderately safe beach when waves are less than 1 m. However watch the rip, rocks and boats.

Surfing: There is usually a low shorebreak at the beach. uring bigger swell, the outside of the seawall produces a right break called *Crumpets*.

Fishing: The seawall is the best spot.

Summary: An artificial beach used primarily for launching boats and fishing the seawall.

543   BLACK NOSE POINT

Unpatrolled
Single bar LTTTBR
Beach Hazard Rating **5**
Length 400 m

Black Nose Point is a 20 m high basalt headland that forms the northern border of a 400 m long, east facing beach that is partially protected by Point anger. A vehicle track leads down the backing bluffs and runs the length of the beach to a car park at the southern end. The beach is composed of basalt cobble and boulders, with a wide, shallow sand bar exposed at low tide, particularly to the south. Waves average about 1 m.

Swimming: This beach is better at low tide, as at high tide the waves break over the rocks. Waves above 1 m produce rips against the headlands.

Surfing: There is usually a low beach break that can be surfed at low tide. *Black Nose* is best known for the long right hand break that forms over the reefs off the point. Along the rocks on the south side of the beach, and below the rifle range, is another right point break called *Rifle Range*.

Fishing: There is good access to this beach that can be fished from the steep beach at high tide or the rocks at each end.

Summary: The boulders on this beach make it generally unsuitable for swimming, with most visitors being fishers or surfers, when the point is working.

544   GRANT BAY

Unpatrolled
Single bar TBR rocks reef
Beach Hazard Rating **8**
Length 150 m

Grant Bay, also known as Crayfish Bay, is located between Point anger and Cape Sir William Grant. It is accessible from Cape Nelson Road, which runs along the 40 m high bluffs above the bay. A car park is located above the beach. The beach, which lies deep inside the open bay, faces east and is 150 m long. It consists of a cobble and boulder beach, fronted by rocks and a sandy surf zone.

The beach is partially protected by the cape and receives waves averaging between 1 and 1.5 m. These break over the 100 m wide sand bar and surge heavily up the steep, rocky beach face. There are two permanent rips draining the beach against the rocks at each end.

Swimming: This beach is unsuitable for swimming, owing to the steep cobble and boulder beach and the presence of the rocks, reefs and permanent rips.

Surfing: uring moderate swell, there are beach breaks over the sand off the beach and a left reef break referred to as *Crayfish Bay*. Caution is required in crossing the rocky swash zone.

Fishing: There are permanent gutters against the rocks at each end, and deep water off the beach at high tide.

Summary: A rocky beach with patches of sand, best suited for fishing.

545, 546, 547   NELSON BAY 1, 2 & 3

Unpatrolled				
	Rating	Inner bar	Outer bar	Length
545 Nelson Bay 1	**8**	RBB	RBBLBT	500 m
546 Nelson Bay 2	**8**	RBB	RBBLBT	150 m
547 Nelson Bay 3	**8**	TBR	RBBLBT	100 m
			Total length	750 m

Nelson Bay is a 4 km wide, south facing, rocky bay lying between Cape Sir William Grant and Cape Nelson. eep inside the bay are three remnants of a once larger beach. Today the bay consists of 50 to 90 m high cliffs, composed largely of dune calcarenite sitting on top of basalt bases and rock platforms. Behind the beaches, the cliffs rise steeply to heights of 30 to 40 m and are capped with a veneer of dune sand. The dunes were deposited when a larger beach and backing sand ramp occupied the back of the bay, probably about 6 000 years ago.

The beaches are difficult to access. Cape Nelson Road runs above the western beach (**Nelson Bay 3**) and there is a car park, walkway, viewing platform and steps down the bluffs to this beach. However, the other two (**Nelson Bay 1 & 2**) must be reached on foot around the rocks at low tide.

All three beaches receive waves averaging about 1.5 m, that break over a 200 m wide surf zone. An inner, rip dominated bar is usually detached from the beach, and the outer bar has larger rips. Rocks and reefs also dominate the inner surf. In addition, the narrow, high tide sand beaches are often awash at high tide. These are backed by cobble beaches at the base of the cliffs.

Swimming: These are three little used and hazardous beaches, that are generally unsuitable for safe bathing. In addition, the Portland sewer outfall is located next to the first beach (Nelson Bay 1).

Surfing: This spot is known as *Yellow Rocks* after the bright yellow calcarenite cliffs. It produces some good beach breaks for experienced surfers. When the surf is running, so are the strong rips.

Fishing: Permanent rip holes and gutters are a feature of these three beaches, which can be fished from the beach or rocks.

Summary: It is worth driving out to view these beaches, but they are only used by fishers and experienced surfers.

Victorian Beaches - Region 9: West (Portland to Nelson)

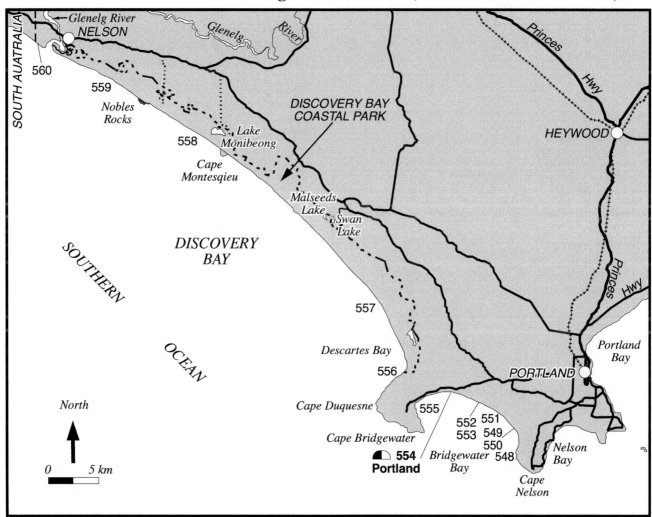

Figure 131. Region 9 (western section): Portland to Nelson (Beaches 548 - 560)

CAPE NELSON STATE PARK

Area   210 ha
Coast length 3 km (rocky)
Beaches   547 to 548 (2 beaches)

Cape Nelson is a basalt headland capped by massive layers of dune calcarenite, which rise to 100 m. The yellow calcarenite cliffs have been eroded by high waves, and often have dark basalt rock platforms at their base. A lighthouse built in 1884 is located on the southern tip of the large cape and stands 75 m above sea level. The park provides vehicle access to the lighthouse, parking and picnic areas, and several cliff-top and nature walking trails.

Information Portland (055) 923 3232

Bridgewater Bay is an open, 13 km wide, south-west to south facing bay. Most of the bay is exposed to persistent, high, south-west waves and swell, and it contains some of the widest, highest and wildest surf in Australia. Originally the bay would have contained one continuous beach. However, loss of sand to the extensive sand dunes (and probably offshore) has seen the beach retreat to the point where it is now broken into eight segments by calcarenite bluffs. Although the beach is segmented, the surf zone forms a continuous line during big swell.

548, 549, 550   BRIDGEWATER BAY 1, 2 & 3

Unpatrolled				
	Rating	Inner bar	Outer bar	Length
548 Bridgewater Bay 1	6	LTTTBR	-	1 000 m
549 Bridgewater Bay 2	8	TBR	RBB	200 m
550 Bridgewater Bay 3	8	TBR	RBB	50 m
			Total length	1 250 m

Bridgewater Bay is located between 50 m high Cape Nelson and 130 m high Cape Bridgewater. Both capes have basalt cores, that are remnants of an extinct volcano, and are capped by extensive dune calcarenite. The bay faces south and receives high waves and strong winds.

The eastern 1.5 km of the bay contains three south-west facing beaches. They are backed by dune-capped calcarenite slopes and bluffs rising to 100 m. There is no direct access to these beaches and a 4W and local knowledge are required to locate them. The first, **Bridgewater Bay Beach 1**, is 1 km long and swings

around to face west against Cape Nelson. It receives some protection from the cape and extensive offshore reefs, with waves averaging between 1 and 1.5 m. These result in a single attached bar, which is more sheltered toward the cape, but has higher waves and rips toward the north.

The second two **Bridgewater Bay Beaches** are smaller and rich in shells. They are also bordered and broken up by low calcarenite bluffs, rocks and reefs and receive waves averaging more than 1.5 m. As a result, they have a 200 m wide surf zone with an inner, rip dominated bar and an outer bar.

Swimming: These beaches are difficult to access and little used, so take care. Bridgewater Bay Beach 1 is the safest of the three, particularly in lee of the cape. The other two are rip, rock and reef dominated and are hazardous.

Surfing: Local surfers use these beaches (also known as _Murrells Beach_), as they offer persistent beach and reef breaks.

Fishing: There are permanent rip channels at the Bridgewater Bay Beaches. These can be fished from the beach or rocks.

Summary: Three out of the way and little used beaches, except by keen fishers and experienced surfers.

551, 552, 553   BRIDGEWATER BAY CENTRE, SHELLY 1 & 2

Unpatrolled				
	Rating	Inner bar	Outer bar	Length
551 Bridgewater Bay Centre	7	TBR	RBBLBT	5 500 m
552 Shelly 1	8	TBR rocks	RBBLBT	1 800 m
553 Shelly 2	8	TBR rocks	RBBLBT	700 m
			Total length	8 000 m

The central portion of **Bridgewater Bay** contains 8 km of south-west to south facing beach, which is broken into three beaches by the occurrence of dune calcarenite. The first section (551) is 5.5 km long and consists of a continuous sand beach. This is fronted by a 300 m wide surf zone, with a rip dominated inner bar and a wide outer bar. This beach is backed by extensive dunes and calcarenite extending up to 3 km inland, and reaching 125 m high at Mount Chaucer. Access is by 4W tracks.

The second two beaches (**Shelly Beach**) are sandy, but dominated by calcarenite bluffs, rocks and inner reefs. The surf zone links with the beaches to either side and continues as a double bar system. These two beaches lie 200 to 300 m from Bridgewater Road and can be reached by a few vehicle tracks leading from the road to the top of the 30 to 40 m high bluffs.

Swimming: These are three exposed, high energy beaches, with rip dominated inner bars, together with permanent rips and rocks on the second two beaches. Not suitable for safe bathing.

Surfing: There are numerous beach breaks along here, with best conditions occurring in a low swell and northerly winds.

Fishing: The inner surf zone tends to be shallow, so look for the rip holes on the long beach. There are many holes around the bluffs, and reefs off the second two beaches.

Summary: Three little used and potentially hazardous beaches.

554 BRIDGEWATER BAY SURF

Portland SLSC
Patrols: late November to Easter holidays **Surf Lifesaving Club:** Saturday, Sunday and Christmas public holidays **Lifeguard:** no lifeguard on duty or weekday patrols Inner bar LTTTBR   Outer bar RBBLBT Beach Hazard Rating **7** Length 2 800 m
For map of beach see Figure 132

Bridgewater Bay is the site of the Portland Surf Life Saving Club, which is located on this western, 2.8 km long section of beach. This beach also has the best public access of the eight Bridgewater Bay beaches. The eastern half of the beach is backed by 20 to 30 m high, dune-capped calcarenite bluffs that extend east to the calcarenite bluff that separates it from the previous beach. The surf lifesaving club is located toward the western end, where the beach faces south-east and is partially sheltered by Cape Bridgewater.

The area around the surf club has a car park, toilets, showers and a kiosk. Apart from a few houses, tea rooms etc. on the other side of the road, these buildings comprise the only development along this large bay. The original Portland Surf Life Saving Club was founded in 1947, with the current club at Bridgewater Bay established in 1971. It averages 3 rescues annually.

The beach at the surf club consists of fine sand, with waves averaging 1.4 m. These produce a wide, flat beach and surf zone. West of the surf club, the waves and surf decrease, however to the east, the waves and surf rapidly build and widen. There are usually widely spaced rips located along the inner bar, with even more widely spaced rips on the outer bar. In addition, rocks and reefs outcrop along the beach, inducing strong, permanent rips.

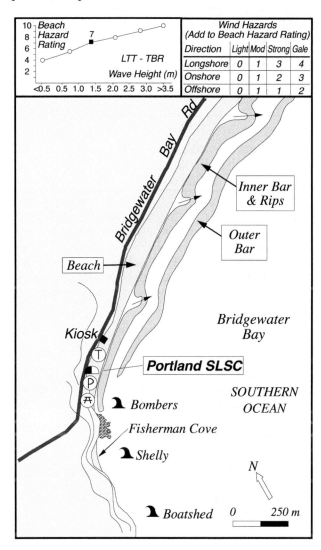

Wind Hazards (Add to Beach Hazard Rating)				
Direction	Light	Mod	Strong	Gale
Longshore	0	1	3	4
Onshore	0	1	2	3
Offshore	0	1	1	2

Figure 132. Portland Surf Life Saving Club is located 18 km west of Portland at the western end of the energetic Bridgewater Bay. The club is fronted by a wide, gently sloping surf, with strong rips east of the clubhouse.

Swimming: The only safe place to bathe on this beach is in the western patrolled area and on the shallow bars. o not bathe at the eastern end, in the rips or near the rocks.

Surfing: The wide surf zone is more popular with wind surfers than board riders, though learners will find the wide, inner surf zone a good place to practise. Bigger swell can generate beach breaks at *Bombers* and *Shelly*, while further out on the rocks is a right hand break called *Boatshed*.

Fishing: There is increasingly high energy beach fishing to the east along the beach, while safer rock fishing is available out along the rocks in Fisherman Cove.

Summary: An attractive, if isolated, bay and beach. It is most popular during summer in low wave conditions and when it is patrolled. Higher waves rapidly produce a hazardous beach.

555   FISHERMAN COVE

Unpatrolled
Single bar LTT
Beach Hazard Rating **4**
Length 100 m

Fisherman Cove is, as the name suggests, an area of quieter water in the energetic Bridgewater Bay. It is located around the rocks about 500 m south-west of the surf lifesaving club. The cove is 100 m long and consists of a low, flat beach, with scattered rocks, backed by grassy slopes that rise to 90 m. The cove faces east and is well protected by Cape Bridgewater, with waves averaging less than 1 m. These produce a shallow, attached bar, usually free of rips.

Swimming: This is the safest beach in the area under normal wave conditions, while higher waves will induce rips against the rocks.

Surfing: sually a low beach break, however big swell generates a heavy break over the reefs south of the beach, called *Boatshed*.

Fishing: The rocks along the shore on either side of this beach are relatively safe, and front moderately deep water.

Summary: This is a picturesque little beach, with grassy slopes and old boatsheds further along the shore.

556   WHITES

Unpatrolled
No bar R reefs
Beach Hazard Rating **6**
Length 200 m

Whites Beach is a small patch of sand lying on the north side of Cape Bridgewater and at the southern end of escartes Bay. It is backed by 50 m high calcarenite

bluffs. The beach can be reached by way of Amos Road, which ends on the bluffs past the southern end of the beach.

The beach faces west and is partially protected by the cape and extensive rock reefs. Waves average about 1 m and surge up a steep beach with no bar. There are reef-controlled rips off the beach and against the rocks at each end.

Swimming: An isolated beach with usually low waves but permanent rips, rocks and reefs.

Surfing: There is a rock-strewn reef break that provides a good left if you are game. Only for experienced surfers.

Fishing: A good spot with permanent rip channels that can be fished from the beach or rocks.

Summary: A difficult to find, isolated beach best suited to fishing.

CAPE BRIDGEWATER COASTAL RESERVE
Coast length   13 km
Beaches   556 (1 beach)
Cape Bridgewater is a basalt headland capped by massive layers of dune calcarenite. On the western side, the dunes have climbed basalt cliffs tens of metres high. Along the south-west section is a reserve that provides a guided walking track around the area, including basalt dykes, blowholes and petrified tree trunks in the calcarenite.
Park Information   Portland   (055) 923 3232

557   DESCARTES/DISCOVERY BAY

Unpatrolled
Inner bar LTTTBR Second bar RBB Outer bar LBT
Beach Hazard Rating **8**
Length 29 000 m

The first beach in **Discovery Bay** runs for 29 km from Cape Bridgewater, where it faces west, up to Nobles Rocks, by which time it is facing south-west. The entire length is backed by massive coastal dune systems up to 4 km wide and 60 m high. There are only four access points to the beach. The first is at the very eastern end in **Descartes Bay**, called Tarragal, where the road from Cape Bridgewater leads to a grassy car park on 50 m high slopes. A walking track leads down the slopes and over the foredune to the beach, where it is known as Blacks Beach. The second is a 2 km long walking track from Swan Lake. The third is a 500 m long track from the camping area at Lake Monibeong, in lee of Cape Montesquieu. The fourth provides vehicle access to a car park located behind the foredune about 500 m east of Nobles Rocks.

The beach is low and flat, with the southern escartes Bay end having a green tinge from the mineral olivine, that is washed out of the basalt of Cape Bridgewater. It is fronted by a 400 m wide surf zone which, during high swell, has three bars working. The inner bar is usually attached to the beach with massive rips cutting it every 500 m. The second bar has more widely spaced rips, while the outer bar only breaks when the swell exceeds 2 m.

Swimming: This is the highest energy beach in Victoria and potentially very hazardous. If swimming here (although it is not advised), stay well inshore on the attached portions of the inner bar and well clear of the rips and rocks at each end.

Surfing: There is always some swell along this beach and it works best with a low swell. A lot of experience is required to handle moderate to high swell.

Fishing: There are rip holes and gutters along the length of the beach, however the inner surf tends to be shallow. At either end are the rocks of Cape Bridgewater and Nobles Rocks, which have permanent rips against them.

Summary: A long, exposed, high energy beach worth visiting and viewing, but be very careful if entering the surf.

558   NOBLES ROCKS

Unpatrolled
Inner bar TBR rocks reef   Second bar RBB
Outer bar LBT
Beach Hazard Rating **9**
Length 2 000 m

Nobles Rocks is a 2 km long section of low, calcarenite bluffs that break up the run of the beach. There is car access to the southern end of the rocks, with a 4W required to travel along the remainder of the beach. The beach faces south-south-west and, like the rest of the bay, receives waves averaging over 1.5 m. It has a 400 m wide, high energy, triple bar surf zone, together with strong permanent rips against the rocks at each end.

Swimming: A hazardous section of beach owing to the wide, high energy surf, coupled with the bordering low, calcarenite bluff that induces strong permanent rips.

DISCOVERY BAY COASTAL PARK	
Area	8 530 ha
Coast length	50 km
Beaches	557 to 560 (4 beaches)

Discovery Bay is an open, 50 km long bay that faces south-west and has some of the highest energy beaches and most active sand dunes in southern Australia. It runs from Cape Bridgewater up to the Glenelg River mouth at Nelson, and then into South Australia. It contains a near continuous, 43 km long beach between Cape Bridgewater and the border, with the calcarenite Nobles Rocks and river mouth dividing it into three beaches.

The iscovery Bay Coastal Park includes all three beaches and much of the backing dune systems. There are access points in the east at Tarragal above escartes Bay, via Swan Lake, Lake Monibeong and Black Swamp; and at Nelson in the west. Camping areas are provided at Swan Lake, Lake Monibeong and Long Swamp. The latter can only be reached on foot along the beach.

The park is part of the 200 km long Great South West Walk, which runs the length of the great beach system.

Park Information	Portland	(055) 923 3232
	Nelson	(087) 938 4051

Surfing: The permanent rips provide an easy access to the outer surf, where there is usually a wide beach break over the inner and outer bars.

Fishing: A relatively popular spot that is accessible by vehicle, with permanent holes against the rocks.

Summary: The rocky section of iscovery Bay Beach.

559   DISCOVERY BAY/NELSON

Unpatrolled
Inner bar TBR   Second bar TBR   Outer bar LBT
Beach Hazard Rating **8**
Length 10 300 m

The second long section of the **Discovery Bay** beach system runs for 10.3 km from the western end of Nobles Rocks Beach to the Glenelg River mouth. This section is backed by active coastal dunes extending a few hundred metres inland and reaching heights of 10 to 20 m. The only access points are at each end at Nobles Rocks and **Nelson**. There is also a walk-in campsite at Long Swamp.

The beach faces south-west and receives waves averaging over 1.5 m, which combine with the fine sand to maintain a 400 m wide, triple bar system. The inner bar is attached and cut by rips every 500 m, while deep troughs separate the outer two bars. In addition, there are permanent rips against Nobles Rocks and several patchy reefs along the beach. Tidal currents and shifting shoals are associated with the Glenelg River mouth.

Swimming: This is a long, isolated and hazardous beach. The safest place to get wet is in close on the inner section of the attached bar and well clear of the rips. Be very careful near Nobles Rocks, the reefs and the river mouth. At the river mouth there is safe bathing in parts of the estuary away from the tidal channels.

Surfing: There are beach breaks along the length of this beach. These are best in a low swell and north-easterly winds. Higher swell requires a long paddle and sufficient experience to handle the wide surf zone.

Fishing: Nobles Rocks is a reasonably accessible and popular beach and rock fishing location, with rip channels along the length of the beach. Channels, currents and shoals are associated with the Glenelg River mouth, as well as the backing estuary.

Summary: A long beach with good access at each end, but care is required if entering the surf, and near the river mouth.

560   DISCOVERY BAY WEST/GLENELG RIVER WEST

Unpatrolled
Inner bar LTTTBR   Second bar RBB
Outer bar LBT
Beach Hazard Rating **8**
Length 2 000 m

The westernmost beach in Victoria is a 2 km long section of **Discovery Bay** lying between the shifting **Glenelg River** mouth and the South Australia border. The beach is the westernmost part of iscovery Bay National Park, however there is no direct access to it. You can walk across the river mouth when it is closed, however when it's open you need a boat to cross the estuary.

The beach is low and wide. It is backed by a 10 to 15 m high foredune and is fronted by a 250 m wide surf zone, with rips cutting the attached inner bar every 500 m. In addition, there are the shifting tidal channels and shoals of the river mouth. Toward the border, offshore reefs begin to lower the waves and the inner bar becomes more continuous as the rips infill.

Swimming: This is a hazardous beach, with the river mouth at one end and rips along the beach into South Australia. If swimming here, stay well clear of the river, on the attached parts of the inner bar and away from the rips.

Surfing: There are beach breaks all the way along, but these are rarely surfed as there are more accessible spots on the Nelson side of the river.

Fishing: The river mouth is the most popular location, with rip holes along the beach.

Summary: An energetic and, at times, difficult to access beach. It is the most western in the state.

GLOSSARY

bar (sand bar) - an area of relatively shallow sand upon which waves break. It may be attached to or detached from the beach, and may be parallel (longshore bar) or perpendicular (transverse bar) to the beach.

barrier - a long term (1 000s of years) shore-parallel accumulation of wave, tide and wind deposited sand, that includes the beach and backing sand dunes. It may be 100s to 1000s of metres wide and backed by a lagoon or estuary. The beach is the seaward boundary of all barriers.

beach - a wave deposited accumulation of sediment (sand, cobbles or boulders) lying between modal wave base and the upper limit of wave swash.

beach face - the seaward dipping portion of the beach over which the wave swash and backwash operate.

beach type - refers to one of three beach types dissipative, intermediate and reflective. Each of these possesses a characteristic combination of hydrodynamic processes and morphological character.

beach types (abbreviations, see Figure 26 and 27)
    R - reflective
    LTT - low tide terrace
    TBR - transverse bar and rip
    RBB - rhythmic bar and beach
    LBT - longshore bar and trough
    - dissipative

beach hazards - elements of the beach environment that expose the public to danger or harm. Specifically water depth, breaking waves, variable surf zone topography, and surf zone currents, particularly rip currents. Also include local hazards such as rocks, reefs and inlets.

beach hazard rating - the scaling of a beach according to the hazards associated with its beach type as well as any local hazards.

berm - the nearly horizontal portion of the beach, deposited by wave action, lying immediately landward of the beach face. The rear of the berm marks the limit of spring high tide wave action.

blowout - a section of dune that has been destabilised and is now moving inland. Caused by strong onshore winds breaching the dune.

cusp - a regular undulation in the high tide swash zone (upper beach face), usually occurring in series with spacing of 10 to 40 m. Produced during beach accretion by the interaction of swash and sub-harmonic edge waves.

foredune - the first sand dune behind the beach. In Victoria it is usually vegetated by spinifex and marram grass, then teatree thickets.

gutter - a deeper part of the surf zone, close and parallel to shore. It may also be a rip feeder channel.

hole - a localised, deeper part of the surf zone, usually close to shore. It may also be part of a rip channel.

Holocene - the geological time period (or epoch) beginning 10 000 years ago (at the end of the last Glacial or Ice Age period) and extending to the present.

lifeguard - in Australia this refers to a professional person charged with maintaining public safety on the beaches and surf area that they patrol. Also known as *beach inspectors.*

lifesaver - an Australian term referring to an active volunteer member of Surf Life Saving Australia, who patrol the beach to maintain public safety in the surf.

megacusp - a longshore undulation in the shoreline and swash zone, with regular spacings between 100 and 500 m, which match the adjacent rips and bars. Produced by wave scouring in lee of rips (megacusp or rip embayment) and shoreline accretion in lee of bars (megacusp horn).

megacusp embayment - see megacusp

megacusp horn - see megacusp

parabolic dune - a blowout that has extended beyond the foredune and has a shape when viewed from above.

Pleistocene - the earlier of the two geological epochs comprising the uaternary Period. It began 2 million years ago and extends to the beginning of the Holocene Epoch, 10 000 years ago.

rip channel - an elongate area of relatively deep water (1 to 3 m), running seaward, either directly or at an angle, and occupied by a rip current.

rip current - a relatively narrow, concentrated seaward return flow of water. It consists of three parts the *rip feeder current* flowing inside the breakers, usually close to shore; the *rip neck*, where the feeder currents converge and flow seaward through the breakers in a narrow 'rip'; and the *rip head*, where the current widens and slows as a series of vortices seaward of the breakers.

rip embayment - see megacusp

rip feeder current - a current flowing along and close to shore, which converges with a feeder current arriving from the other direction, to form the basis of a rip current. The two currents converge in the rip (megacusp) embayment, then pulse seaward as a rip current.

rock platform - as per shore platform.

rock pool - a wading or swimming pool constructed on a rock platform and containing sea water.

sea waves - ocean waves actively forming under the influence of wind. sually relatively short, steep and variable in shape.

set-up - rise in the water level at the beach face resulting from low frequency accumulations of water in the inner surf zone. Seaward return flow results in a *set-down*. Frequency ranges from 30 to 200 seconds.

shore platform - a relatively horizontal area of rock, lying at the base of sea cliffs, usually lying above mean sea level and often awash at high tide and in storms. The platforms are commonly fronted by deep water (2 to 20 m)

swash - the broken part of a wave as it runs up the beach face or swash zone. The return flow is called *backwash*.

swell - ocean waves that have travelled outside the area of wave generation (sea). Compared to sea waves, they are lower, longer and more regular.

trough - an area of deeper water in the surf zone. May be parallel to shore or at an angle.

tidal pool - a naturally occurring hole, depression or channel in a shore platform, that may retain its water during low tide.

wave (ocean) - a regular undulation in the ocean surface produced by wind blowing over the surface. While being formed by the wind it is called a *sea* wave; once it leaves the area of formation or the wind stops blowing it is called a *swell* wave.

wave refraction - the process by which waves moving in shallow water at an angle to the seabed are changed. The part of the wave crest moving in shallower water moves more slowly than other parts moving in deeper water, causing the wave crest to bend toward the shallower seabed.

wave shoaling - the process by which waves moving into shallow water interact with the seabed causing the waves to refract, slow, shorten and increase in height.

wave bore - the turbulent, broken part of a wave that advances shoreward across the surf zone. This is the part between the wave breaking and the wave swash and also that part caught by bodysurfers. Also called *whitewater*.

REFERENCES AND FURTHER READING

Bird, E C F, 1993, The Coast of Victoria. Melbourne niversity Press, 324 pp. An excellent reference and description of the nature, geology, geomorphology and evolution of the entire Victorian coast.

a Costa, G, 1988, Car Touring and Bush Walking in East Gippsland. Australian Conservation Foundation, Melbourne, 188 pp. A complete guide to seeing East Gippsland by car or foot.

Land Conservation Council, 1993, Marine and Coastal Special Investigation - escriptive Report. Land Conservation Council, Melbourne, 254 pp. Contains a wealth of scientific information on the Victorian coast.

Loverridge, R, 1987, A Guide to the Surf Beaches of Victoria. Lothian Publ. Co., Melbourne, 126 pp. The best surfing guide to the Victorian coast.

Readers igest, 1986, Guide to the Coast of Victoria, Tasmania and South Australia. Readers igest, Sydney, 208 pp. Excellent aerial photographs and information on the more popular spots along the Victorian coast.

Ross, J (editor), 1995, Fish Australia. Viking, Melbourne, 498 pp. An excellent coverage of all Victorian coastal fishing spots.

Short, A , 1993, Beaches of the New South Wales Coast. Australian Beach Safety and Management Program, Sydney, 358 pp. The New South Wales version of this book.

Surf Life Saving Australia, 1991, Surf Survival; The Complete Guide to Ocean Safety. Surf Life Saving Australia, Sydney, 88 pp. An excellent guide for anyone using the surf zone for swimming or surfing.

Warren, M, 1988, Mark Warren's Atlas of Australian Surfing. Angus Robertson, Sydney, 232 pp. Covers many Victorian surfing spots.

Young, N, 1980, Surfing Australia's East Coast. Horwitz, Sydney, 112 pp. Provides maps and a description of most east coast surfing spots.

GENERAL INDEX

Also see BEACH INE page 290
and SRF INE page 297

lagoon
 Flynn, 147
 McHaffie, 147
LBT, 50
lifeguard
 Anglesea, 220
 Cowes, 149
 Front (Torquay), 216
 Gunnamatta, 157
 Jan Juc, 217
 Lakes Entrance, 101
 Lorne, 227
 Point Lonsdale, 209
 Point Roadknight, 221
 Port Fairy, 266
 Portsea, 161
 Seaspray, 106
 Smith Beach, 143
 Torquay, 216
 Venus Bay, 128
 Warrnambool, 263
 Woolamai, 141
lifesaver, 281
lighthouse
 Aireys Inlet, 224
 Cape Nelson, 276
 Cape Otway, 241, 242
 Cape Schanck, 157
 Gabo Island, 81
 Point Hicks, 92
 Point Lonsdale, 209
 Wilsons Promontory, 117
Lorne, 227
LTT, 43

Mallacoota, 81
Marine and Coastal Park
 Bunurong, 129
Marine Park
 Corner Inlet, 109
 Nooramunga, 109
 Shallow Inlet, 122
 Wilsons Promontory, 113
Marine Reserve
 Harold Holt, 162, 166, 167
 Point Cook, 197
 Wilsons Promontory, 113
megacusp, 281
 embayment, 281
 horn, 281
megarip, 57
minirip, 43
modified beach
 Henty Park, 273
 Nunns, 273
 Pivot, 273
 Surf Beach, 272
Mornington Peninsula, 138, 150

Narrawong, 272
National Park
 Otway, 238
 Mornington Peninsula, 155, 163, 166, 167
 Port Campbell, 248
 Wilsons Promontory, 113
Nature Walk
 Wingan, 89
Nelson, 279

ocean
 currents, 28
 processes, 18
 salinity, 31
Ocean Grove, 5, 211
optional dress beach
 Addiscot, 219
 Campbells Cove, 198
 Point Impossible, 215
 Southside, 219
 Sunnyside North, 175
Otway Basin, 5

parabolic dune, 281
penguin colony, 253
Peterborough, 253, 254
Phillip Island, 5, 138
pier
 Chelsea, 180
 Half Moon Bay, 187
 Hampton, 188
 Mordialloc, 183
 Princes, 195
 Seaford, 177
 Williamstown, 196
platform beach
 Beach 272, 163
 Beach 273, 163
 Beach 274, 163
 Beach 275, 163
 Beach 371, 234
 Beach 373, 234
 Beach 389, 238
 Beach 401, 240
 Blanket Bay, 239
 Blanket Bay North, 239
 Bridgewater Bay, 160
 Browns Creek, 234
 Carisbrook Creek, 234
 Caryard, 226
 Cheviot, 163
 Coal Point East, 134
 Coal Point West, 135
 Cowrie Bay, 146
 Crayfish Point, 241
 immicks 1, 159
 immicks 2, 159
 Eagles Nest, 131
 Eric the Red, 240

BEACH INDEX

(SLSC) = Surf Life Saving Club
(LSC) = Life Saving Club
(LG) = Lifeguard

BAY BEACHES

SURF INDEX

This index lists most recognised surfing breaks, including many point and reef breaks. Most beach 'breaks' are listed under the beach name in the BEACH INE and not in this index.

BEACHES OF THE AUSTRALIAN COAST

Published by the Sydney University Press for the
Australian Beach Safety and Management Program
a joint project of

Coastal Studies Unit, University of Sydney and Surf Life Saving Australia

by

Andrew D Short
Coastal Studies Unit, University of Sydney

BEACHES OF THE NEW SOUTH WALES COAST
Publication: 1993   **ISBN:** 0-646-15055-3
358 pages, 167 original figures, including 18 photographs; glossary, general index, beach index, surf index.

BEACHES OF THE VICTORIAN COAST & PORT PHILLIP BAY
Publication: 1996   **ISBN:** 0-9586504-0-3
298 pages, 132 original figures, including 41 photographs; glossary, general index, beach index, surf index.

BEACHES OF THE QUEENSLAND COAST: COOKTOWN TO COOLANGATTA
Publication: 2000   **ISBN** 0-9586504-1-1
369 pages, 174 original figures, including 137 photographs, glossary, general index, beach index, surf index.

BEACHES OF THE SOUTH AUSTRALIAN COAST & KANGAROO ISLAND
Publication: 2001   **ISBN** 0-9586504-2-X
346 pages, 286 original figures, including 238 photographs, glossary, general index, beach index, surf index.

BEACHES OF THE WESTERN AUSTRALIAN COAST: EUCLA TO ROEBUCK BAY
Publication: 2005   **ISBN** 0-9586504-3-8
433 pages, 517 original figures, including 408 photographs, glossary, general index, beach index, surf index.

Order online from **Sydney University Press** at

http://www.sup.usyd.edu.au/marine

Forthcoming titles:

BEACHES OF THE TASMANIAN COAST AND ISLANDS (publication 2006)   1-920898-12-3

BEACHES OF NORTHERN AUSTRALIA: THE KIMBERLEY, NORTHERN TERRITORY AND CAPE YORK (publication 2007)   1-920898-16-6

BEACHES OF THE NEW SOUTH WALES COAST (2nd edition, 2007)   1-920898-15-8

SYDNEY UNIVERSITY PRESS